AQA GCSE biology

Authors

Simon Broadley

Sue Hocking

Mark Matthews

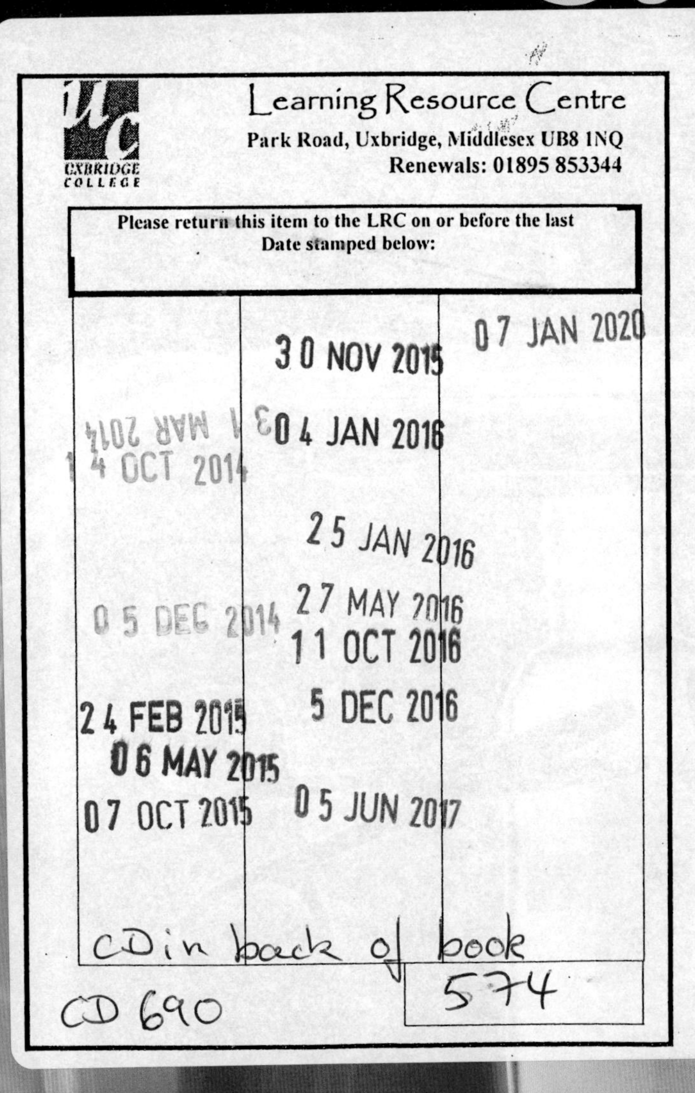

Contents

Welcome to your AQA GCSE Biology course. This book has been specially written by experienced teachers and examiners to match the 2011 specification.

On these two pages you can see the types of pages you will find in this book, and the features on them. Everything in the book is designed to provide you with the support you need to help you prepare for your examinations and achieve your best.

Unit openers

Specification matching grid: This shows you how the pages in the unit match to the exam specification for GCSE Biology, so you can track your progress through the unit as you learn.

Why study this unit: Here you can read about the reasons why the science you're about to learn is relevant to your everyday life.

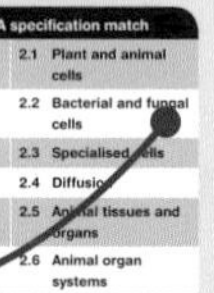

B2 Part 1 Cells and the growing plant

AQA specification match		
2.1	2.1	Plant and animal cells
2.1	2.2	Bacterial and fungal cells
2.1	2.3	Specialised cells
2.1	2.4	Diffusion
2.2	2.5	Animal tissues and organs
2.2	2.6	Animal organ systems
2.2	2.7	Plant tissues and organs
2.3	2.8	Photosynthesis
2.3	2.9	The leaf and photosynthesis
2.3	2.10	Rates of photosynthesis
2.3	2.11	Controlling photosynthesis
2.4	2.12	Sampling techniques
2.4	2.13	Handling environmental data
Course catch-up		
AQA Upgrade		

Why study this unit?

Photosynthesis is one of the most important biological processes. It is through photosynthesis that energy is trapped into the living world. Once trapped, this energy is used to power the entire living world in all its glory.

In this unit you will study photosynthesis as a process, and look at where the process occurs. You will also look at processes for sampling the distribution of living organisms in the environment.

The cell is often considered to be the basic building block of living things. This unit looks at the structure of plant and animal cells. You will observe how cells develop special structures to perform special functions. The principles of how cells work together to form organs, and how organs work together in organ systems, will form part of your study of cells. Finally, the problem of how molecules can get into and out of cells will be explored.

You should remember

1. You are made of cells that are organised into tissues, organs, and systems – such as the reproductive system.
2. Plant and animal cells both have a membrane, cytoplasm, nucleus, mitochondria, and ribosomes, but plant cells also have a cell wall and a large vacuole.
3. Cell structure and function.
4. The environment can be studied by sampling the distribution of organisms.
5. Plants make food by the process of photosynthesis.
6. Photosynthesis occurs in the leaves.
7. Diffusion is the movement of particles from a high concentration to a low concentration.

Have you ever considered how garden centres and nurseries get plants ready for sale at just the right time? Just how do poinsettia plants reach their stunning best in time for Christmas? It's all down to controlling plant growth.

The poinsettia has become a symbol of Christmas. In 2009 over 4 million British grown plants were sold, plus many more imported plants. The production process for these plants starts in late August. Large greenhouses are used to grow them in. Shoots are pinched out to keep the plants short and full. To achieve the characteristic red colour, the plants are exposed to shortened days – less than 12 hours' light with no light during the night. To produce strong, healthy plants, the daylight must be very bright, and the temperature must be kept above 10 °C, to promote the maximum rate of photosynthesis. The result will be strong, bushy plants, with glorious red colour.

You should remember: This list is a summary of the things you've already learnt that will come up again in this unit. Check through them in advance and see if there is anything that you need to recap on before you get started.

Opener image: Every unit starts with a picture and information on a new or interesting piece of science that relates to what you're about to learn.

Main pages

Learning objectives: You can use these objectives to understand what you need to learn to prepare for your exams. Higher Tier only objectives appear in pink text.

Key words: These are the terms you need to understand for your exams. You can look for these words in the text in bold or check the glossary to see what they mean.

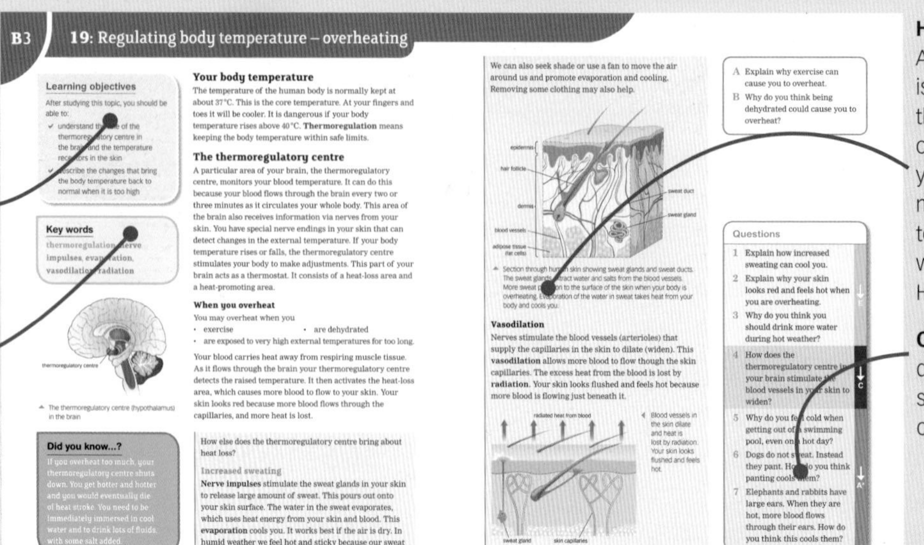

Higher Tier content: Anything marked in pink is for students taking the Higher Tier paper only. As you go through you can look at this material and attempt it to help you understand what is expected for the Higher Tier.

Questions: Use the questions on each spread to test yourself on what you've just read.

Summary and exam-style questions

Every summary question at the end of a spread includes an indication of how hard it is. These indicators show which grade you are working towards. You can track your own progress by seeing which of the questions you can answer easily, and which you have difficulty with.

When you reach the end of a unit you can use the exam-style questions to test how well you know what you've just learnt. Each question has a grade band next to it.

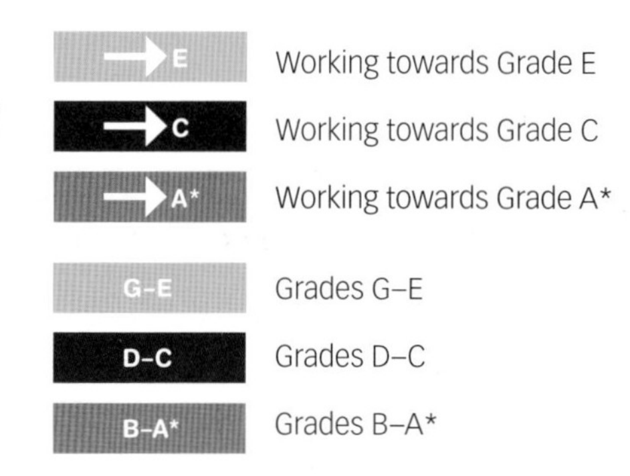

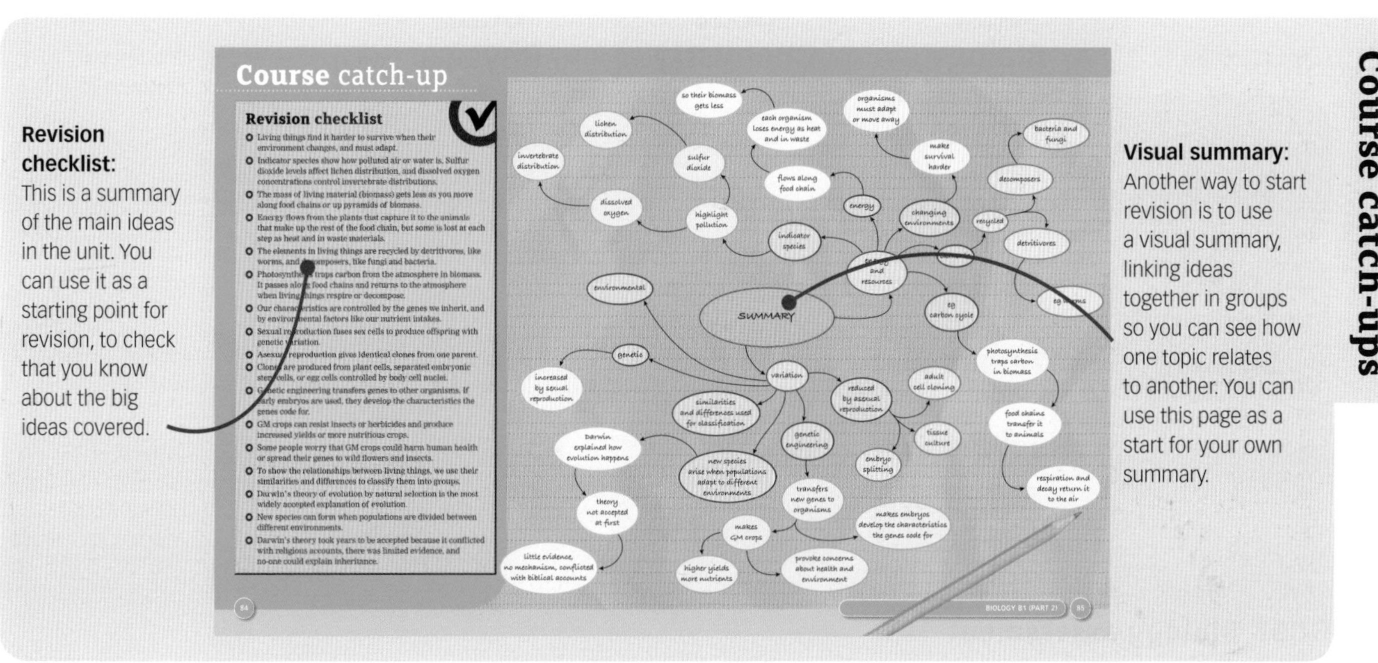

Revision checklist: This is a summary of the main ideas in the unit. You can use it as a starting point for revision, to check that you know about the big ideas covered.

Visual summary: Another way to start revision is to use a visual summary, linking ideas together in groups so you can see how one topic relates to another. You can use this page as a start for your own summary.

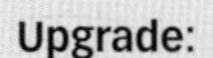

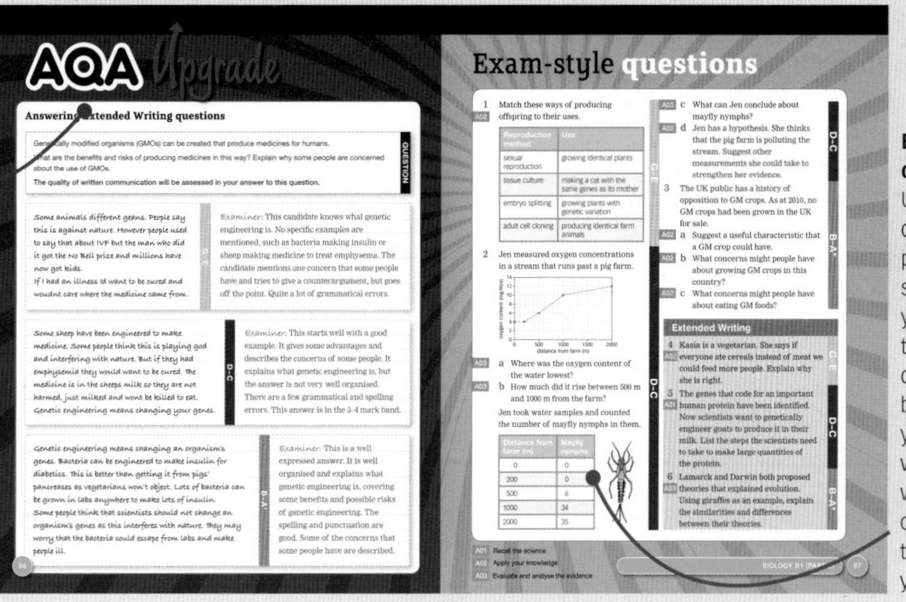

Upgrade: Upgrade takes you through an exam question in a step-by-step way, showing you why different answers get different grades. Using the tips on the page you can make sure you achieve your best by understanding what each question needs.

Exam-style questions: Using these questions you can practice your exam skills, and make sure you're ready for the real thing. Each question has a grade band next to it, so you can understand what level you are working at and focus on where you need to improve to get your target grade.

Routes and assessment

Matching your course

The units in this book have been written to match the specification for **AQA GCSE Biology**.

In the diagram below you can see that the units and part units can be used to study either for **GCSE Biology** or as part of **GCSE Science** and **GCSE Additional Science** courses.

	GCSE Biology	GCSE Chemistry	GCSE Physics
GCSE Science	B1 (Part 1)	C1 (Part 1)	P1 (Part 1)
	B1 (Part 2)	C1 (Part 2)	P1 (Part 2)
GCSE Additional Science	B2 (Part 1)	C2 (Part 1)	P2 (Part 1)
	B2 (Part 2)	C2 (Part 2)	P2 (Part 2)
	B3 (Part 1)	C3 (Part 1)	P3 (Part 1)
	B3 (Part 2)	C3 (Part 2)	P3 (Part 2)

GCSE Biology assessment

The units in this book are broken into two parts to match the different types of exam paper on offer. The diagram below shows you what is included in each exam paper. It also shows you how much of your final mark you will be working towards in each paper.

Unit		%	Type	Time	Marks available
Unit 1	B1 (Part 1) B1 (Part 2)	25%	Written exam	1 hr	60
Unit 2	B2 (Part 1) B2 (Part 2)	25%	Written exam	1 hr	60
Unit 3	B3 (Part 1) B3 (Part 2)	25%	Written exam	1 hr	60
Unit 4	Controlled Assessment	25%		1 hr 30 mins + practical	50

Understanding exam questions

When you read the questions in your exam papers you should make sure you know what kind of answer you are being asked for. The list below explains some of the common words you will see used in exam questions. Make sure you know what each word means. Always read the question thoroughly, even if you recognise the word used.

Calculate
Work out your answer by using a calculation. You can use your calculator to help you. You may need to use an equation; check whether one has been provided for you in the paper. The question will say if your working must be shown.

Describe
Write a detailed answer that covers what happens, when it happens, and where it happens. The question will let you know how much of the topic to cover. Talk about facts and characteristics. (Hint: don't confuse with 'Explain')

Explain
You will be asked how or why something happens. Write a detailed answer that covers how and why a thing happens. Talk about mechanisms and reasons. (Hint: don't confuse with 'Describe')

Evaluate
You will be given some facts, data or other information. Write about the data or facts and provide your own conclusion or opinion on them.

Outline
Give only the key facts of the topic. You may need to set out the steps of a procedure or process – make sure you write down the steps in the correct order.

Show
Write down the details, steps or calculations needed to prove an answer that you have been given.

Suggest
Think about what you've learnt in your science lessons and apply it to a new situation or a context. You may not know the answer. Use what you have learnt to suggest sensible answers to the question.

Write down
Give a short answer, without a supporting argument.

Top tips

Always read exam questions carefully, even if you recognise the word used. Look at the information in the question and the number of answer lines to see how much detail the examiner is looking for.

You can use bullet points or a diagram if it helps your answer.

If a number needs units you should include them, unless the units are already given on the answer line.

Controlled Assessment in GCSE Biology

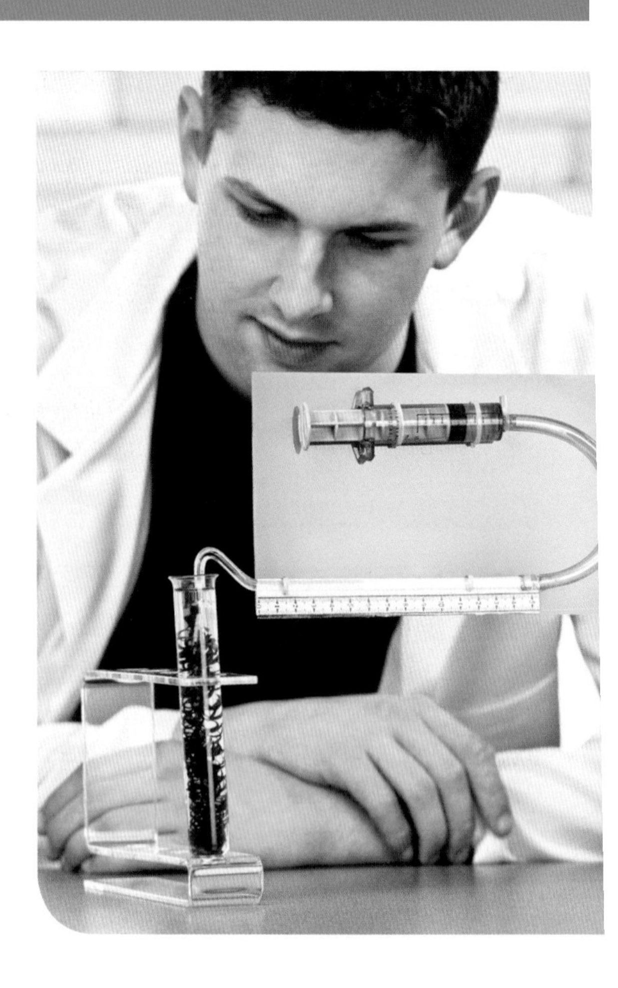

As part of the assessment for your GCSE Biology course, you will undertake a Controlled Assessment task.

What is Controlled Assessment?

Controlled Assessment has taken the place of coursework for the new 2011 GCSE Science specifications. The main difference between coursework and Controlled Assessment is that you will be supervised by your teacher when you carry out your Controlled Assessment task.

What will my Controlled Assessment task look like?

Your Controlled Assessment task will be made up of four sections. These four sections make up an investigation, with each section looking at a different part of the scientific process.

	What will I need to do?	How many marks are available?
Research	• Independently develop your own hypothesis. • Research two methods for carrying out an experiment to test your hypothesis. • Prepare a table to record your results. • Carry out a risk assessment.	
Section 1	• Answer questions relating to your own research.	20 marks
Practical investigation	• Carry out your own experiment and record and analyse your results.	
Section 2	• Answer questions relating to the experiment you carried out. • Select appropriate data from data supplied by AQA and use it to analyse and compare with your hypothesis. • Suggest how ideas from your investigation could be used in a new context.	30 marks
	Total	**50 marks**

How do I prepare for my Controlled Assessment?

Throughout your course you will learn how to carry out investigations in a scientific way, and how to analyse and compare data properly.

On the next three pages there are Controlled Assessment-style questions matched to the content in B1, B2, and B3. You can use them to test yourself, and to find out which areas you want to practise more before you take the Controlled Assessment task itself.

B1 Controlled Assessment-style questions

Hypothesis: Light affects the distribution of woodlice. More woodlice are found in dark areas.

Download the Research Notes and Data Sheet for B1 from **www.oxfordsecondary.co.uk/aqacasestudies**.

Research

*Record your findings in the **Research Notes table**.*

1. Research two different methods that could be used to test the hypothesis.
2. Find out how the results of the investigation might be useful in determining whether animals are well adapted to survive in their normal environment, and that organisms may sense and react to stimuli.

Section 1 — Total 20 marks

Use your research findings to answer these questions.

1. **(a)** Name two sources that you used for your research.
   (b) Which of these sources did you find more useful, and why? [3]
2. **(a)** Identify one control variable.
   (b) Briefly describe a preliminary investigation to find a suitable value for this variable, and explain how the results of this work will help you decide on the best value for it. [3]
3. *In this question you will be assessed on using good English, organising information clearly, and using specialist terms where appropriate.* Describe how to carry out an investigation to test the hypothesis. Include the equipment needed and how to use it, the measurements to make, how to make it a fair test, and a risk assessment. [9]
4. Use your research to outline another possible method, and explain why you did not choose it. [3]
5. Draw a table to record data from the investigation. You may use ICT if you wish. [2]

Section 2 — Total 30 marks

*Use the **Data Sheet** to answer these questions.*

1. **(a)** State the independent and dependent variables, and one control variable. [3]
   (b) Think about the time intervals for your data measurements. The smallest scale division on a measuring instrument is called its resolution. What was the resolution of the instrument you used, and was this resolution suitable for your experiment? [3]
   (c) Display the ***Group A data*** on a bar chart or line graph. *This data has been provided for you to use instead of data that you would gather yourself.* [4]
   (d) Does the ***Group A data*** support the hypothesis? Explain how. [3]
   (e) Describe the similarities and differences between the ***Group A data*** and the ***Group B data***. Suggest one reason why the results of the two groups may be different. *The **Group B data** has been provided for you to use instead of data that would be gathered by others in your class.* [3]
   (f) Suggest how your method might have helped Group A to achieve results that show a clear pattern. [3]
2. **(a)** Sketch a graph of the results in ***Case study 1***. [2]
   (b) Explain to what extent the data from ***Case studies 1–3*** support the hypothesis. [3]
   (c) Use ***Case study 4*** to describe the relationship between light and movement in woodlice. Explain how well the data supports your answer. [3]
3. The context of this investigation (the topic it relates to) is that animals are well adapted to survive in their normal environment, and that organisms may perceive and react to stimuli. Describe how your results may be useful in this context. [3]

Controlled Assessment-style questions

Overview: An increase in temperature may affect the rate of respiration. **You must develop your own hypothesis to test.** You will be provided with a culture of yeast cells, sugar solution, carbon dioxide and pH sensors, data loggers, temperature-controlled water baths, timers, rulers, and common laboratory glassware.

Download the Research Notes and Data Sheet for B2 from **www.oxfordsecondary.co.uk/aqacasestudies.**

Research

*Record your findings in the **Research Notes table**.*

1. Research two methods to find out how temperature affects the rate of respiration.
2. Find out how the results might be useful in selecting the yeast used by breweries.

Section 1 — Total 20 marks

Use your research findings to answer these questions.

1. **(a)** Name two sources that you used for your research.
   (b) Which of these sources did you find more useful, and why? [3]
2. **(a)** Write a hypothesis about how temperature may affect the rate of respiration in yeast cells.
   (b) Using information from your research, explain why you have developed this hypothesis. [3]
3. Describe how to carry out an investigation to test your hypothesis. Include the equipment needed and how to use it, the measurements to make, how to make it a fair test, and a risk assessment. [9]
4. Use your research to outline another possible method, and explain why it was not chosen. [3]
5. Draw a table to record data from the investigation. You may use ICT if you wish. [2]

Section 2 — Total 30 marks

*Use the **Data Sheet** to answer these questions.*

1. Display the ***Group A data*** on a graph. *This data has been provided for you to use instead of data that you would gather yourself.* [4]
2. **(a)** What conclusion can you draw from the ***Group A data*** about a link between temperature and rate of respiration? Use any pattern you can see in the ***Group A data*** and quote figures from it. [3]
   (b) (i) Compare the ***Group A*** and ***Group B data***. Do you think the ***Group A data*** is reproducible? Explain why. *The **Group B data** has been provided for you to use instead of data that would be gathered by others in your class.* [3]
   (ii) Explain how you could use the repeated results from Group B to obtain a more accurate answer. [3]
   (c) Look at the ***Group A data***. Are there any anomalous results? Quote from the data. [3]
3. **(a)** Sketch a graph of ***Case study 1*** results. [2]
   (b) Explain to what extent the data from ***Case studies 1–3*** support or contradict your hypothesis. [3]
   (c) Compare the ***Group A data*** to the data shown on the ***Case study 4*** graph. Explain how far the ***Case study 4*** data supports or contradicts your hypothesis. [3]
4. In brewing, yeast respires anaerobically and produces ethanol as well as carbon dioxide. A company has developed and tested a new strain of yeast, strain X, for brewing strong (high alcohol content) beer. Their hypothesis is that strain X is better than strains Y and Z.
   (a) Does the data from ***Case study 4*** support their hypothesis? Quote from the data. [3]
   (b) The tests show that strain X has a different optimum temperature for growth than the other two strains. Heating yeast uses fuel and costs money. Show how ideas from the ***Group A data*** and the ***Case studies*** could be used by brewers. [3]

B3 Controlled Assessment-style questions

Overview: There is a link between exposure of plants to direct sunlight and the number of stomata found on the undersides of their leaves. **You must develop your own hypothesis to test.** You will be provided with a microscope, microscope slides, clear nail polish, sticky tape, fine forceps, cover slips, a dropper pipette and distilled water, leaves from plants (eg ivy) grown in shade conditions, and leaves from the same plants grown in full sunlight.

Download the Research Notes and Data Sheet for B3 from **www.oxfordsecondary.co.uk/aqacasestudies.**

Research

*Record your findings in the **Research Notes table**.*

1. Research two methods to explore the link between the amount of direct sunlight to which a plant is exposed and the number of stomata found on the underside of its leaves.
2. Find out how the results of the investigation might be useful for garden centres when advising customers about their houseplants.

Section 1 — Total 20 marks

Use your research findings to answer these questions.

1. **(a)** Name the two most useful sources that you used for your research.
   (b) Explain why these sources were the most useful. [3]
2. Write a hypothesis about how exposure of plants to light or shade affects numbers of stomata on the lower surfaces of their leaves. Use your research findings to explain why you've made this hypothesis. [3]
3. Describe how to carry out an investigation to test your hypothesis. Include the equipment needed and how to use it, the measurements to make, how to make it a fair test, and a risk assessment. [9]
4. Use your research to outline another possible method, and explain why it was not chosen. [3]
5. Draw a table to record data from the investigation. You may use ICT if you wish. [2]

Section 2 — Total 30 marks

1. Display the ***Group A data*** on a graph. *This data has been provided for you to use instead of data that you would gather yourself.* [4]
2. **(a)** What conclusion can you draw from the ***Group A data*** about a link between the amount of direct sunlight to which a plant is exposed and the number of stomata found on the underside of its leaves? Use any pattern you can see in the ***Group A data*** and quote figures from it. [3]
   (b) (i) Compare the ***Group A*** and ***Group B data***. Do you think the ***Group A data*** is reproducible? Explain why. *The **Group B data** has been provided for you to use instead of data that would be gathered by others in your class.* [3]
   (b) (ii) Explain how you could use the repeated results from Group B to obtain a more accurate answer. [3]
   (c) Look at the ***Group A data***. Are there any anomalous results? Quote from the data. [3]
3. **(a)** Sketch a graph of the results in ***Case study 1***. [2]
   (b) Explain to what extent the data from ***Case studies 1–3*** support or contradict your hypothesis. Use the data to support your answer. [3]
   (c) What change could you make to the way the investigation was carried out in ***Case study 3*** to obtain data that may support the results of ***Case study 2***? [3]
4. Look at ***Case study 4***.
   (a) State three environmental conditions that the scientists kept constant when measuring the mass of the detached leaves over 48 hours. [3]
   (b) Explain how the ideas from ***Case study 4*** and from the ***Group A*** and ***Group B data*** could be used by a garden centre in customer advice on watering houseplants. [3]

Diet, exercise, hormones, genes, and drugs

AQA specification match		
B1.1	1.1	Diet and exercise
B1.1	1.2	Diet and health
B1.1	1.3	Infectious diseases
B1.1	1.4	Antibiotics and painkillers
B1.1	1.5	Immunity and immunisation
B1.2	1.6	The nervous system
B1.2	1.7	Hormones and control of the body
B1.2	1.8	How hormones control the menstrual cycle
B1.2	1.9	Using hormones to control fertility
B1.2	1.10	Controlling plant growth
B1.3	1.11	Drugs and you
B1.3	1.12	Testing new drugs
Course catch-up		
AQA Upgrade		

Why study this unit?

To stay healthy, you need to understand how your body works so you can adopt the behaviours that keep you healthy.

In this unit you will learn what you need to eat and how to exercise to stay healthy. You will find out about infectious diseases, and how your immune system and medicines can deal with them. You will also find out how your hormones play a key role in your growth and development and in helping your body to function properly.

You will learn how drugs affect your health, and what makes us different from each other.

You should remember

1. You are made of cells which are organised into tissues, organs, and systems.
2. You started as one cell which formed stem cells that then became specialised to do different jobs.
3. Your skin and stomach acids try to stop microorganisms from entering the body.
4. Your immune system tries to deal with any microorganisms that do enter the body.
5. Healthy eating, exercise, and using medicines wisely can keep you healthy.
6. Some drugs are harmful.
7. Genes in your body control your characteristics; they are inherited from your parents.
8. Genes are found on your chromosomes.

Scientists have been developing pill cameras – tiny cameras that fit into a pill capsule swallowed by a patient – since the year 2000. These give doctors incredible internal views of the entire digestive tract. The tiny devices shown here represent the next step. They are pill-sized miniature robots able to travel through the body and perform functions helping doctors with screening and diagnosis, drug delivery, and even therapeutic procedures.

Devices like those pictured can propel themselves through the body under remote control, and some even have extendable 'legs' to push aside tissue, giving doctors a better view in tight spaces within the digestive system. One day a patient may be able to swallow a number of these pills and they could join together in the stomach, forming a powerful robot that doctors could control to perform surgery wirelessly!

Learning objectives

After studying this topic, you should be able to:

- understand the body's metabolic rate and things that affect it
- know the benefits of regular exercise

Different ways of taking exercise

School athletics events give children opportunities for exercise and fun

Metabolism and metabolic rate

You are made of lots of cells, and each cell carries out lots of chemical reactions. These reactions keep you alive. They include things like respiration (which releases energy from food) and making proteins.

All of these chemical reactions in cells are collectively called **metabolism**. The rate at which they go on is your **metabolic rate**.

Measuring your oxygen consumption (use) tells you about your metabolic rate. This is because aerobic respiration, which releases the energy you need for metabolism, uses oxygen.

Your metabolic rate varies with the amount of activity you do and the proportion of muscle to fat in your body, and it may also depend on inherited factors.

A What is metabolism?

B What is metabolic rate?

Activity and metabolic rate

When you exercise:

- Your muscles contract more to move your limbs.
- Muscles need energy to contract.
- So muscle cells need to respire more to release more energy from glucose.
- To provide this glucose, you need to eat more food.
- Exercise increases the amount of energy expended by the body.

People who exercise regularly and often are rarely overweight. They are usually healthier than people who take little exercise. They do not have the health risks linked to being overweight.

C Why does exercise increase the body's need for energy?

The proportion of muscle to fat in your body

Fat cells store fat. They are inactive and have a low rate of respiration to ensure that they do not use their stored fat.

Muscle cells are active. They need energy from respiration to contract. Their rate of respiration is high.

If you have more muscle and less fat, then your body has a higher metabolic rate than someone of the same size who has less muscle and more fat. You need to eat more food to supply the energy needed for the muscle cells.

Males have more muscle and less fat in their bodies. Females have less muscle and more fat. This is largely determined by genes, although females can increase their muscle mass by exercise.

Most males need to eat more than most females.

D Explain why men need to eat more food than women.

Inherited factors and metabolic rate

- Some people have higher metabolic rates than others.
- Tall people have higher metabolic rates, as they lose more heat from their body surface.
- Contrary to popular belief, overweight people have a higher metabolic rate than slimmer people, as their bodies are larger and need more energy to work.
- People with an underactive thyroid gland have lower metabolic rates.

Questions

1 Describe six ways that regular exercise can improve your health. (E)

2 People under the age of 20 years have higher metabolic rates than those over 20. Why do you think this is? (C)

3 List other inherited factors, not mentioned in the list above, that can affect metabolic rates. (A*)

Woman doing weight training in a gym. This increases her muscle mass.

Key words

metabolism, metabolic rate

Did you know...?

People who exercise use up more food and are usually slim. They are at less risk of heart disease, cancer, diabetes, and arthritis. They also make chemicals in the brain that make them feel good and happy. Exercise also strengthens the immune system so they do not get many infections. However, too much exercise can harm joints and weaken the immune system.

Exam tip AQA

✔ You need to mention 'and often' as well as 'regularly' when describing how people should take exercise. After all, once a year is regular!

Learning objectives

After studying this topic, you should be able to:

- know what makes a healthy diet
- understand that inherited factors also affect health

Key words

balanced diet, deficiency disease, obese, statins

In a healthy, balanced diet the largest part should be carbohydrates. The next layer of the pyramid is fruit and vegetables. These provide many vitamins and minerals and fibre. The next layer is protein foods. Finally, at the top of the pyramid are fats and oils, and sweets, which should be eaten sparingly.

What is a balanced or healthy diet?

A **balanced diet** contains the right amount of different foods and the right amount of energy to keep you healthy.

Food/nutrient	Why you need to eat it
carbohydrates	for energy
fats	for energy
proteins	to build cells and repair tissue
mineral ions and vitamins	needed in small amounts to keep the body functioning healthily

You also need water and fibre.

- Your cells contain about 70% water and you lose it in sweat, breath, tears, faeces, and urine.
- Fibre helps prevent constipation and bowel cancer.

A What is a balanced diet?
B Why do you need to eat carbohydrates, fats, and proteins?
C Why do you need to eat vitamins and minerals?
D Why do you need to drink water?

Malnourishment

Some people do not eat a balanced diet. They eat the wrong amount or wrong types of food. A person is malnourished if their diet is not balanced.

Being malnourished (*mal* means 'bad') may lead to being too fat or too thin. It may also lead to a **deficiency disease**. Rickets is a deficiency disease caused by a lack of vitamin D in the diet. Children with rickets have soft bendy bones that can become deformed. They may have a curved spine or enlarged skull.

Underweight

Some people do not eat enough, or exercise too much. If the energy content of the food you eat is less than the energy you use, you lose body mass. This makes people underweight and can cause health problems. Very thin girls and women have irregular periods.

Overweight

If people eat too much they become overweight. If they are very overweight they are **obese**. Being overweight causes health problems such as type 2 diabetes.

However, good health is influenced by many factors, and diet is just one of them. Inherited factors also play a part. Genes (inherited factors) can affect cholesterol levels.

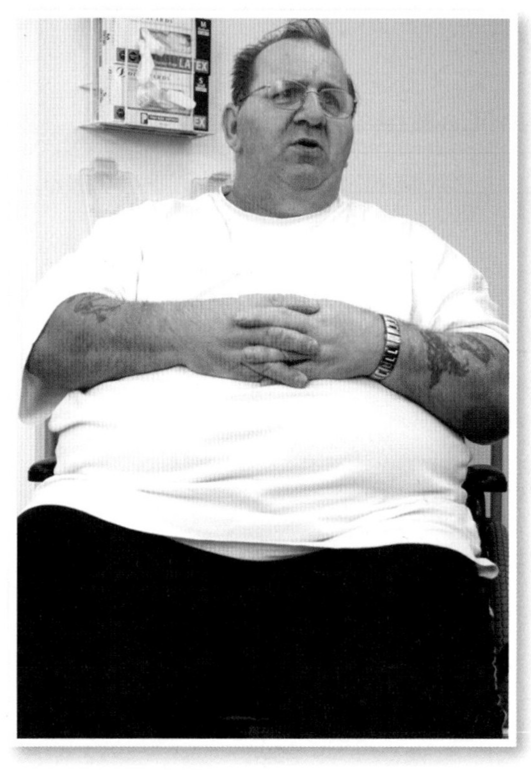

▲ Obesity

Cholesterol also affects our health

Cholesterol is a waxy substance. It is only present in animal tissues. You need some because:

- It helps strengthen your cell membranes.
- Your body uses it to make your sex hormones, and to make vitamin D.

However, having too much cholesterol in your blood increases your risk of heart disease.

You get some cholesterol directly from eating animal products like meat, prawns, and eggs. Your liver also makes it from saturated fats that you eat, like butter, chocolate, and fatty meat.

Some people have a lot of cholesterol in their blood because of their genes. Their livers make too much cholesterol. They can take drugs called **statins** to reduce the cholesterol made by the body. They take the statins in the evening because the body makes cholesterol during sleep. Statins inhibit an enzyme involved in making cholesterol.

Did you know...?

There are one billion obese people in the world. There are also one billion people who do not get enough to eat.

Questions

1 State one health problem linked to being underweight.
2 What does 'obese' mean?
3 What are statins?

E

4 Why do you need some cholesterol in your body?
5 How do you get cholesterol?

C

6 Explain how statins work.

A*

Learning objectives

After studying this topic, you should be able to:

- state that microorganisms that cause infectious diseases are called pathogens
- know that bacteria and viruses reproduce rapidly inside the body
- know that some bacteria produce toxins

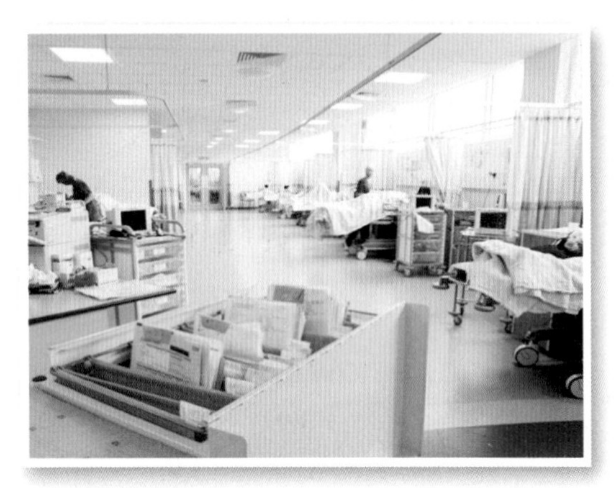

In the mid-twentieth century antibiotics were used to kill bacteria. Unfortunately some bacteria developed resistance to the antibiotics so hospital-acquired infections are now difficult to treat.

Bacteria. Some are rod shaped (shown here as grey), some are round (red), and some are spiral (cream).

Stopping infections spreading

Before the nineteenth century, people did not know about **microorganisms**. Diseases were often passed on from one person to another because doctors did not wash their hands.

A Hungarian doctor called Ignaz Semmelweis was the first to realise that washing hands helps prevent the spread of disease. He noticed that the number of deaths in one hospital maternity ward was higher than in another. At this time nobody knew about germs, but when he investigated he found that on the ward with more deaths the doctors went straight from dissecting bodies to delivering babies. He asked doctors to wash their hands between tasks and the number of deaths reduced significantly.

Hospitals today have codes of practice and strict hygiene regimes for staff, patients and visitors. There are alcohol-based hand gels at the entrance to each ward and by each bed.

- Visitors should clean their hands before entering a ward and when leaving a ward.
- Nurses should clean their hands after dealing with one patient and before going to the next patient.
- Patients should also regularly clean their hands.

Hands spread many microorganisms because we touch:

- food
- door knobs
- surfaces
- each other.

However, some people are not as conscientious as they should be about using the hand gels.

A Why should visitors clean their hands before entering a ward and when leaving a ward?

B Why should nurses clean their hands after dealing with a patient and before dealing with the next patient?

Pathogens

Any microorganism that can cause an infectious disease is called a **pathogen**. Some **bacteria** are pathogens. All **viruses** are pathogens.

Bacteria

Not all bacteria are pathogens. We have millions of them on our skin and in our gut and we could not live without them. However, if they get through our gut wall or skin and into our blood or cells they can make us very ill. They reproduce rapidly inside our warm bodies and some produce **toxins** (poisons) that make us feel ill. Some bacterial toxins may cause death.

Viruses

It is debatable whether viruses are living or not. They are not made of cells. They cannot carry out any life processes. They have to insert themselves into a host cell and hijack that cell's parts to make copies of the virus. These new virus particles can then burst out of the cell and infect many other host cells in a 'chain reaction'. Viruses damage and destroy our cells when they infect us in order to reproduce. Viruses are much smaller than bacteria.

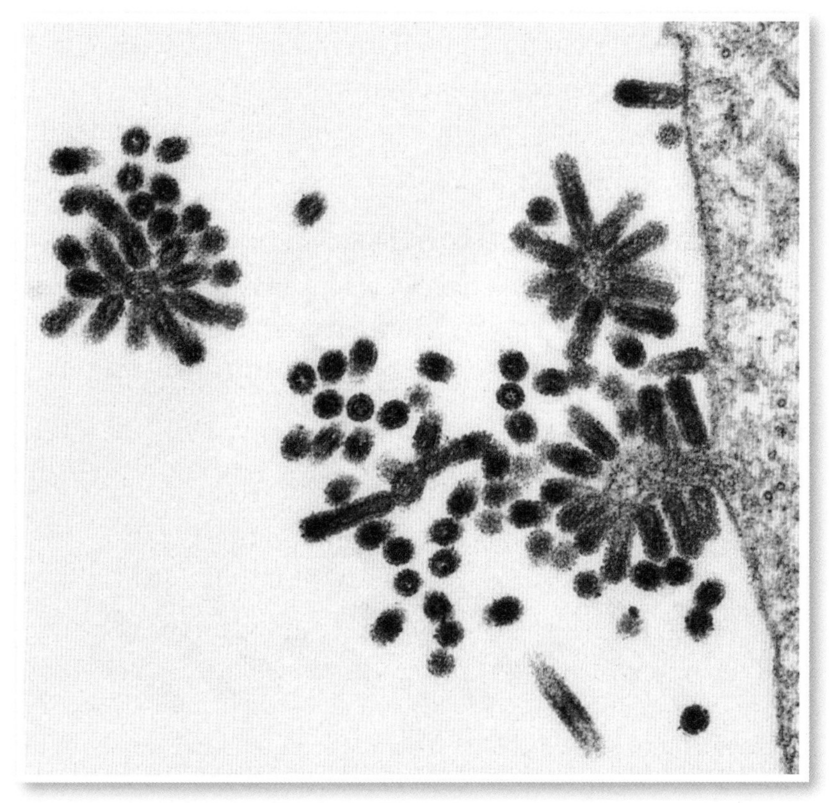

▲ Coloured transmission electron micrograph (TEM) of influenza viruses, shown in red, that have reproduced inside a cell of the respiratory tract and are breaking out of the cell. They have damaged the host cell. (× 150 000).

Key words

microorganism, pathogen, bacteria, virus, toxin

Did you know...?

Although all viruses are pathogens, they don't all infect humans. Some infect animals or plants and some live in the sea and infect algae. This is useful to the environment as it keeps the algal growth in check. Some viruses attack bacteria and enter their cells.

Exam tip AQA

✔ Remember that viruses cannot produce toxins. They are not really alive and do not have any metabolic processes or structures for making proteins.

Questions

1 What are pathogens?
2 Do viruses infect other species besides humans?
E
3 How do viruses damage our cells?
4 Why is it hard to say whether viruses are truly living?
C

Learning objective

After studying this topic, you should be able to:

- ✔ understand the roles of painkillers and antibiotics in treating diseases

A Why is pain useful?

B What are the symptoms of many infectious diseases?

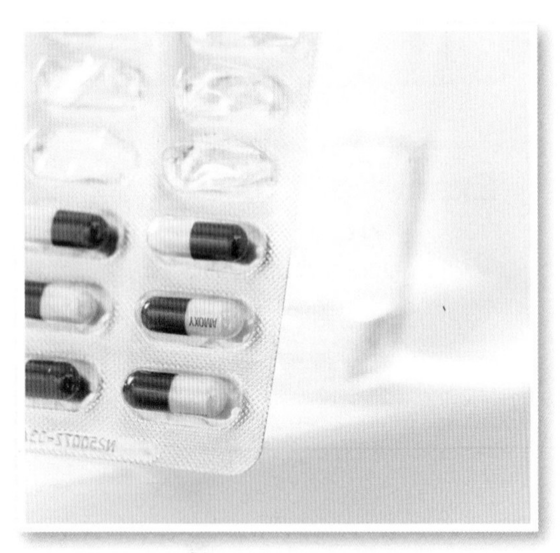

▲ Amoxicillin antibiotic pills. Amoxicillin is a type of penicillin.

C What are antibiotics?

D Name the process by which strains of bacteria develop resistance to antibiotics.

Why do you take painkillers when you have an infection?

Pain is useful because it tells you that something is wrong. Many infections give us a headache and aching muscles. We may also get a fever (high temperature) or feel shivery. However, we all want to carry on with our lives while we get better, so relieving the painful **symptoms** is useful. **Painkillers** such as paracetamol or codeine can do this. They do not kill the pathogens but make us feel better while our immune system or antibiotics (or both) kill the pathogens.

Antibiotics

Antibiotics have been available since the 1940s and have saved millions of lives. Antibiotics are medicines that can kill infective bacteria inside the body. They can help cure diseases caused by bacteria, such as chest or ear infections, TB, or blood poisoning. Penicillin is an antibiotic. It does not kill all types of bacteria. Since it was discovered, scientists have found many more types of antibiotics. Specific antibiotics target specific bacteria. Using the appropriate antibiotic can help reduce the spread of antibiotic resistance.

Resistance to antibiotics

Many strains of bacteria have developed **resistance** to various antibiotics. This resistance has developed as a result of chance (spontaneous) mutation and **natural selection**.

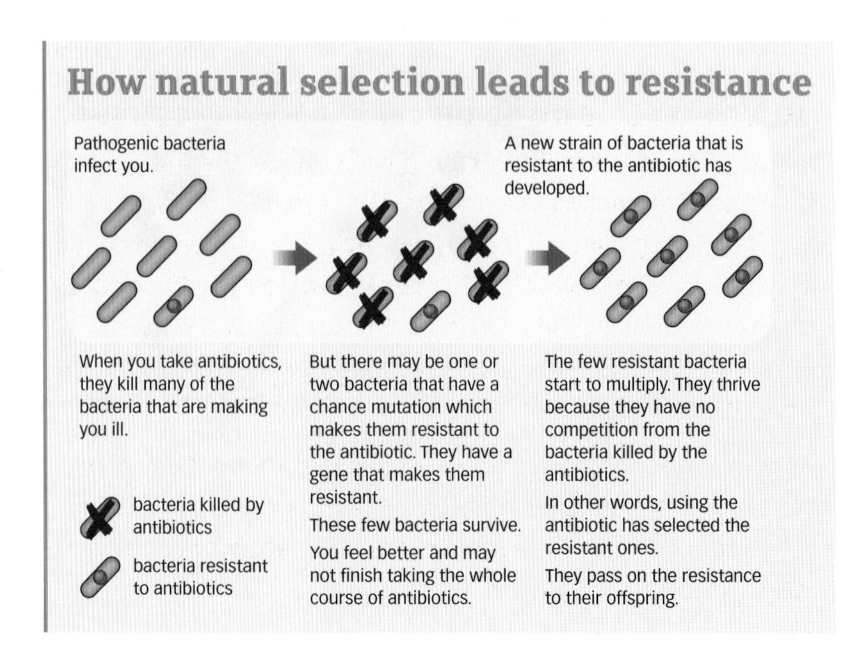

How resistance to an antibiotic develops in bacteria. A new strain of bacteria that is resistant to the antibiotic has developed. Some of these bacteria may infect other people and, when these people are given the antibiotic, it does not work. ▶

MRSA

One of the most troublesome strains of bacteria that has developed resistance to antibiotics is MRSA. Its full name is methicillin-resistant *Staphylococcus aureus*. Methicillin is a very strong antibiotic only used in bad cases of infection, but this bacterium is resistant to even that. It causes many hospital-acquired infections each year and many deaths.

This bacterium lives on our skin and in our noses where it does no harm. But if your skin is not clean when you have an operation and the bacterium gets into a wound, it can cause a dangerous infection which could kill you. However, this bacterium can easily be killed with antiseptics and disinfectants, so it is important that hospitals practise good hygiene.

How to prevent further resistance arising

In the 1940s and 1950s, when antibiotics were new, many doctors prescribed them when they were not really needed. Patients did not always take the whole course of the drugs, but stopped when they began to feel better. This overuse of antibiotics increased the rate at which antibiotic resistance developed.

Nowadays, doctors are aware of the dangers of overusing antibiotics. They do not prescribe them for non-serious infections like sore throats. This is to slow down the rate of resistant strains of bacteria being produced.

Antibiotics cannot be used to kill viruses

Each type of antibiotic interferes with one of the life processes that bacteria carry out. For example, some antibiotics stop bacteria making their cell walls and some stop them making proteins. However, antibiotics do not kill viruses because viruses do not carry out any life processes. Viruses get inside your cells and force your cells to make more copies of them.

It is difficult to make a drug that kills viruses without also damaging your cells, because viruses have to be inside your cells to reproduce. There are drugs called antivirals that stop your cells making copies of viruses.

▲ Wound infected with MRSA. The spread of MRSA can be prevented by doctors and nurses washing their hands between treating patients.

Key words

symptoms, painkiller, antibiotic, resistance, natural selection

Exam tip

✔ Remember that it is the bacteria that can become resistant to antibiotics, not the patient.

Questions

1. Why do people take painkillers when they have an infection? (E)
2. Why is MRSA such a problem?
3. How can doctors and patients prevent further antibiotic resistance happening? (C)
4. Explain why antibiotics cannot be used to kill viruses.
5. Why is it difficult to make drugs which cure viral infections? (A*)

Learning objectives

After studying this topic, you should be able to:

- ✔ describe how the immune system deals with pathogens
- ✔ know that people can be immunised against some diseases

Key words

immune system, **antibody**, **antigen**, **immunity**, **immunisation**, **epidemic**, **pandemic**

A How does your body stop pathogens entering it?

Did you know...?

Your immune system also protects you from cancer. It can recognise and kill cancerous cells.

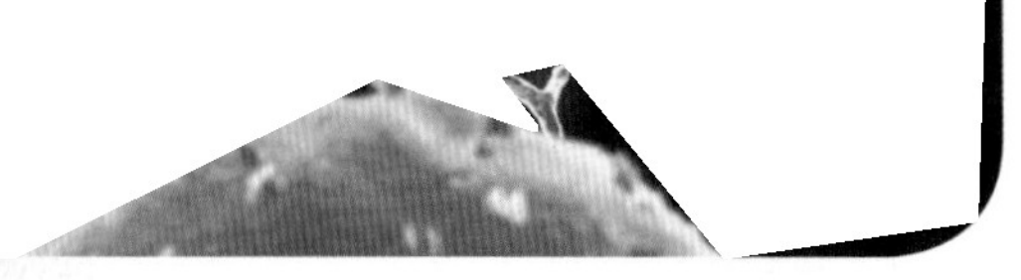

Antibodies surrounding a virus particle

How your immune system deals with pathogens

Your body has barriers to stop pathogens entering it. These include

- skin
- tears
- blood clotting when you cut yourself.
- stomach acid
- mucus in the airways

However, sometimes some pathogens manage to enter your body. When they do, your **immune system** may deal with them. This involves your white blood cells.

White blood cells

You have different sorts of white blood cells.

- Phagocytes engulf (ingest) the pathogens.
- Lymphocytes produce antibodies or antitoxins.

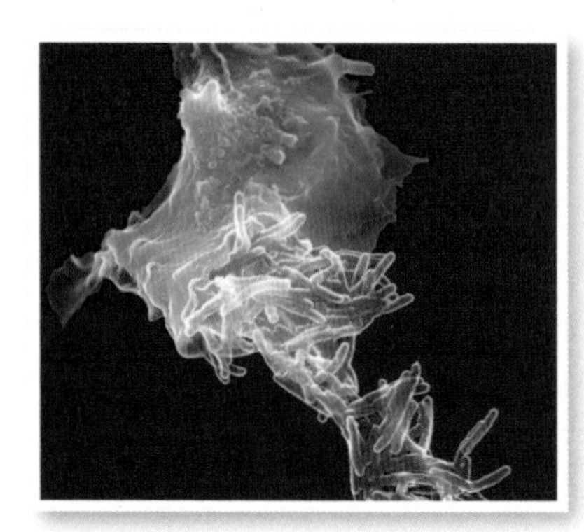

Coloured scanning electron micrograph (SEM) of a phagocyte ingesting TB bacteria (× 3500)

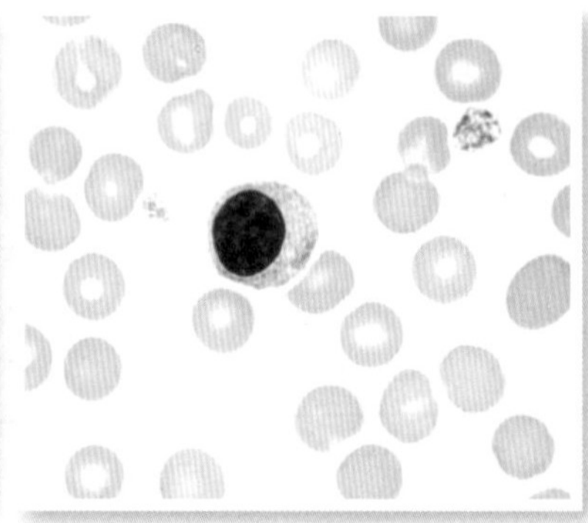

Light micrograph of blood showing lymphocyte (stained purple) and red blood cells. Lymphocytes are white blood cells that make antibodies (× 400).

Antibodies

Antibodies are proteins. Each type of antibody can destroy a particular type of bacterium or virus. This is because:

- Each type of pathogen has particular **antigens** (proteins) with a specific shape on its surface.
- Each type of antibody (also a protein) has a particular shape and can lock on to a particular antigen.
- Your immune system makes the right sort of antibodies to lock on to the antigens of the particular pathogen that is in your body.
- Once the pathogen is coated with antibodies, white blood cells can ingest and kill the pathogens.

Once you have recovered from an infection, you have **immunity** to it. If that same pathogen enters your body again, your body makes antibodies so quickly that the pathogen is destroyed before it makes you feel ill.

B Describe two ways that white blood cells can kill pathogens.

C Explain how you become immune to a disease such as measles.

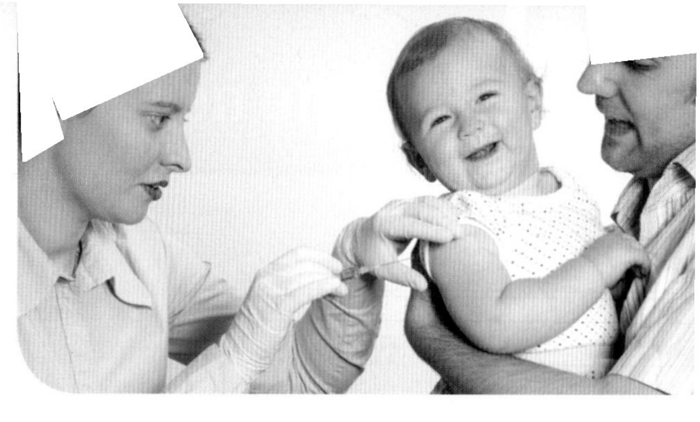

▲ This baby is having an MMR vaccination to protect her against measles, mumps, and rubella

D Draw a flow diagram to show how immunisation works.

Immunisation

Many people used to die from infections like TB and smallpox. In the developing world today many die from measles, malaria, and cholera. **Immunisation** can make people immune to a disease, without them having the disease. Widespread immunity in a population reduces the spread of infection within it.

Here's how it works:

- A small amount of dead or inactivated pathogen is introduced into your body – usually by injection. This is called being vaccinated. The dead or inactivated pathogens still have the antigens on their surface.
- Some of your white blood cells recognise these antigens on the pathogens and respond to them by making antibodies.
- If, later on, the live pathogens get into your body, your white cells quickly make the right sort of antibodies.
- These antibodies destroy the pathogens before they can make you ill.

How do mutations affect vaccines?

Some viruses, like the flu virus, mutate often. This causes them to have slightly different antigens. Your immune system does not recognise these viruses, so they can make you ill again with flu, even though you may have had flu before. So every year new vaccines are made for the new strains of flu that are likely to infect people that year.

In 2009 many people in many countries were immunised against swine flu to prevent the disease sweeping across countries and causing an **epidemic**, or across continents and causing a **pandemic**.

Exam tip AQA

✔ Do not confuse the terms 'immunity' and 'resistance'. People become immune to infectious diseases because they have an immune system. Bacteria (not people) may be resistant to antibiotics.

Questions

1 Describe one way that pathogens can enter your body.

2 What is (a) an antigen, and (b) an antibody?

↓ E

3 Describe two ways that you can become immune to mumps.

4 Explain how mutations in viruses can lead to epidemics.

↓ C

5 Healthcare workers are vaccinated each year against flu. Why do you think this is?

↓ A*

Learning objectives

After studying this topic, you should be able to:

- ✔ understand the role and organisation of the nervous system
- ✔ know that receptors detect stimuli
- ✔ know that nerve impulses pass from receptors along neurones
- ✔ recall how reflex actions come about

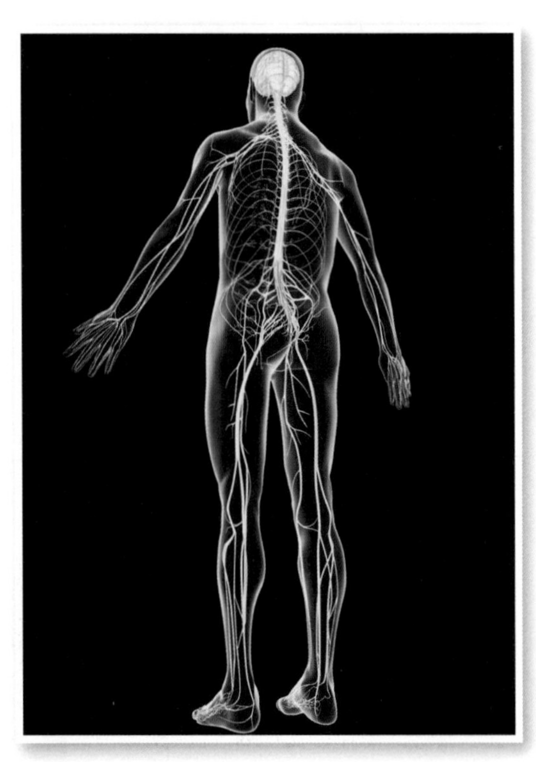

▲ The nervous system is made up of the central nervous system (CNS) and the peripheral nervous system.

Key words

stimulus, central nervous system, peripheral nervous system, receptor, neurone, reflex action, synapse, effector

Why you need to respond to change

You, and all living things, need to be able to respond to changes in the environment. These changes are called **stimuli**. If you could not detect and respond to stimuli you would not be able to find food or avoid danger. You also need to learn from your experiences and coordinate your behaviour.

The structure of the nervous system

There are two main parts to the nervous system:

- The **central nervous system** (CNS) – the brain and spinal cord.
- The **peripheral nervous system** – nerves taking information from sense organs into the CNS, and nerves taking information from the CNS to effectors (muscle or glands).

A What is a stimulus?

B Why do we need to be able to detect stimuli?

C List four stimuli that your skin can detect.

Sense organs or receptors

Receptor cells are special cells adapted to detect stimuli. Like most animal cells they have

- a nucleus
- a cell membrane
- cytoplasm.

Information from the receptors passes as electrical impulses. It travels along nerve cells called **neurones** to the brain. The brain then coordinates the response. Some responses are voluntary – they are consciously controlled by the brain. For example you may hear part of a song on the radio and decide whether to listen and turn up the volume or to switch off the radio.

Reflex actions

Sometimes you need to respond to a potentially dangerous situation very quickly. For example, if you touch a hot object, you need to quickly withdraw your hand before it burns.

There is no time to think, so the brain does not need to be involved. Instead the response is coordinated by the other part of the CNS, the spinal cord. These responses are called **reflex actions**. They are fast, automatic, and protective.

Each reflex action follows the pathway: stimulus → receptor → sensory neurone → relay neurone → motor neurone → synapses → effector → response.

This pathway is described as a reflex arc.

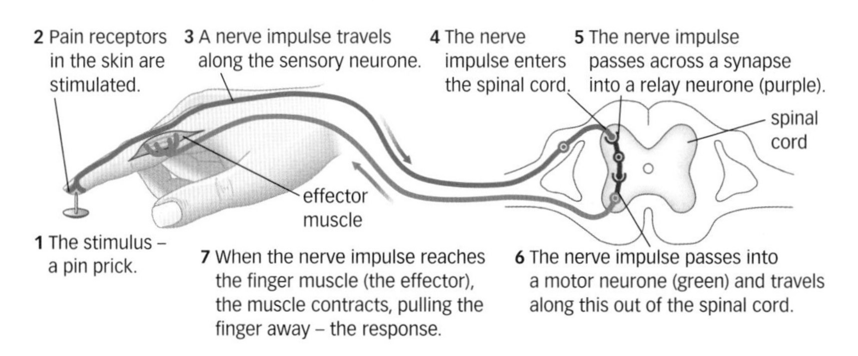

▲ A reflex arc. The impulse goes from receptor to CNS and then to effector to bring about the response. The relay neurone inside the spinal cord coordinates the response by connecting the sensory neurone to an appropriate motor neurone. The information travels from one neurone to another across a small gap called a **synapse**.

Effectors

Effectors are glands or muscles that carry out a response.

- A muscle responds by contracting.
- A gland responds by secreting chemical substances.

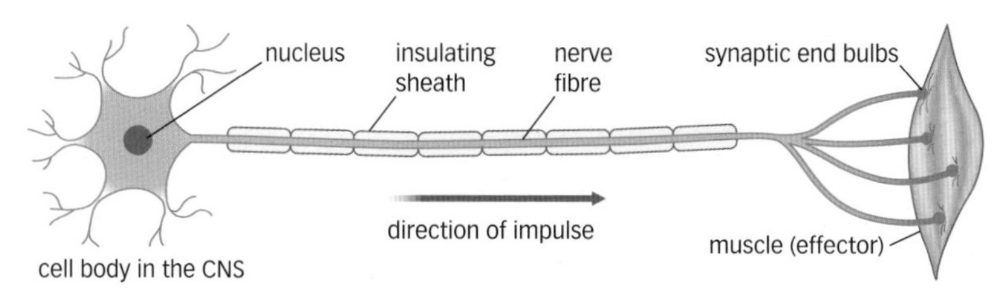

▲ Structure of a motor neurone. The nerve impulse is carried along the nerve fibre.

> Neurones are adapted to their functions because they are long, have an insulating sheath to prevent impulses leaking away, and have branched endings so they can communicate with many other neurons.

Examples of reflex actions include:

- the knee-jerk reflex
- withdrawing a hand from a hot plate.

▲ The knee-jerk reflex test. When the leg is tapped just below the knee, the leg straightens. This reflex is used when we walk.

Did you know...?

The pupil-shrinking reflex does involve the brain, but not at the conscious level, so it is still very quick and automatic.

Exam tip AQA

✔ When you are writing about nerve transmission, always use the technical term 'impulses'. Don't write about 'signals' or 'messages' or you will lose marks.

Questions

1 What are voluntary responses?

2 Why are reflexes automatic and rapid?

E

3 How does the pupil-shrinking reflex protect the eye?

4 Which part of the neurone does the impulse travel along?

C

5 What is the function of the insulating sheath?

A*

Learning objectives

After studying this topic, you should be able to:

- ✔ know that your internal conditions are controlled
- ✔ know that many processes in the body are coordinated by hormones

Key words

hormone, secrete, gland, target organ, ion

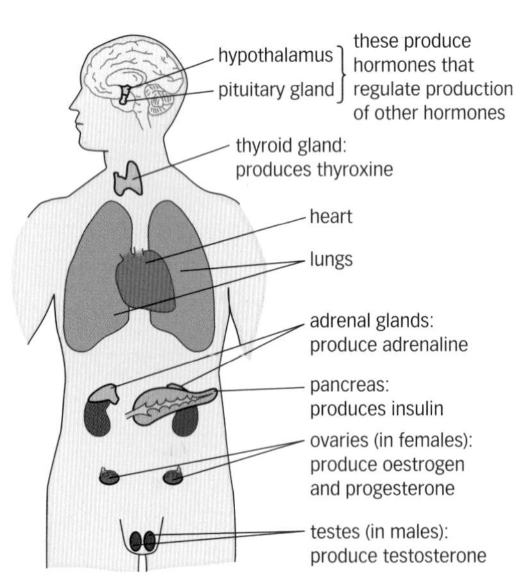

▲ Some of the main glands and the hormones they produce

> **A** Write down four internal conditions that have to be controlled.
>
> **B** How does your body lose water?

Internal conditions are controlled

Many things inside your body are controlled. Their levels are kept within a narrow acceptable range. In this section you will learn about

- the water content of the body
- the ion content of the body
- body temperature
- blood sugar levels.

Hormones

Hormones are chemicals. They

- regulate the functions of many organs and cells
- coordinate many processes in the body
- are produced (**secreted**) from **glands** into the bloodstream
- travel in the blood to **target organs**.

The body usually reacts slowly to hormones. Hormones coordinate long-term body changes such as maturing (growing up).

However, some hormones act quickly. One hormone that acts quickly is made in part of the brain and acts on the kidneys to regulate the water content of the blood.

How the body loses and gains water

You, like all living things, contain about 70% water in your cells. Your cells need this water, otherwise their chemical reactions cannot go on properly. Although your skin is waterproof, you lose water from the:

- skin in sweat
- lungs by breathing out
- kidneys in urine.

The amount of water in your body has to be balanced. You gain water from:

- drinks containing water
- food that contains water
- respiration of digested food.

Your kidneys help regulate the water content of your body. A hormone, made in part of the brain, helps the kidneys to only pass out water when you have too much.

Blood sugar level

When you eat and digest food, the products of digestion pass into your bloodstream. They are then carried to cells. Sugar is delivered to cells so they can get a constant supply of energy. Your blood sugar level has to be controlled so there is never too much or too little. Hormones regulate the blood sugar level.

The ion content of the body

Your blood and the watery fluid in between all your cells contain **ions** (sometimes called electrolytes). These include sodium ions, potassium ions, magnesium ions, calcium ions, and hydrogen ions. They are all very important in helping your nerves to work and keeping the body fluids at the right pH. Hormones work with the kidneys to make sure the balance of these ions is right. You may pass out unwanted ions in urine.

Temperature

Your body temperature has to be maintained at around 37 °C. This is warm enough for the chemical reactions in cells to happen quickly enough to keep you alive. The enzymes that control these chemical reactions work well at this temperature. If your body temperature was much higher, some enzymes and many of the other proteins in your body would stop working.

The heat made during respiration keeps your body warm. When you exercise, your muscle cells are working harder and need to respire more. This makes more heat. Your body gets rid of the heat by sweating. You also lose heat when you breathe out warm air, and you lose it from the warm blood flowing near the skin surface.

Did you know...?

Some animals, like desert rats, are well adapted to living in dry places. They do not drink but get all of their water from food. They do not sweat but keep cool by finding shade.

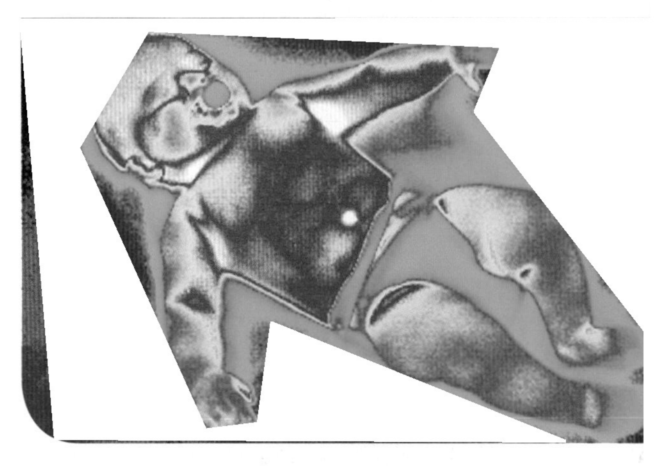

▲ Thermal image of a baby. The hottest parts look white. Then the scale goes from red, yellow, green, blue, to purple (coolest).

Exam tip AQA

✔ If you are asked what a hormone is, give as much information as possible. Don't just say it is a chemical. Say where hormones are made (glands) and that they travel in the blood to target organs, to coordinate body processes.

Questions

1 What are hormones?

2 Name four ions in your body fluids.

E

3 Why do you need to have the right amounts of ions in your body fluids?

4 How does your body get rid of unwanted (excess) ions?

C

5 How does your body (a) generate heat; (b) lose heat?

6 Explain why your body temperature needs to be maintained within narrow limits at around 37 °C.

A*

8: How hormones control the menstrual cycle

Learning objectives

After studying this topic, you should be able to:

- know that several hormones are involved in the menstrual cycle
- describe the role of FSH and oestrogen in the menstrual cycle
- explain how oestrogen and progesterone can be used as contraceptives

Did you know...?

At puberty males produce small amounts of oestrogen and progesterone and females produce small amounts of the male sex hormone, testosterone. In females testosterone causes pubic hair and hair under the arms to grow.

A Where is FSH made in the body?

B What does FSH do?

C When during the menstrual cycle is the level of FSH highest?

Growing up

At puberty, children's bodies begin to change into those of sexually mature adults. This change takes several years. The first thing that happens is that the ovaries in females and testes in males develop and begin to produce the **sex hormones**.

In females the sex hormones **oestrogen** and **progesterone** are made in the ovaries. These hormones are involved in the **menstrual cycle**.

The menstrual cycle

At puberty females begin to have a menstrual period each month. Several hormones help coordinate the menstrual cycle:

- The pituitary gland in the brain produces a hormone called follicle stimulating hormone (**FSH**).
- FSH causes eggs in the ovaries to mature, one each month.
- It also stimulates the ovaries to make oestrogen**.**
- Oestrogen stimulates the pituitary gland to make luteinising hormone (**LH**), which triggers the release of an egg (**ovulation**) from the ovary.
- Oestrogen also prevents more FSH being secreted, and it repairs the uterus lining after menstruation.
- Progesterone maintains the uterus lining, and works with oestrogen to prevent secretion of FSH.

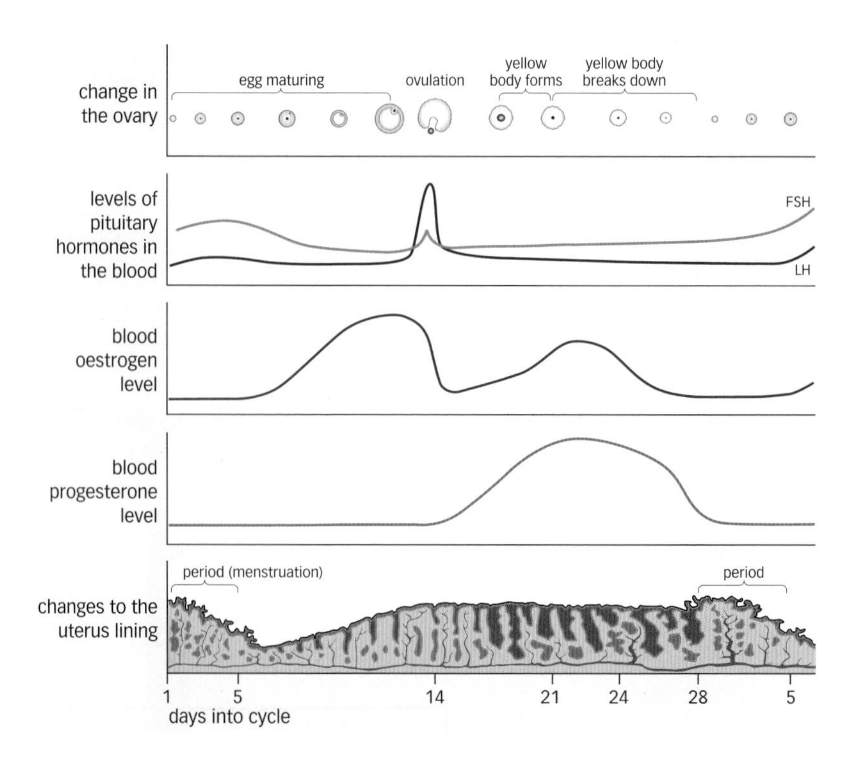

Events in the menstrual cycle

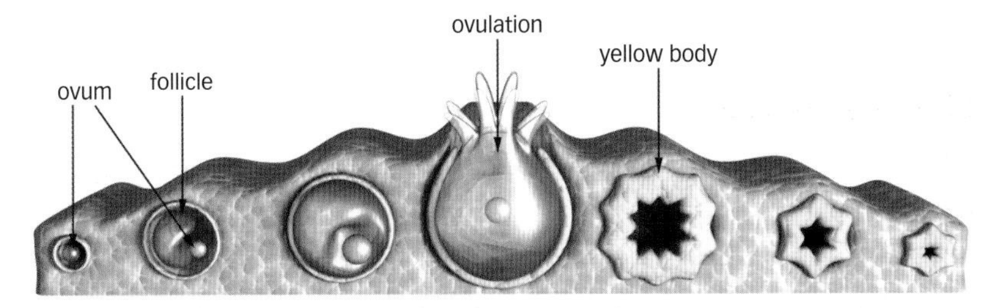

Changes in the ovary during the menstrual cycle. The egg develops in a follicle. It then bursts out of the follicle. The empty follicle develops into a yellow body which makes a hormone, progesterone. This stops menstruation.

If the egg is not fertilised, at the end of the cycle the uterus lining passes out of the body. This is the period. If the egg is fertilised then the uterus lining stays so that the baby can develop.

How the female sex hormones control fertility

The female sex hormones can be used to make a woman less fertile, if she does not want to become pregnant.

During pregnancy both oestrogen and progesterone levels are high and they inhibit FSH production from the pituitary gland. This prevents the development and release of any more eggs.

Scientists realised that if women took these hormones in a daily pill, the high levels in the body would prevent ovulation. Without ovulation women cannot become pregnant. These hormones are used in **contraceptive** pills.

The first birth-control pills contained high amounts of oestrogen. They prevented ovulation but many women suffered from side-effects. Birth-control pills now contain a much lower dose of oestrogen and some progesterone to inhibit egg production. These give fewer side-effects. Some birth-control pills contain only progesterone.

Some women may not conceive easily and want to improve their fertility. FSH may be used to help them conceive.

Key words

sex hormones, oestrogen, progesterone, menstrual cycle, FSH, LH, ovulation, contraceptive

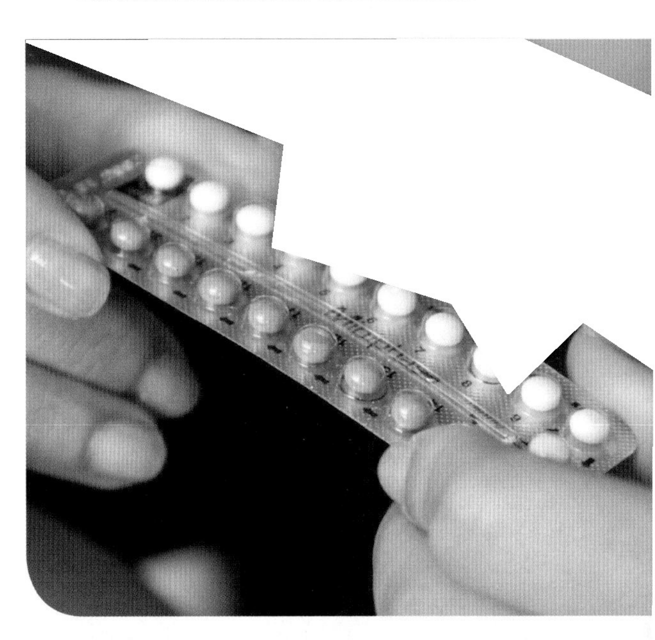
Each pack contains enough contraceptive pills for one month. They are usually taken for 21 days of each month and then not taken for 7 days, so the woman has a period.

Questions

1 State two functions of oestrogen in controlling the menstrual cycle.

E

2 What is the function of progesterone?

3 During which part of the menstrual cycle is a woman most likely to conceive?

C

4 Explain how the oestrogen in contraceptive pills prevents pregnancy.

A*

Learning objectives

After studying this topic, you should be able to:

- know that hormones can be used in in vitro fertilisation (IVF) to control fertility
- evaluate the benefits and problems of using hormones to control fertility

Robert Edwards, British IVF pioneer

Key words

in vitro fertilisation

Test tube baby

You have seen how female hormones are used in contraceptive pills. They have also been used to help infertile women have babies.

In 1978 in the UK, the first 'test tube baby' in the world was born. She is Louise Brown. Her parents, John and Lesley, had been trying to conceive a baby for nine years, but Lesley had blocked oviducts. The procedure that helped them conceive was developed by two British doctors, gynaecologist Patrick Steptoe and physiologist Robert Edwards.

Louise developed inside her mother's womb, not in a test tube. However, the egg was fertilised in a glass dish. The Latin word for glass is '*vitro*' so this is where the term **in vitro fertilisation** (IVF) comes from.

How IVF works

Doctors collect eggs from the woman. Before they do that they inject her with FSH and LH. This causes her ovaries to make more than the usual one egg during her menstrual cycle. The eggs are then collected from the mother and fertilised by mixing them with the father's sperm in a glass dish. To make the procedure more likely to work, sperms are selected and one is injected into each egg. This is done under a powerful microscope.

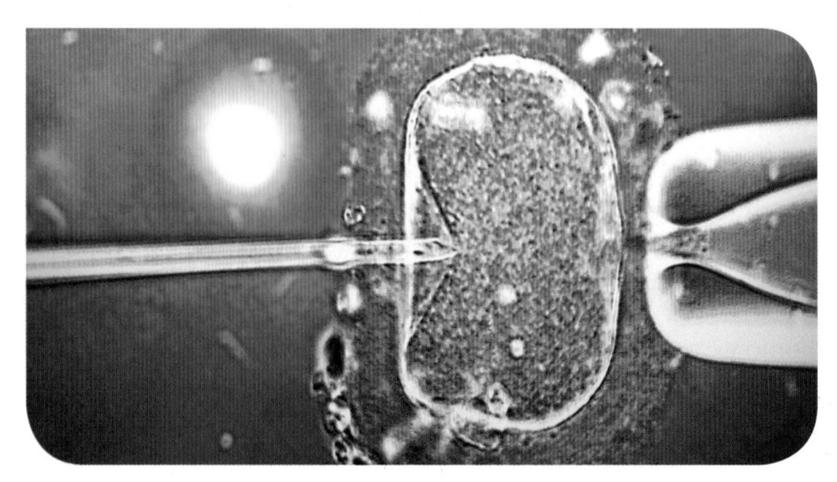

Human sperm being injected into a human egg

The fertilised eggs begin to develop into embryos. When they are tiny balls of cells, two are chosen and inserted into the mother's womb.

Benefits and problems of using hormones to control fertility

Benefits	Problems
• Couples who are infertile can be treated and have their own children. • The embryo can be tested for any genetic disorders before being implanted. • People can choose when to have children. • They may wish to delay having children whilst developing their career. • They may want to wait until they can better afford to look after the children. • People can limit the size of their family. • Some people may want to choose not to have children.	• Some religions feel that it is wrong for humans to exercise control over their fertility. • Some people feel IVF is wrong, as some of the embryos made are not allowed to develop into people. • Some people think that infertile couples should adopt instead, as there are many orphaned children without homes. • IVF is expensive and is not always successful. • People may delay having their families for too long and then find that it is difficult to conceive. • Some women may suffer side-effects of taking contraceptive pills, such as mood swings, weight gain, or feeling sick.

▲ A human embryo produced by IVF, before being implanted into the mother's uterus

Questions

1 Describe how FSH and LH can be used to increase fertility in women who cannot conceive.

2 Why do you think two embryos are implanted into the mother's uterus?

↓ E

3 Evaluate the advantages and disadvantages of IVF.

4 Some other fertility drugs contain a chemical that inhibits oestrogen. Suggest how this can increase fertility.

↓ C

5 Some families have a history of a particular genetic disorder. In these cases, fertilisation can be in vitro and the embryos can be tested for genetic defects. Only healthy embryos will be implanted. Do you think this is a good or bad idea? Give reasons for your answer.

↓ A*

Exam tip AQA

✓ If you are using abbreviations, like IVF, give the full name first, with the abbreviation after it in brackets. Then use the abbreviation.

10: Controlling plant growth

Learning objectives

After studying this topic, you should be able to:

- ✔ know that plants are sensitive to light, moisture, and the force of gravity
- ✔ understand that plants produce hormones to control and coordinate growth
- ✔ recall how plant hormones can be used in agriculture

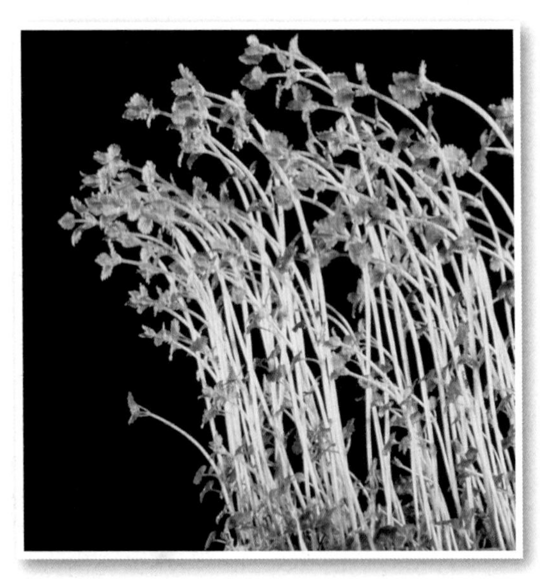

▲ Seedlings growing towards light. They are showing a positive phototropic response.

Key words

tropism, phototropic, auxin, geotropic

A What do plant growth hormones control in plants?

B What is (a) phototropism, and (b) geotropism?

Plants respond to their environment

Plants as well as animals respond to stimuli, namely changes in their environment. Plants make chemicals called plant hormones (plant growth substances) that control and coordinate

- the growth of shoots and roots
- flowering
- ripening of fruits.

You may have noticed that plants growing on a windowsill tend to bend towards the light source. If you want them to grow straight you have to keep turning them. A plant's response to a stimulus is called a **tropism**.

Phototropism

Plant shoots grow towards light. They are positively **phototropic**.

- Plant shoot tips are a growing point. They make a plant hormone called **auxin** and this moves down to other parts of the stem.
- When light strikes one side of the shoot tip, more auxin builds up on the *other* side of the shoot tip (furthest from the light).
- This causes the shoot to bend over, towards the light.
- This is useful to the plant, as it needs light to make food.

How does auxin make the shoot bend?

Auxin is unevenly distributed in the shoot tip. This auxin then moves down the stem and causes cells on the side of the shoot furthest from the light to elongate more than those nearest the light.

Geotropism

Geotropism is the response of a plant to gravity. Roots grow downwards in response to the pull of gravity. Auxins may be involved in this response, but other chemicals that inhibit growth may also play a part.

Plant hormones in agriculture and horticulture

Plant hormones can be applied to plants to either speed up or slow down their growth.

Venus flytrap

This sensitive plant's leaves close up after the plant has been touched

Weedkillers

Auxins are used as selective weedkillers. Agent Orange was used in the Vietnam War. It made trees lose their leaves. Without leaves, the trees cannot make food and they die.

Rooting powder

If cuttings are dipped into auxin powder, this helps the cuttings make new roots. With new roots the cuttings can anchor in the soil and take up water and minerals.

Fruit ripening

Auxin can be sprayed on to fruit trees to prevent the ripe fruit from dropping. Then it can all be harvested at the same time and the fruit is not bruised by falling to the ground. If a high dose of auxin is sprayed later, it makes all the fruits fall.

Control of dormancy

Seeds are usually dormant when they are shed from the parent plant. This stops them germinating at the end of the summer, as new plants may not survive the harsh winter. However, if commercial growers want to germinate seeds in greenhouses during winter, applying auxin can break the dormancy.

Questions

1. Why is it an advantage to a plant if the shoots are positively phototropic?
2. Why is it an advantage to a plant if the roots are positively geotropic?

E

3. Describe four commercial uses of auxins.

C

4. Explain how auxin causes shoot tips to bend towards the direction of the light source.

A*

Did you know...?

Plants also respond to other stimuli. Plant roots grow towards water.

Some plants respond to touch. Climbing plants put out tendrils and cling to walls or canes.

Some plants move their leaves downwards very quickly when touched. This startles animals trying to eat them and protects the plant. The Venus flytrap responds to flies touching hairs on its leaves. It snaps the leaves shut, trapping the flies so it can digest them.

Exam tip AQA

✔ Remember that although plant growth substances like auxin are called plant hormones, they do not work in the same way as animal hormones.

Horticulturist dipping a cutting in rooting powder

11: Drugs and you

Learning objectives

After studying this topic, you should be able to:

- know about different types of drugs
- evaluate why some people use illegal drugs for recreation

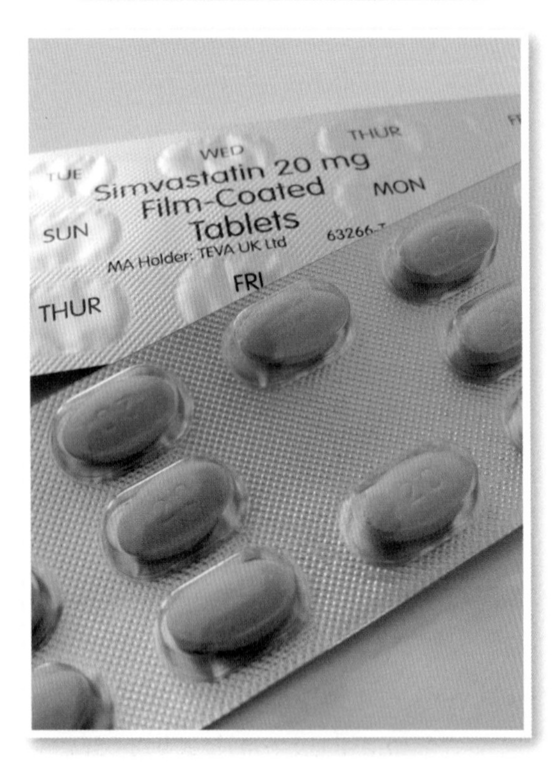

Statins lower blood cholesterol. Some statins cause muscle pain in some people, and a doctor will need to find the best type for each patient.

A Why do you need to have a prescription from a doctor to get certain drugs, like strong painkillers?

B Why do legal drugs have a greater impact overall on people's health than illegal drugs?

Drugs may be beneficial or harmful

A **drug** is a chemical that alters the way your body or brain works. Drugs may alter your behaviour as well as altering your metabolism.

Beneficial drugs are medicines like painkillers, antibiotics, and statins. Some drugs have to be prescribed by a doctor. This is because they may

- have side-effects
- interfere with another medicine the patient is taking
- be harmful for a particular patient if they have another condition
- be harmful if taken too often.

Using drugs for recreation

Legal recreational drugs

Some drugs are legal and used for recreation. These include caffeine, nicotine in tobacco, and alcohol. Caffeine is usually not harmful. Nicotine makes people **addicted** to tobacco, and that causes cancer. Alcohol can harm the nervous system. It alters people's behaviour and may lead to violence or accidents.

Illegal recreational drugs

Some athletes use performance-enhancing drugs like anabolic steroids. These can have harmful side-effects. It may be unethical to use them as it gives some athletes an unfair advantage. The athletes may also suffer side-effects from taking the anabolic steroids.

Progression to hard drugs

Many young people experiment briefly with some types of drugs. Unfortunately some of them may go on to take hard drugs like heroin and cocaine. Both these drugs are very addictive. When users try to stop taking them they get **withdrawal symptoms**.

Impact on health

All drugs have an effect on your health. However, the legal drugs, like alcohol and tobacco, have a greater overall impact and cause more harm. This is because more people use them so more people are harmed.

Field of opium poppies in Dorset. Opium is obtained from the seed heads of this poppy. Opium contains morphine and codeine. Opium can also be refined to make the illegal drug heroin.

Cannabis

Some people believe cannabis is a very good painkiller. People with multiple sclerosis find it relieves their symptoms. However, some people have concerns that the chemicals in cannabis smoke may

- lead to mental health problems in some people
- lead the user on to addiction to hard drugs like heroin and cocaine
- increase the risk of heart attacks and strokes.

In the UK cannabis is illegal and cannot be prescribed. However, it is used illegally for recreation by some people.

Testing new drugs

New drugs have to be rigorously tested before being licensed. They are tested on laboratory animals and human tissue to see if they are toxic. Then they are trialled on human volunteers.

Questions

1 Name three drugs that can be obtained from opium poppies. ↓E

2 Explain the following terms: drug; addiction; withdrawal. ↓C

3 Explain why new drugs have to be tested before they are licensed for use as medicines. ↓A*

Did you know...?

Some animals self-medicate. They eat certain leaves that they do not normally eat, to treat parasitic infections.

Key words

drug, addiction, withdrawal symptoms

Cannabis products: seeds, a leaf, dried parts, and marijuana

C What are the medical benefits of cannabis?

D Why is cannabis not available on prescription in the UK?

Exam tip

✔ Do not fall into the trap of saying that because something is made from natural substances it is bound to be good for you. Many strong poisons come from plants.

Learning objectives

After studying this topic, you should be able to:

- ✔ understand why new drugs have to be tested
- ✔ describe how a double blind trial is carried out

Key words

Thalidomide, clinical trial, placebo, double blind trial

Did you know...?

Sometimes clinical trials do not have a placebo. If the patients are very ill it would not be ethical to give them a placebo, because they need some treatment. In this case the new drug is tested against the best current treatment. Once again, the trial is double blinded.

A Why was Thalidomide given to pregnant women?

B What is a side-effect?

Thalidomide

Between 1957 and 1961 a drug called **Thalidomide** was developed as a sleeping pill. It was prescribed to pregnant women, as it also prevented morning sickness. However, it had not been properly tested on animals, or in humans in **clinical trials**. Unfortunately it had side-effects. These are effects of the drug on the body other than the beneficial effects it is designed for. Many side-effects are minor, but thalidomide caused birth defects. The babies of women who took the drug in pregnancy had very short limbs.

Later research found that the drug interfered with genes. It prevented the normal development of limbs in the fetus.

The drug was then banned. However, since the 1980s it has been used in some countries to relieve the side-effects of the drugs used to treat leprosy. Unfortunately doctors have not always checked that patients are not pregnant. As a result, in those countries, children have more recently been born with very short limbs.

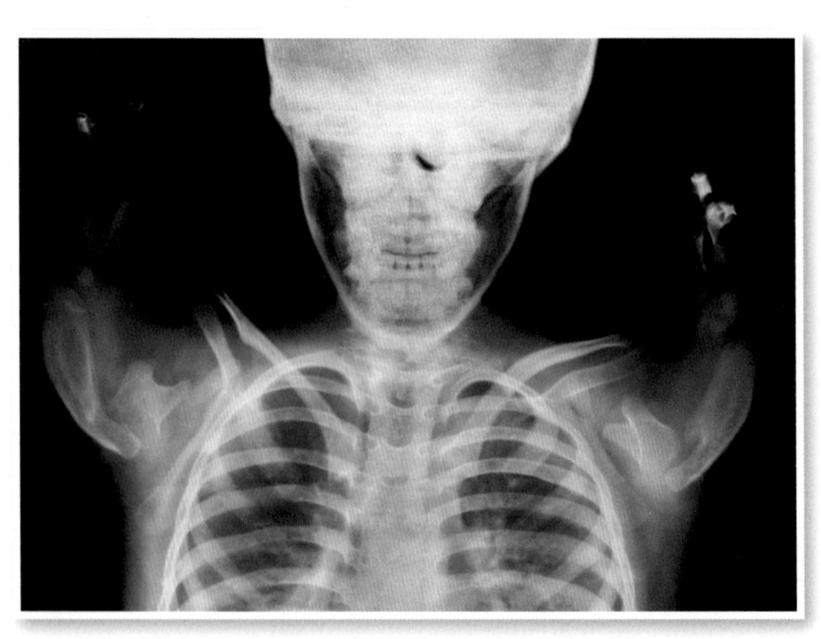

▲ X-ray of upper chest and arms of a baby. The baby's mother was given Thalidomide when she was pregnant. This caused the baby to have very short arms because the arm bones did not develop properly.

Because of this tragedy, new drugs have to be rigorously tested in clinical trials before being licensed.

How new drugs are tested

- New drugs are tested in laboratories, on human tissue and animals, to see if they work and to find out how toxic they are.
- If they pass these tests the drugs are tested on humans in clinical trials.
- At first very low doses of the drug are given to volunteers.
- Then doses are increased to find the dose that works best.
- Volunteers are divided into two groups. The control group is given a **placebo** (dummy pill) and the experimental group is given the real drug.
- Neither the doctors nor the patients know who is getting the placebo and who gets the real thing, until the end of the trial. This is called a **double blind trial**. It makes the trial fair.
- At the end of the trial the two groups are compared to see if there is any real difference between them.
- If the drug makes a real difference and causes no harm it is licensed for use.
- However, if some new side-effects occur, some drugs are recalled even when they have been used on many people.

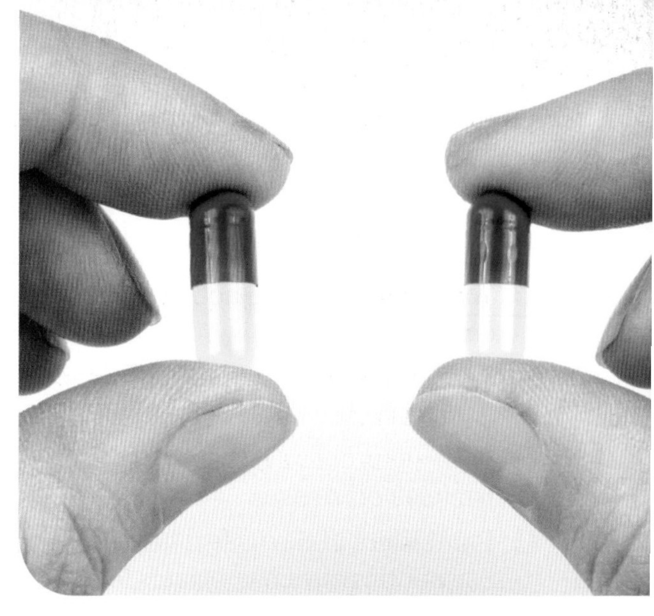

One of these capsules contains real medicine. The other is a placebo. It looks like the real thing but does not contain medicine. Sometimes people feel better when taking a placebo, because they believe they are being treated.

This boy is taking a tablet of omega-3 oil. He is part of a clinical trial to see if omega-3 can improve children's brain function. One group took the real tablets twice a day for eight weeks. The control group took a placebo twice a day for eight weeks. Members of each group had memory tests. Their reaction times and attention spans were also measured.

Questions

1 What were the effects of Thalidomide when given to pregnant women?

2 Explain why new drugs have to be tested before they are licensed for use as medicines.

3 How are new drugs tested?

4 Discuss why in a double blind trial (a) the patient and (b) the doctor does not know whether they are being treated with the real drug or a placebo.

Course catch-up

Revision checklist

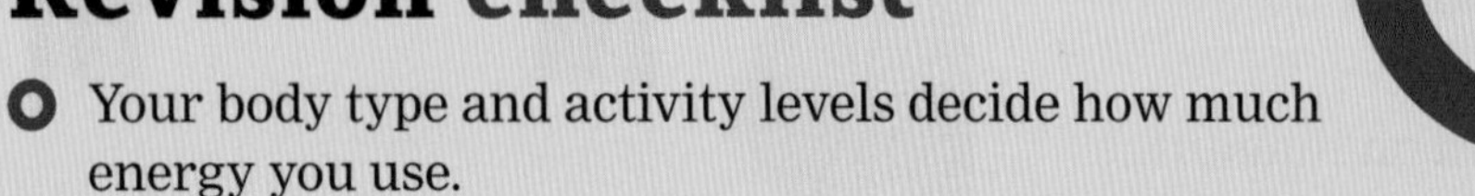

- Your body type and activity levels decide how much energy you use.

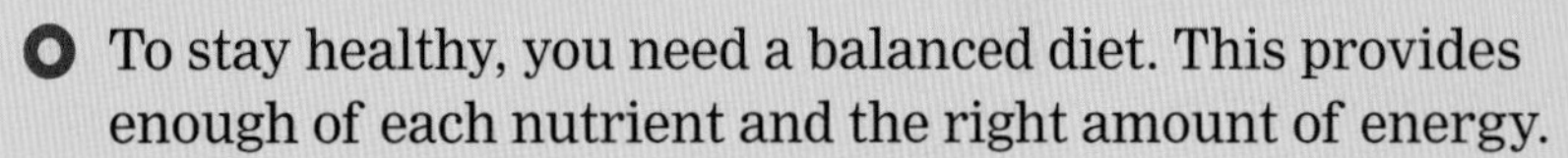

- To stay healthy, you need a balanced diet. This provides enough of each nutrient and the right amount of energy.
- Pathogens cause disease by multiplying inside us. They can be bacteria or viruses.
- Painkillers make us feel better, but do not kill pathogens. Antibiotics cure bacterial infections until mutation and natural selection let resistant strains develop.
- Our bodies keep most pathogens out. White blood cells fight infections by ingesting pathogens, producing antibodies or making antitoxins. Vaccines provide immunity by preparing white blood cells to produce antibodies faster.
- Our brains process signals from sensory cells and decide how to respond. We also have fast, automatic reflex actions which protect us from danger.
- We keep our temperature, water content, ion content, and blood glucose steady. Chemical messengers called hormones help by coordinating the actions of different cells.
- Hormones regulate a woman's menstrual cycle and are used as contraceptives.
- Hormones also make IVF possible. Controlling fertility brings benefits and problems.
- Plants also use hormones to respond to change.
- Drugs alter the way your body or brain works. Some people use recreational drugs to alter their mood, but drugs can be harmful and/or addictive.
- New medicinal drugs need testing to show they work and aren't harmful. Double blind trials are used, which compare drugs with placebos. Doctors and patients aren't told who took which, so the results cannot be biased.

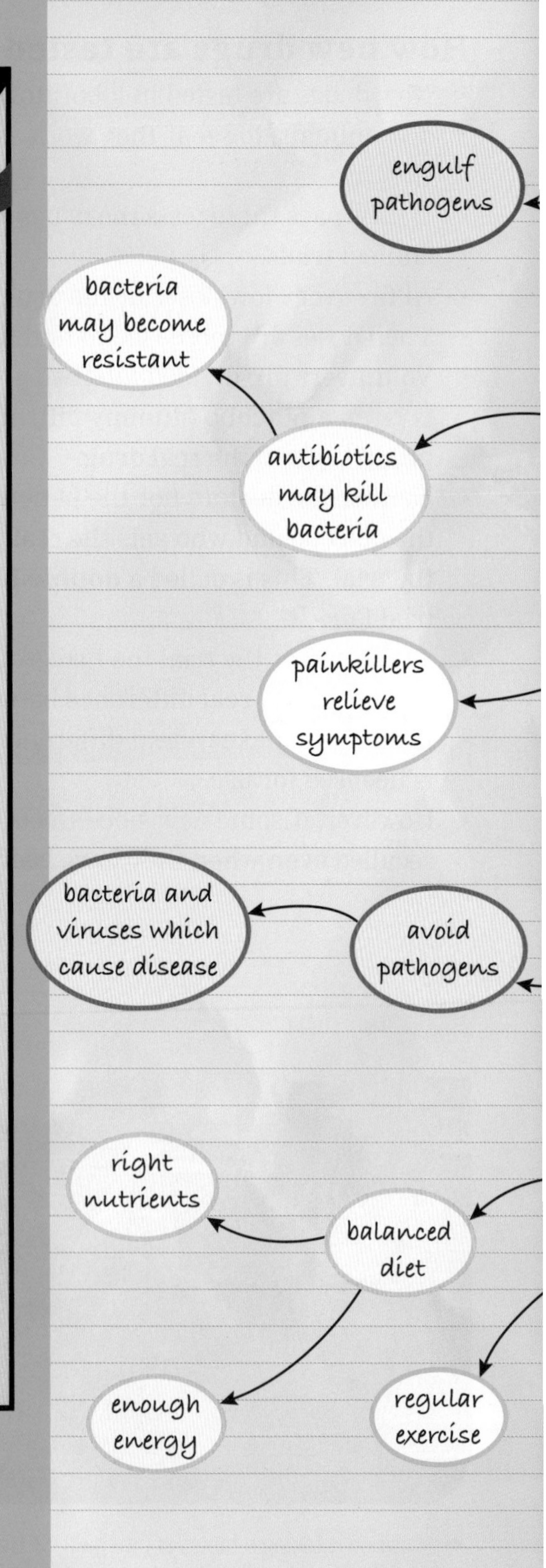

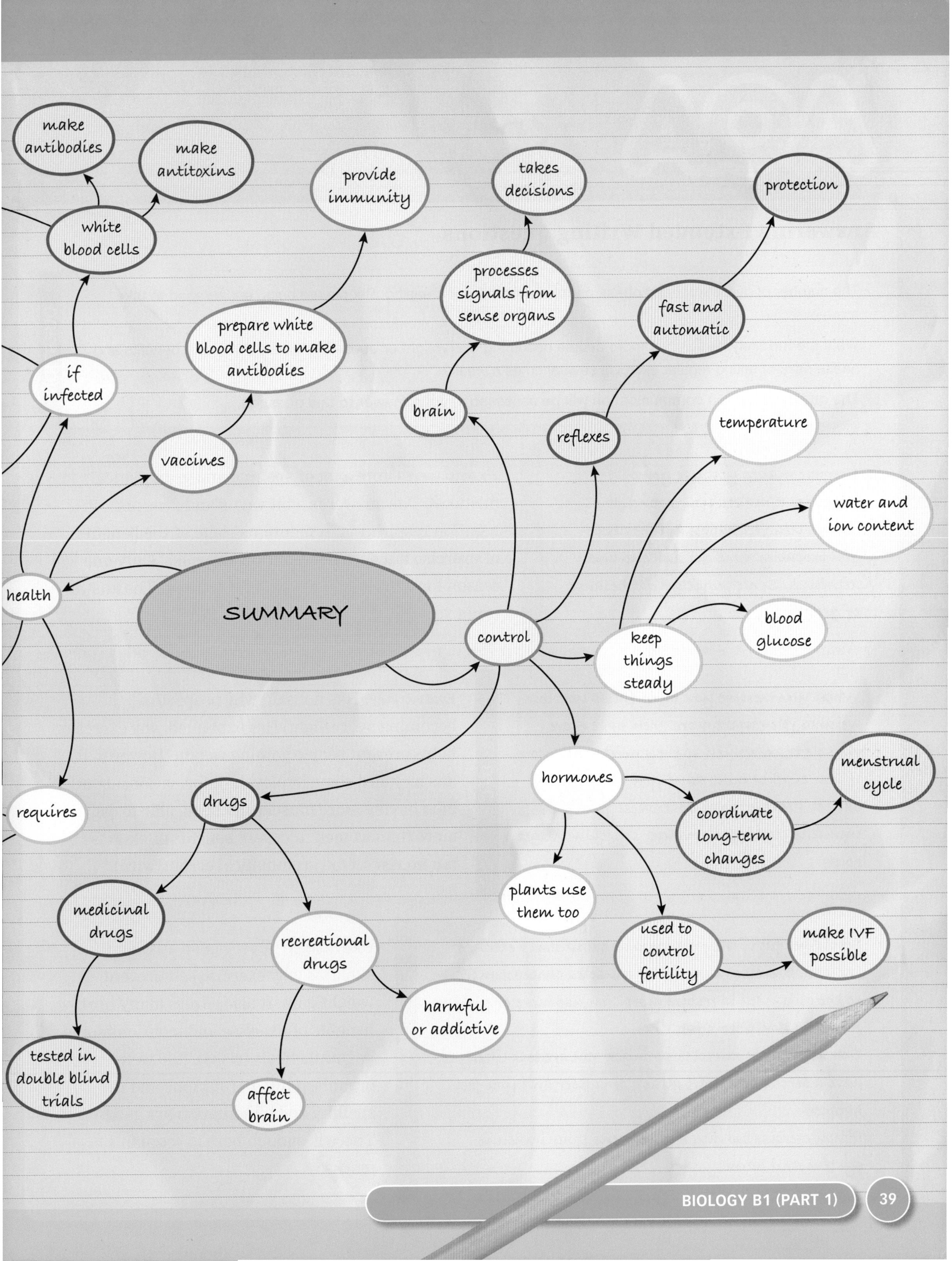

SUMMARY
health
if infected
white blood cells
make antibodies
make antitoxins
vaccines
prepare white blood cells to make antibodies
provide immunity
requires
control
brain
processes signals from sense organs
takes decisions
reflexes
fast and automatic
protection
keep things steady
temperature
water and ion content
blood glucose
hormones
coordinate long-term changes
menstrual cycle
plants use them too
used to control fertility
make IVF possible
drugs
medicinal drugs
tested in double blind trials
recreational drugs
harmful or addictive
affect brain

Answering Extended Writing questions

QUESTION

The number of overweight and obese people in the UK is increasing. People are being encouraged to take more exercise.

What are the likely reasons for more people becoming overweight or obese? What are the health benefits and health risks of taking regular exercise several times a week?

The quality of written communication will be assessed in your answer to this question.

G–E

Exercise makes you use your mussels. It makes you strong. It makes you feel happy and healthy. You probably wont get diabetic and arthritus. Some people get fat because of genetics.

Examiner: Quite a lot of spelling and grammatical mistakes. The answer is a bit vague and does not mention respiration. It mentions some of the benefits of exercise but does not mention any risks. Only one reason for why people gain weight is given. The answer is not very well organised.

D–C

When you exercise you burn fat so you lose weight. You respire more. Men have more muscle than women so they need to eat more. If you exercise most days you wont get cancer, heart attacks or strokes. Your immune system is stronger so you wont get colds.

Examiner: Has not said why people are becoming overweight. Has explained why exercise helps prevent people gaining weight. However, it has included some irrelevant information, about men having more muscle and being able to eat more than women. It covers benefits of exercise but no risks. One grammatical error (twice).

B–A*

People get fat if they eat too much and don't exercise enough. Exercise makes your muscles contract more so the cells need to respire more. You use more of your food and don't store fat.

Exercise makes muscles stronger. You are less likely to have a stroke or a heart attack or cancer and you feel happier.

If you exercise too much it weakens your immune system and harms joints.

Examiner: A very good answer. It explains why exercise can prevent weight gain. It also covers many health benefits of exercise and mentions some harmful effects (risks). It is well organised into paragraphs and the spelling and punctuation are good. There is enough here to score full marks.

Exam-style questions

1 The diagrams show two organisms that can cause disease.

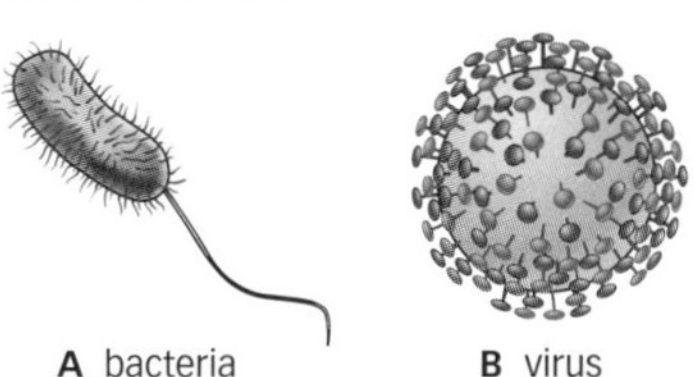

A bacteria B virus

Answer the following questions by writing A, B, neither, or both.

A01 a Which could be called a pathogen?

A01 b Which is destroyed by antibiotics?

A01 c Which is destroyed by painkillers?

A01 d Which would be able to multiply rapidly inside you?

A01 e Which multiplies inside cells?

G–E

2 Cannabis changes how people see and hear things, and makes them feel relaxed. But could it be doing long-term harm? Scientists used an MRI machine to compare the brains of users and non-users. They were all the same age. The 15 users had smoked cannabis five times a day for ten years. The brain area that deals with emotions was 12% smaller in users' brains.

A02 a Why did the scientists choose long-term cannabis users for their study?

A02 b Why was it important to compare users and non-users of the same age?

A03 c Can we be sure that the cannabis has changed their brains?

A03 d Suggest a way to get stronger evidence that cannabis shrinks parts of the brain.

A03 e Suggest one ethical problem you might have if you carried out this experiment.

D–C

3 Many women use hormones to control their fertility.

A01 a Explain what a hormone is.

A01 b Why do hormones circulate in your blood?

A01 c Explain why oestrogen and progesterone are used in oral contraceptives.

A01 d Give two advantages and two disadvantages of using oral contraceptives.

A02 e Suggest why the composition of the pills has changed since they were first used.

B–A*

Extended Writing

4 A02 Dev cut his hand in the garden. His finger is red and swollen. It's infected. Explain how Dev's white blood cells will try to combat the microbes.

G–E

5 A02 Kevin is taking part in a drug trial. It is a double blind trial of a new acne cream. Twenty people are taking part and will stay at the test centre for three weeks. They all have acne. Write a set of instructions for the nurses running the trial.

D–C

6 A02 A03 Mary and John want to have children using IVF. Explain how and why IVF is used, and state arguments against its use.

B–A*

A01 Recall the science

A02 Apply your knowledge

A03 Evaluate and analyse the evidence

Surviving and changing in the environment

AQA specification match	
B1.4	1.13 Animal adaptations
B1.4	1.14 Plant adaptations
B1.4	1.15 Extreme adaptations
B1.4	1.16 Competition for resources
B1.4	1.17 Changes in distribution
B1.4	1.18 Indicating pollution
B1.5	1.19 Pyramids of biomass
B1.5	1.20 Energy flow in food chains
B1.6	1.21 Recycling in nature
B1.6	1.22 The carbon cycle
B1.7	1.23 Variation
B1.7	1.24 Reproduction
B1.7	1.25 Cloning
B1.7	1.26 Genetic engineering
B1.7	1.27 GM crops
B1.7	1.28 Concerns about GM crops
B1.8	1.29 Classification
B1.8	1.30 Surviving change
B1.8	1.31 Evolution in action
B1.8	1.32 Evolutionary theory
Course catch-up	
AQA Upgrade	

Why study this unit?

You have seen how organisms are adapted to their environment. In this unit you will continue to look at how organisms survive. Organisms gradually change – better characteristics making them more suited to their environment.

In this unit you will look at how the distribution of organisms will change, as natural conditions change. These changes in conditions may be the result of human pollution. The movement of energy through the living world, and the cycling of elements between the living and, non-living world, will also be looked at.

You will also look at the process of reproduction in animals, and the ways in which humans can manipulate reproduction. This will lead on to the controversial issues surrounding cloning. Finally, the ideas of groups within the variety of life, and an explanation of how nature selects characteristics most suited to their environment via natural selection, will be investigated.

You should remember

1. Organisms depend on each other for their survival.
2. Feeding relationships are shown in food chains and webs.
3. Energy is passed through the food chain by animals feeding.
4. Plants and animals are sorted into groups based on their characteristics. This is called classification.
5. There is variation between individuals of the same species.
6. Plants and animals reproduce to produce new individuals.

One of the most majestic bears on Earth is the polar bear. The polar bear and its closest relative, the brown bear, diverged about 150 000 years ago. Males can be 2.4 metres long and can weigh up to 680 kg, making it the largest land carnivore alive today. Our increasing demand for energy has led to global warming which has caused the ice the polar bear lives on to melt. Their numbers are now in decline because of this change in habitat, and also due to hunting. In the past the polar bear has adapted to changes in climate, as natural climate change occurs slowly. But biologists are worried that the bear will not have time to adapt to the change this time.

13: Animal adaptations

Learning objectives

After studying this topic, you should be able to:

- know that adaptations of organisms help them survive
- identify and explain adaptations of animals

Did you know...?

Some Amazonian tribes use the poison from the skin of poison dart frogs to tip their arrows. This is one of the most deadly poisons known. They use the arrows to kill monkeys.

What are different environments like?

Conditions in different environments can vary greatly. Deserts are hot and dry environments; the Arctic is cold. There are animals that can survive in each of these environments, but they must be adapted to survive. An **adaptation** is a feature of an animal's body which helps it to live in its environment.

Adapted to compete

Animals have to be adapted not only to the physical environment, but also to cope with other organisms in their environment. Some animals are **camouflaged** to blend in with their environment, so predators or prey do not notice them. Other animals are poisonous and have bright warning colours to prevent them being eaten. An example is poison dart frogs. The giraffe's tongue is adapted to stretch out and pull leaves from between the thorns of the acacia tree.

Adaptations to cold environments

Adaptation	How this aids survival
small ears	the surface area of the ear is reduced, and so less heat is transferred to the environment
thick white fur	insulates the body against the cold, and camouflages the bear
sharp teeth	to kill prey
strong legs long legs	contain large muscles which contract so the bear can run on land or swim in water
big feet with fur on the soles	spread the load of the animal on the snow or ice; fur helps grip and helps insulate against the snow
claws	for killing and holding prey
blubber below the skin	a thick layer of fat which insulates against heat transfer to the environment; the stored fat can also be used for respiration to generate heat
large body size	reduces the relative surface area, and so reduces heat loss
fins	balance the whale during swimming
muscular tail	contains large muscles which contract to generate movement during swimming

A State three adaptations of the polar bear that only help it survive if conditions are cold.

B Penguins live in the Antarctic. Describe five features of a penguin that help it to survive there.

Adaptations to hot environments

Adaptation	How this aids survival
hump of fat	fat is stored in one place, which reduces all-round insulation; fat can be broken down to release water
bushy eyelashes	stop sand entering the eyes
nostrils which close	prevent breathing in of sand
body tolerance to temperature changes	does not need to sweat so much when hot
long legs	lift body off hot sand
large feet	spread the load, stop the camel sinking into the sand
thin fur	less warm air is trapped, reducing insulation
large ears	lose heat by radiation, and are used to fan the body
large body size	can knock over plants and shrubs for food
wrinkled skin	increases the surface area from which to lose heat
trunk	allows the elephant to suck up water to drink, and to spray water over the body to cool itself
large feet	spread the load and stop the elephant sinking into the mud

C Sort the adaptations of the elephant and camel into two lists – those that help them survive hot conditions, and those that are not related to the heat.

D Explain how the difference in ear size between elephants and polar bears helps the two animals survive in their different environments.

Key words

adaptation, camouflage

Questions

1. A close relative of the elephant is the woolly mammoth. It lived in cold environments. Suggest two features that differ from those of the elephant and helped the mammoth survive. (E)
2. Explain how the features you identified in Question 1 would aid the mammoth's survival.
3. The distribution of camels is limited to the desert. Explain why the camel is well adapted to life in the desert. (C)
4. Explain why slow-moving camels are not well adapted to live in the community of animals on the African savannah. (A*)

Exam tip AQA

✔ When looking for adaptations, notice particular features of an animal, and try to suggest how the animal uses these features to survive.

Learning objectives

After studying this topic, you should be able to:

- know that plants show adaptations to hot and cold environments
- understand that the adaptations of a plant determine where it can grow

Key words

pore, needles, surface area, spines

A What is the advantage to the oak tree of its large green leaves?

B Name the advantage to the pine tree of reducing its leaves to needles.

C Suggest a disadvantage to the pine tree of reduced leaves.

Exam tip AQA

- Think about how a plant is adapted to its environment, and how its adaptations restrict its distribution.

Plants are adapted too

Plants are found in all sorts of environments. They need to be adapted just as well as animals to survive. Plants show a range of adaptations, most of them to do with absorbing light and retaining water. Leaves absorb light, and water is lost from the surface of the leaves through **pores**.

Adaptations to hot and cold environments

Northern pine forests	Temperate forests
Pine tree	Oak tree
Conditions	**Conditions**
For much of the winter the water freezes in the soil.	Cold and not much light in winter; sunny and moist in summer.
Adaptations	**Adaptations**
Leaves are reduced to **needles** to reduce the **surface area**; this reduces water loss.	Leaves fall from the trees in autumn; there is not much light for photosynthesis, so leaves have no function in the winter. Leaves have a larger surface area to make the most of the summer sun.
Thick wax on the surface of the leaf also reduces water loss.	Wax is thinner on the leaf surface, as there is plenty of water in the soil for most of the year.

Adaptations to dry environments

Beaches and sand dunes	Deserts
Marram grass	Cactus
Conditions	**Conditions**
Rainwater quickly drains through sand. Can be hot and windy.	Very little water in the sandy soil; very low rainfall. Air is hot and dry.
Adaptations	**Adaptations**
Leaves are long thin spikes, which reduces their surface area; this reduces water loss.	Leaves are **spines**, which reduces their surface area; this reduces water loss.
Leaves are rolled; leaves lose water through pores, and rolling keeps these pores on the inside of the roll.	Spines are less likely to be eaten by animals.
Waxy layer on the outside of the leaf to reduce water loss.	Wax on stem reduces water loss.
Deep root system to absorb water, and to anchor the plant against the wind.	Shallow root system to cover great areas, and to absorb water when it does rain. Stems are swollen to store water. Many stems have grooves which can expand and flatten out when the stem fills with water.

The acacia tree is common on the African savannah. These trees are heavily grazed by animals such as giraffes. In order to reduce the damage caused by grazing, acacias have a number of adaptations.

- They have developed large thorns which discourage the giraffes.
- The thorns provide a home to colonies of stinging ants.
- The leaves release a nasty poisonous chemical which giraffes dislike.

A giraffe browsing an acacia tree

D The pine tree and the cactus both show similar adaptations of their leaves. Explain why the same adaptation helps plants survive in two different environments.

E Beaches are often windy environments. Explain how the adaptations of marram grass help it survive the windy conditions.

Questions

1. State the adaptations that help marram grass live in dry soil. ↓ E
2. Explain why the dandelion isn't well adapted for living in the desert. ↓ C

3. Suggest a reason why flowering plants tend to reproduce more in the warmer summer months.
4. The acacia tree has a very long tap root. Explain how this adaptation helps it to survive on the African savannah. ↓ A*

Learning objectives

After studying this topic, you should be able to:

- know that many extreme conditions are found on Earth
- know that organisms are found in most environments, and show adaptations to those extreme environments

Key words

extremophiles

A red mangrove tree growing in an estuary in the Caribbean

A What is an extremophile?

B Explain why plants which live on salt marshes in estuaries need to be adapted to deal with salty conditions.

Living on the edge

There are many places on Earth with very difficult conditions – too extreme for humans to live there. But some organisms do live in these places. They are called **extremophiles**. These organisms are often fascinating because they have interesting and unusual adaptations.

Coping with salt

Mangrove trees are common plants at the water's edge on tropical coasts. Their roots are permanently in sea water, with salt levels that would kill most land plants. So how do they survive? Water is lost from their leaves and, as in most plants, this causes the plant to draw in large amounts of water through the roots. But mangroves have special roots that do not allow salt to enter as the water is absorbed.

Under pressure

Many marine animals, such as whales, are able to dive to astonishing depths. Deep in the ocean, the pressure increases. It would crush a human, but whales can survive because their bodies are highly adapted to cope. Their lungs are smaller than expected for an animal so big, and they do not fully inflate them before diving. This makes them less buoyant. Their muscles can store large amounts of oxygen, which allows them to remain underwater for long periods.

A sperm whale is adapted to life deep in the ocean

Microbes far and wide

Microorganisms have such a wide range of adaptations that there are microbes of one kind or another living in almost all the conditions found on Earth. Bacteria are found in the depths of the ocean and on the peaks of mountains. They span cold environments such as Arctic wastes to the hot springs of Yellowstone Park in North America.

▲ This lake in Yellowstone Park contains many thermophilic bacteria which colour the water

Thermophiles

Thermophiles are bacteria that are adapted to living in very high temperatures. They have been found living in environments above 80 °C, such as in volcanic areas, hot springs, and geysers. They often have bright colours.

Thermophilic bacteria are adapted to survive in these hot conditions by having special enzymes. Most enzymes are destroyed at high temperatures, but these particular enzymes are able to withstand the hot temperatures. These bacteria are useful in industry. The heat-resistant enzymes are used in genetic engineering.

Bacteria at the bottom of the sea

Biologists have discovered bacteria living under great pressure at the bottom of the ocean. These bacteria are adapted by having an equally high pressure inside their cells, to balance the pressure of the water around them. They are biologically important because they can carry out reactions which release energy for them to survive.

When scientists look for life on other planets, they tend to look for bacteria, because these are the organisms that are most likely to survive the extreme conditions.

▲ Thermophilic bacteria are often brightly coloured

Questions

1. What is a thermophile? (E)
2. Describe what adaptations are shown by bacteria that live in hot-water geysers. (C)
3. Explain why thermophilic bacteria are of use to scientists who carry out cell reactions at high temperatures. (A*)

16: Competition for resources

Learning objectives

After studying this topic, you should be able to:

- ✔ know that organisms compete with each other for resources, and that this can affect their distribution

Key words

resource, competition, population

Did you know...?

Humans are probably the greatest competitors of all time. This has led to many other species losing out. Dodos and Tasmanian tigers have both become extinct because of humans.

A Why do farmers and gardeners remove weeds from around their plants?

B Explain why bluebells only grow in the spring in a British woodland.

What do plants and animals compete for?

There is a limited supply of **resources** for plants and animals. Nearly all living things are locked in a battle to get enough materials and other resources to survive. This fight for resources is called **competition**. The availability of resources can affect the distribution of an organism – that is, where it lives.

Plants compete for:

- light
- space
- water
- minerals
- carbon dioxide.

Animals compete for:

- food
- space or territory
- mates
- water.

A population is the number of organisms of a particular species in a named area. For example, the woodlice living under a stone are a **population**. If conditions are good and the woodlice reproduce, the population gets larger. If conditions are bad and some woodlice die out, the population gets smaller. Competition affects the size of populations.

Competition between plants

Carbon dioxide: plants need this for photosynthesis. The air contains only 0.03% carbon dioxide. The massive canopy of tree leaves absorbs carbon dioxide, so there is less available under the tree for ground plants.

Light: the energy supply for photosynthesis. The tree leaves absorb some light, and not much light passes through, so it is too shady under a tree for many plants to grow.

Space: the tree roots take up most of the space in the ground, leaving little room for ground plants.

Water: used in photosynthesis and to cool the plant. The large tree will absorb most of the water in the soil, leaving little for the smaller ground plants.

Soil minerals: needed to keep the plant healthy. These are absorbed in water through the roots. Again, the tree can absorb far more minerals than the small plants.

▲ Plants do not move around, so they can only live in places where resources are available. If they cannot compete with other plants for these resources, they will not be able to survive there.

Competition between animals

Animals usually compete with each other for food. For example, some coastal birds compete for the same food supply, and they must find a solution for the different species to survive together.

C Name the most common resource for which animals compete.

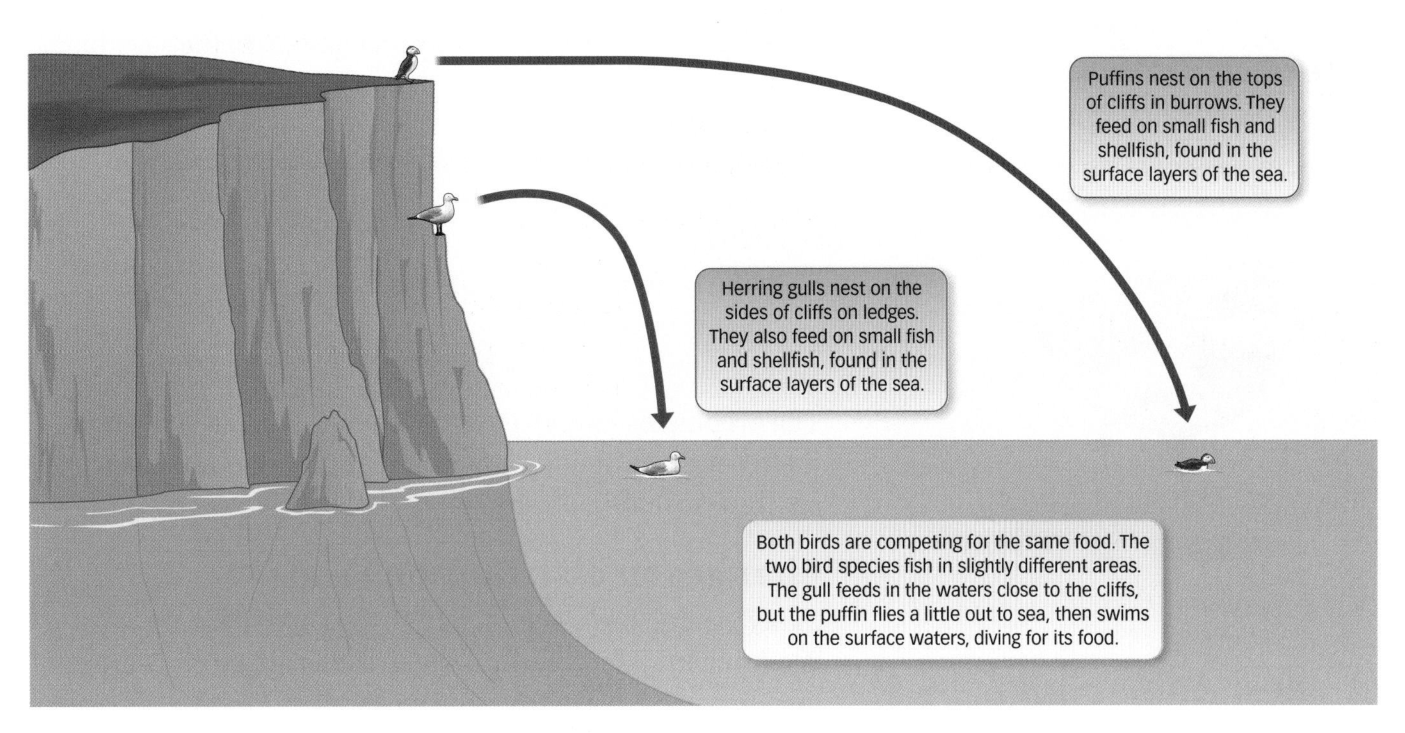

▲ Competition between the puffin and the herring gull

Questions

1. Explain why plants need light to survive, but animals do not. (E)
2. Explain what will happen to the ground plants in a woodland when a large tree dies and falls. (C)
3. Red squirrels are a native species in the UK. Around 1900 a close relative, the grey squirrel, was introduced. The grey squirrel is a better competitor. Describe the effect the grey squirrel's introduction had on the red squirrel population. (A*)

D If humans fished the coastal surface waters, explain what would happen to the population of puffins.

Exam tip

✔ When two populations compete with each other, the number of individuals in each population is reduced.

Learning objectives

After studying this topic, you should be able to:

- know that the distribution of organisms is affected by the environment
- discuss the change in distribution of named species

Key words

distribution

Ringed plover

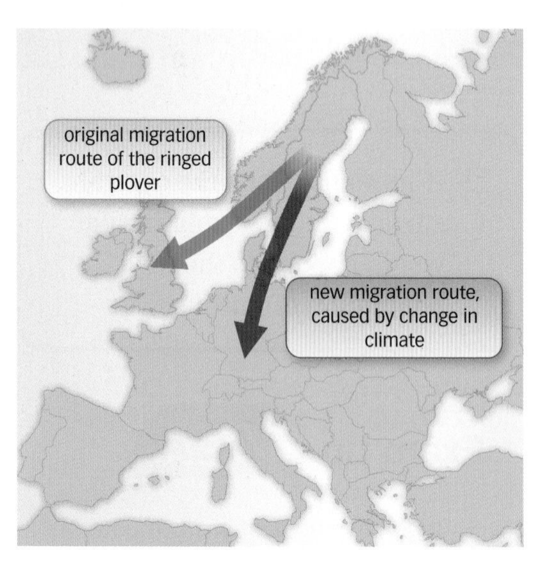

Climate change has resulted in lower numbers of ringed plover in the UK

Fit for the job

A well-adapted animal survives well in its environment. It is fit for the job.

Elephants are well suited to the grassy African savannah, so there are large numbers of elephants on the savannah. They cannot necessarily survive well elsewhere. Elephants are not adapted to survive further north in Africa, where there are deserts, because they cannot survive long periods without water, and there would not be sufficient food for them. So the distribution of African elephants is limited to the savannahs.

Moving home

If the environment changes suddenly, the organisms there may not be well suited to these changes. Such changes might result from:

- living factors, such as humans, predators, or disease-causing microbes
- non-living factors, such as rainfall and temperature.

The organisms have to move to an environment that suits them better, or they will die. Many fast environmental changes are caused by humans.

Changes in distribution: birds

Many wading birds common to British estuaries are declining in numbers. Birds such as the ringed plover used to migrate to Britain from the north for the milder winters. However, as the climate has become warmer throughout Europe during the winter, they have been moving to mainland Europe where the conditions are better. The result is much lower numbers of ringed plover in Britain.

Biologists working for the Royal Society for the Protection of Birds (RSPB) have been concerned about this trend. They have used computers to predict bird **distributions** in the future. The results have suggested that the average bird species distribution will shift nearly 550 km north-east by the end of this century. This is equivalent to the distance from Plymouth to Newcastle. The distribution is also likely to be reduced. This may result in some birds declining in numbers or even dying out altogether.

Human developments, such as the flooding of bays like Cardiff Bay, result in the loss of natural homes for wading birds. During such developments, the planners make provision for the creation of nature reserves close to the development, where bird species can be relocated.

A What factors restrict the distribution of an organism to a particular area?

B Suggest a reason why beech trees are restricted to growing mainly in the southern half of the UK.

C What might be the impact of global warming on the distribution of beech trees?

Changes in distribution: bees

Bees are another species that have been affected by the changing environment. They are declining in number. Biologists believe that there are four reasons for the decline:

- a virus that attacks the bee larvae
- summers are cooler and damper
- the use of agricultural chemicals
- an increase in air pollution.

Questions

1 Describe what might cause the distribution of an animal species to change. (E)

2 State what might happen to a population of birds if they did not move in response to changes in the environment.

3 The ice caps are melting due to global warming. Explain what this means for the distribution of animals such as the polar bear. (C)

4 Why is it important that developers create nature reserves for species when developing an area? (A*)

An artificial wetland habitat created as a nature reserve at Cardiff Bay, Wales

The honey-bee is declining in the UK

18: Indicating pollution

Learning objectives

After studying this topic, you should be able to:

- know that organisms are affected by pollution
- know that organisms can be used to indicate levels of pollution

Key words

pollutant, indicator species

A biologist collecting water samples for testing

Pollution and biodiversity

Pollutants are harmful substances that humans add to the environment. Pollutants have an impact on the number and type of organisms that can survive. Generally, in more polluted areas, fewer species can survive. In some cases particular species will be the main or only survivors, and may even be adapted to cope quite well with the pollution. The presence of these species show biologists that the area is polluted. They are called **indicator species**.

Measuring pollution levels

Biologists use the following two methods to measure the levels of pollution:

1. Non-living indicators – biologists measure a variable such as the temperature, pH, or oxygen level in the environment.
2. Living indicator species – the presence of species adapted to survive in polluted conditions shows that the area is polluted.

Biological oxygen demand

Water in streams and lakes contains oxygen, which allows fish and other organisms to survive. Pollution can lead to lower oxygen levels. These are generally caused by bacteria, which use up the oxygen. The amount of oxygen being used by the bacteria is called the biological oxygen demand (BOD). Biologists can measure the levels of oxygen in water with an oxygen meter. In clean water the BOD is low, but in polluted waters the BOD is higher. The BOD is a measure of pollution.

Other common variables that biologists use to measure pollution in different habitats include

- temperature, measured with a thermometer
- rainfall levels, measured by collecting rainwater and recording the volume.

An indicator species in water

Rat-tailed maggots are invertebrate animals that are adapted to survive in water with very little oxygen (a high BOD). These maggots have a long, tail-like tube which is hollow. It acts like a snorkel, allowing the maggot to take in air containing oxygen from above the polluted water. This is why the maggot thrives in these conditions, and is an indicator species of polluted water.

Rat-tailed maggot

Lichens: another indicator species

Burning fossil fuels releases many chemicals into the air, including sulfur dioxide. This causes air pollution, which reduces the variety of organisms that can survive in the area. Lichens are one group of living things that act as indicator species of air pollution.

A Explain the importance of indicator species.

B The National Rivers Authority (NRA) is responsible for keeping fishing rivers clean. Suggest what they might look for in water samples from rivers.

Some lichens can cope with high levels of pollution, and are found in cities. Other lichens cannot grow there, and are only found in areas with clean air away from cities and motorways. Lichens are great indicators of the level of air pollution in an area.

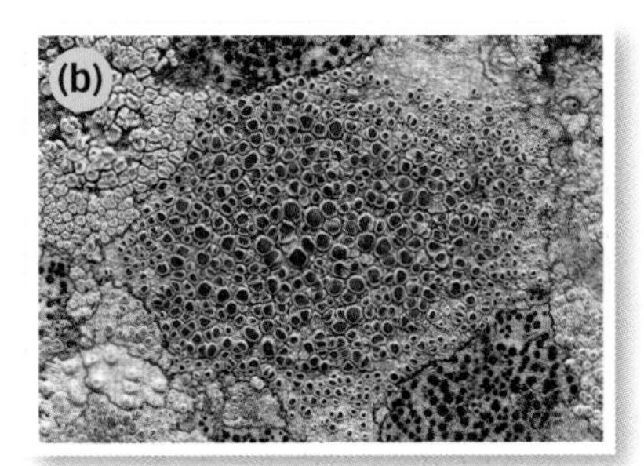

These lichens are indicators of
(a) polluted air
(b) moderate pollution
(c) clean unpolluted air

Questions

1 If a water sample was tested and showed a high BOD, what would this tell you about the water? (E)

2 Explain how you could measure the average rainfall in your school playground per day durng December.

3 Rat-tailed maggots are poor competitors. Suggest why they do not survive well in clean water. (C)

4 Describe an experiment you could carry out to show how lichens can be used to indicate the levels of pollution as you move out of a city. (A*)

19: Pyramids of biomass

Learning objectives

After studying this topic, you should be able to:

- ✔ understand how food chains and pyramids of biomass show the feeding relationships between organisms
- ✔ appreciate the ways in which scientists work

Food chains

A **food chain** shows how the organisms living in an environment are linked by which organisms eat which others. The food chain shows the flow of food and energy from one organism to the next. Each organism is linked by an arrow which shows the direction of flow of food and energy. Here is a food chain for organisms living on the African savannah.

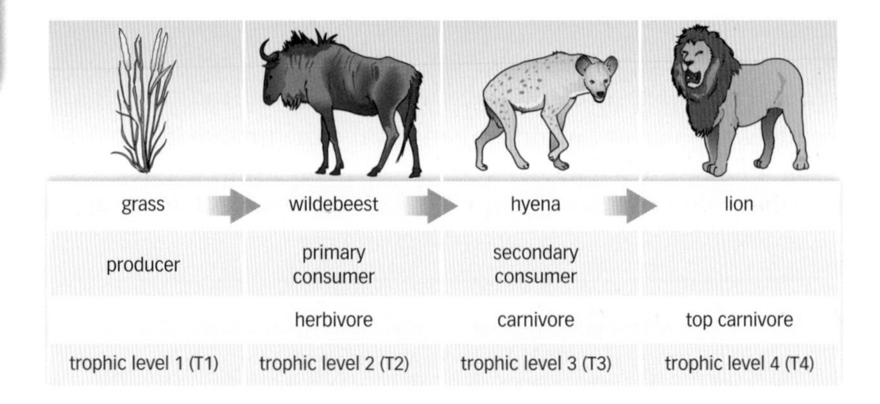

▲ Pyramid of numbers for the food chain on the African savannah

Biologists often study how many organisms there are in each link of a food chain. For example, there are large numbers of small grass plants at the start of the chain shown on the right, but only a few lions at the end of the food chain. These numbers can be plotted in a pyramid of numbers. You can see in the pyramid on the left that the number of organisms decreases at each link in the food chain.

Pyramids of biomass

In a **pyramid of biomass**, biologists plot the biomass of the organisms at each link of the food chain rather than the number of organisms. To calculate the biomass, they multiply the number of organisms at each link of the food chain by the dry mass of one organism. The biomass for each link in the chain can now be plotted as a pyramid of biomass.

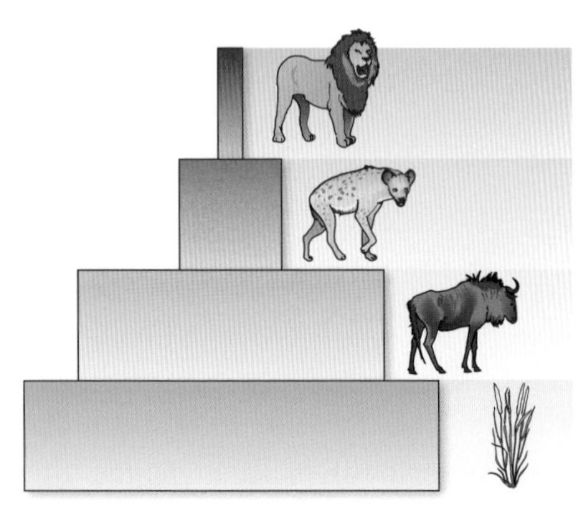

▲ Pyramid of biomass for the food chain on the African savannah

A Describe what happens to (a) the number and (b) the size of organisms as you pass along a food chain.

B What does a pyramid of biomass show?

C Sketch a likely pyramid of biomass for the following food chains:
grass → caterpillar → blue tit → hawk
grass → impala → cheetah → fleas

How scientists work

A case study: lions under threat

Pyramids give us a picture of the state of the environment. If the pyramids change shape over time, this might suggest a problem in the environment. Over the last 60 years, scientists have been concerned about the falling numbers of lions on the African savannah. They wanted to know the cause.

Step 1: analyse the data

Scientists have collected data over the last 60 years for lion food chains and plotted pyramids.

The scientists realised that all the organisms have fallen in number, but particularly the lions. There were 500 000 lions in 1950 but only 20 000 in 2010.

Step 2: interpret the data

The scientists identified a relationship between the falling numbers and human activity on the savannah. They suggested that the fall in numbers might be due to

- habitat destruction by humans to build towns, which reduces the numbers of all species
- hunting, which has a specific effect on the lion population.

Step 3: use the information to inform decisions

Following these findings, people around the world who care about wildlife need to develop plans to protect the lion before it becomes endangered.

Questions

1. State what happens to the biomass of organisms as you move along the food chain. (E)
2. Describe how the scientific community makes use of food pyramids. (C)
3. Construct a pyramid of biomass for the following food chain. Use graph paper where one small square will represent 100 kg. (A*)

grass	→	zebra	→	lion
1 000 000		100		4
0.1 kg		300 kg		250 kg

Key words

food chain, pyramid of biomass

Did you know...?

The wild lion population of Kenya could become extinct by 2030 if nothing is done to stop its decline.

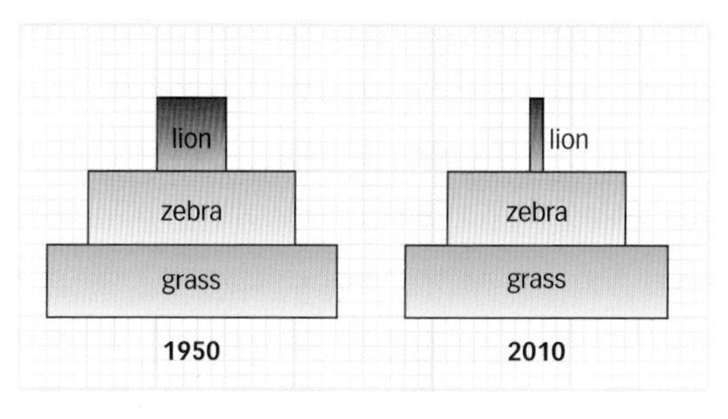

▲ Pyramids of numbers for an African savannah food chain, 1950 and 2010

Exam tip

✔ You can think of a food pyramid as a graph turned on its side.

20: Energy flow in food chains

Learning objectives

After studying this topic, you should be able to:

- know that energy is lost at every link in the food chain
- appreciate that farming methods try to reduce energy loss

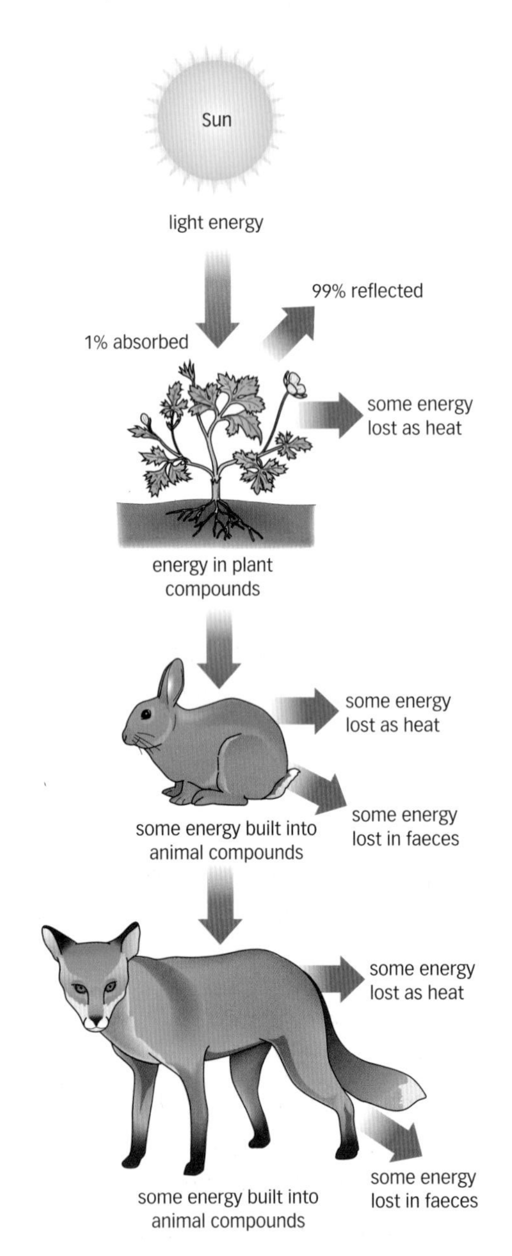

▲ Energy flows through a food chain. Some energy is transferred out at each stage.

Energy

Food chains show not only the flow of materials in biomass from one organism to another, but also the flow of **energy**. All living things need energy to stay alive.

Energy enters the food chain when green plants or algae capture sunlight energy, converting it into chemical energy by means of photosynthesis. Energy flows through the food chain, and leaves it as heat or in waste materials. There must be a continual supply of energy into the food chain to maintain life. However, some of the energy leaves the food chain at each link.

How is energy transferred out of the food chain?

- Not all the food eaten by animals is digested. Some passes straight through the body undigested and comes out in droppings. This transfers energy out of the food chain.
- Every organism in the food chain uses some energy for respiration. This process releases energy for the animal's movement, and heat to warm the body. Birds and animals are 'warm blooded' – they keep their bodies warm at a constant core temperature. It takes a lot of energy from food to maintain this temperature, so they need to eat more food than cold-blooded animals.

Because energy leaves at each link in the food chain, there is very little energy left for the organisms at the end of the chain. This keeps food chains short. Most land food chains do not have more than five organisms. This is also why food pyramids tend to have fewer organisms and less biomass at the top, because there is less energy than at the base.

A State three ways in which energy is transferred out of the food chain.

B Name the source of energy for
(a) a plant
(b) a herbivore.

Energy efficiency in farming

Modern farming needs to produce food as cost-effectively as possible. Farmers need to minimise energy losses from the food chain to get the best yield, maximising energy **efficiency**. Science tells us the following, which helps the farmer:

- The shorter the food chain, the less energy is transferred out. Farm animals tend to be herbivores. This is more energy efficient. If farmers raised carnivores for us to eat, they would need to farm large numbers of herbivores to feed them.
- Less energy is transferred out from animals if they use less heat to keep themselves warm. In intensive farming, larger animals may be kept indoors in a barn, or smaller animals like chickens are often reared in cages. However, it is important to consider animal welfare issues; some battery conditions are inhumane.

▲ Battery farming reduces energy losses

- Energy can also be kept in the food chain by reducing animals' movement. Again, this is achieved by keeping animals indoors or in cages.
- Farmers and growers need to minimise the energy that is transferred out to pests if they eat crops, or to weeds that compete with their plants for light.

Key words

energy, efficiency

Exam tip

✔ Remember that energy can't be created or destroyed. Never talk about energy being made or used up, simply say that it is transferred.

Questions

1. Name the process by which light energy is converted into chemical energy by plants.

2. Explain why fast-moving predators need a large amount of food.
3. Explain why there are far more zebras than lions on the African savannah.
4. On free-range farms the animals are allowed to roam freely outdoors. Explain why this results in produce which is more expensive.

Learning objectives

After studying this topic, you should be able to:

- ✔ know that nature recycles by the decay of dead material
- ✔ know that **microbes** play an important part in the process of decay

Key words

microbe, decay

▲ A millipede eating leaf litter – a detritivore

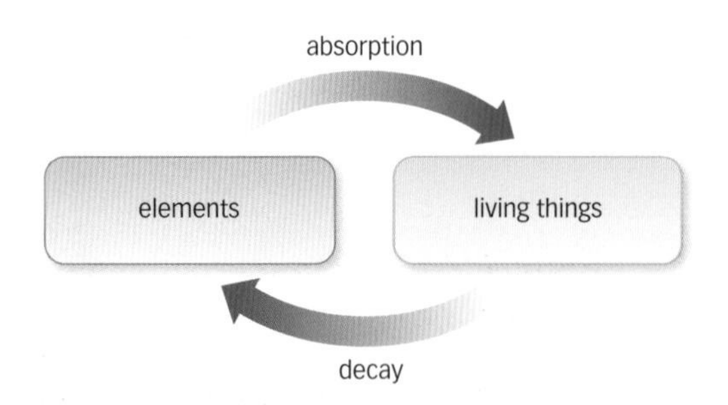

▲ Elements cycle between the living and non-living world

Did you know...?

Some of the carbon in your body is made up of the bodies of people in the past. Maybe some of it was part of King Henry VIII!

Round and round

Elements pass between the living world and the non-living world – air, water, soil, and rocks – in a constant cycle. Plants absorb elements, including carbon and nitrogen, and build them into useful molecules which help the plant grow. When an animal eats a plant, the plant's molecules become part of the animal.

Eventually, all plants and animals die. Their bodies **decay** and this decay process releases the elements back into the environment for plants to reuse. And so the cycle continues.

This is a kind of natural recycling process. Nature breaks down the remains of plants and animals to return the elements to the environment so they can be used again.

In a stable natural community like a woodland, this cycle keeps turning at a steady rate. The processes that remove materials from the environment and lock them up in plants and animals are balanced by the processes of decay, which return the materials to the environment.

▲ Fungi on dead wood – decomposers

Breaking down the dead

A number of organisms play a role in the process of decay. They break down dead plants and animals, and animal waste. There are two main groups of decay organisms:

- Detritivores, such as earthworms, maggots, millipedes, and woodlice, eat small parts of the dead material, which they digest and then release some as waste. This activity increases the surface area of the dead remains for decomposers to act on.
- Decomposers such as bacteria and fungi chemically break down dead material, releasing ammonia into the soil.

> **Saprotrophic feeding**
> Most decomposers feed by releasing enzymes on to the dead animal or plant. The enzymes digest the dead material, and the decomposers then absorb the digested chemicals. This process is called saprotrophic feeding.

Conditions for decay

Decay happens faster at certain times of the year, especially during autumn. This is because conditions are right for the bacteria and fungi:

- plenty of food (dead plants or fallen leaves)
- oxygen
- a suitable temperature
- moisture.

> **C** Use your knowledge of microbes to suggest why the carbon cycle slows during the winter.
> **D** Why is autumn a good season for the process of decay?

Organic gardening

Gardeners make use of natural recycling. They gather garden waste like grass cuttings, leaves, and twigs. They pile them up or put them into a compost bin, and allow them to decay. The result is a nutrient-rich compost which can be used to grow plants. Some local councils collect garden waste and put it in massive compost heaps. This reduces the use of landfill sites, and produces useful compost for parks and gardens.

> **A** Name two important elements that plants need in order to grow.
> **B** Describe why bacteria are important in natural recycling.

▲ Compost is a product of natural recycling

Questions

1. Name two types of organism that cause decay. (E)
2. Describe the difference between a decomposer and a detritivore. (C)
3. Describe how an organic farmer, who does not want to use manufactured fertilisers on his farm, could produce compost to help his crops grow. (A*)

Learning objectives

After studying this topic, you should be able to:

- know that elements are cycled between the living and non-living world
- understand the steps in the carbon cycle
- be aware of processes that cause an imbalance in the cycle

Key words

carbon cycle, **photosynthesis**, **respiration**

Carbon in the living and non-living world

Carbon is the one of the most common elements in living things. All of our major molecules, including carbohydrates, proteins, fats, and DNA, contain carbon. The process of carbon moving between the living and non-living world is called the **carbon cycle**.

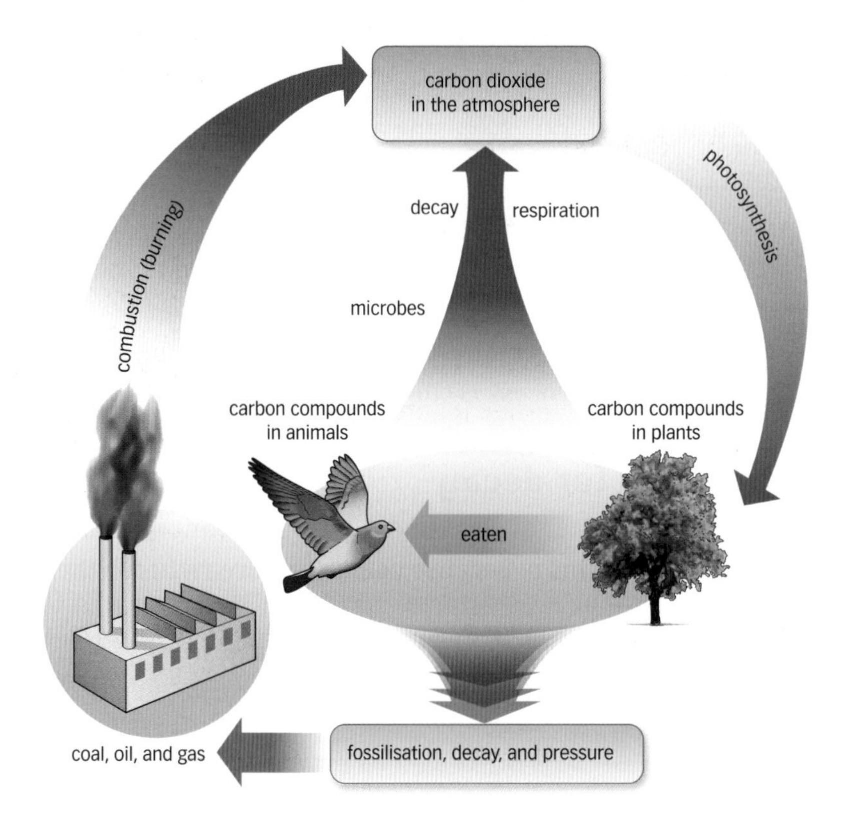

▲ The major steps in the carbon cycle

- Carbon is present in the atmosphere as carbon dioxide.
- Carbon dioxide is absorbed by green plants and algae and built into carbohydrate molecules such as sugars. This happens during **photosynthesis**.
- The plant uses some of the sugars to make other molecules such as cellulose, fats, and proteins, which it uses to grow.
- Plants are eaten by animals and so these carbon compounds can pass into animals and become part of their bodies.
- Both plants and animals **respire**. This returns some carbon dioxide back into the atmosphere.
- When plants and animals die, microbes digest their bodies in the process of decay and carbon dioxide is released back into the atmosphere.

- Not all plant and animal bodies will decay. Some are buried under layers of silt and over millions of years begin to fossilise.
- This forms fossil fuels such as coal, oil, and gas.
- Humans extract fossil fuels and burn them to release energy.
- Burning fossil fuels releases the carbon that was stored millions of years ago as carbon dioxide.
- In a stable environment, the amount of carbon dioxide released should approximately equal the amount absorbed.

A Humans burn vast amounts of fossil fuels. What is the impact of this activity on the carbon cycle?

B How might planting trees balance the carbon given out by your activities in daily life?

Locked-up carbon

Carbon often gets built into the bodies of plants or animals and stays there for millions of years.

There are giant redwood trees in the forests in California which have been growing for more than 2000 years. They have built carbon into the molecules that make up the wood in the tree. Carbon has been locked up in the tree trunk for all that time. Even when these trees die, they are very difficult for microbes to break down. Detritivores are small animals that eat bits of dead plant remains, but they can only eat very small amounts of the dead redwood tree.

When humans cut down forests, the carbon locked up in the trees is released by decay, and there are fewer trees to absorb carbon dioxide from the air. Burning the trees releases the carbon more quickly. Deforestation upsets the balance of the carbon cycle.

▲ Giant redwood trees in the forests of California contain locked-up carbon

Questions

1 (a) Name the process by which carbon enters living organisms.
(b) Name two processes by which carbon is released from living organisms.

E

2 Describe the key role of microbes in the carbon cycle.

C

3 Describe how the actions of humans are leading to an imbalance in the carbon cycle.

A*

Learning objectives

After studying this topic, you should be able to:

- state that most body cells contain chromosomes, which carry information in the form of genes
- state that genes control the characteristics of the body
- know that differences in characteristics may be due to differences in genes or the environment or both

Everyone in the picture is human. However, within this large group there is a lot of variation. There are differences in the characteristics of different individuals.

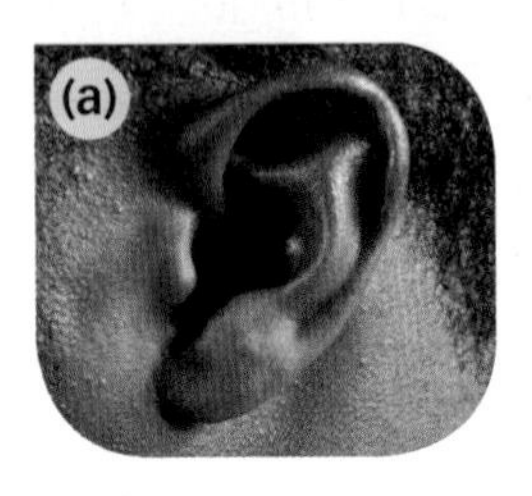

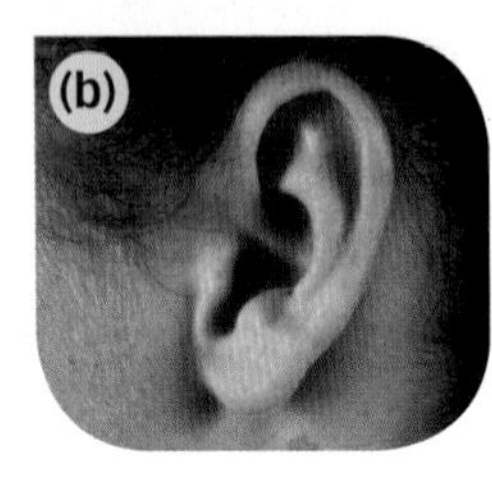

The shape of your earlobe is determined by genes. (a) Attached earlobe; (b) free hanging earlobe.

Did you know...?

Chimpanzees have 24 pairs of chromosomes in each body cell and dogs have 39 pairs of chromosomes in each body cell.

Differences in characteristics

Although all humans are similar to each other, we are also different in many ways. We share characteristics like eye colour; we all have coloured eyes. However, some of us have brown eyes, some blue, some green, and so on. Different individual people show differences in their characteristics.

How human characteristics are determined

Some characteristics are determined by the environment, some by **genes**, and some by both genes and environment.

Characteristics determined by the environment

These include:

- scars
- learning to speak a language.

You get scars after you have injured yourself. Your children will not be born with the same scars.

At birth you cannot speak; you have to learn by listening to your parents and copying them. If you never hear anyone speak, you will not be able to develop language.

Characteristics determined by genes

These characteristics may be inherited. They include:

- eye colour
- earlobe shape
- nose shape.

Characteristics determined by both genes and environment

Some characteristics depend on both genes and suitable environmental factors, such as:

- intelligence
- height
- body mass.

For example, you may inherit genes which control your brain development in such a way that you will be able to have high intelligence. However, if you are not fed properly whilst growing up, or if no one talks to you or reads to you, and you are not given opportunities for stimulating play, you will not develop your full intelligence potential. You may have a genetic potential to be very tall, but you will not reach it if you are undernourished.

A State two human characteristics that are determined by the environment.

B State three human characteristics that are determined by genes.

C State three human characteristics that are determined by both genes and the environment.

Genes and chromosomes

Inside the nucleus of every cell there are thread-like structures called **chromosomes**. All of your body cells have the same number of chromosomes: 23 matching pairs of chromosomes. One member of each pair came from your father, in a sperm, and the other member of each pair came from your mother, in an egg.

Chromosomes contain genes, so half your genes came from your mother and half from your father. This is why you have some characteristics similar to each parent. Humans have 23 pairs of chromosomes, but organisms in different species have different numbers.

- Each chromosome contains many genes.
- Each gene has coded information that controls a particular characteristic.
- Different genes code for different characteristics of the body.

Each gene carries information in the form of coded instructions. These coded instructions control the activity of the cell and the characteristics of the organism.

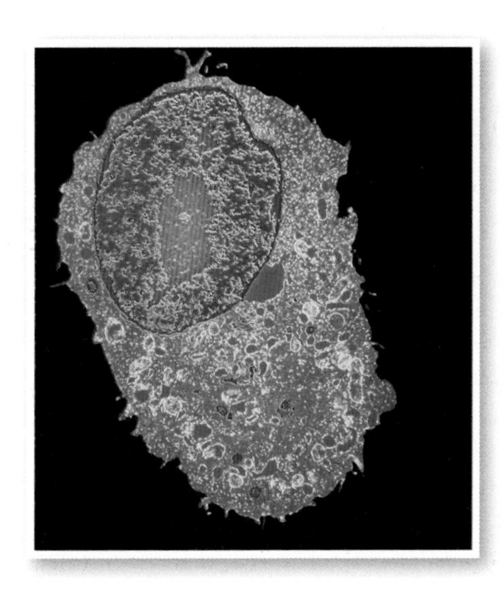

A mammalian cell (× 300). The cytoplasm is coloured blue and the large orange and green structure is the nucleus. Inside the nucleus are chromosomes containing genes.

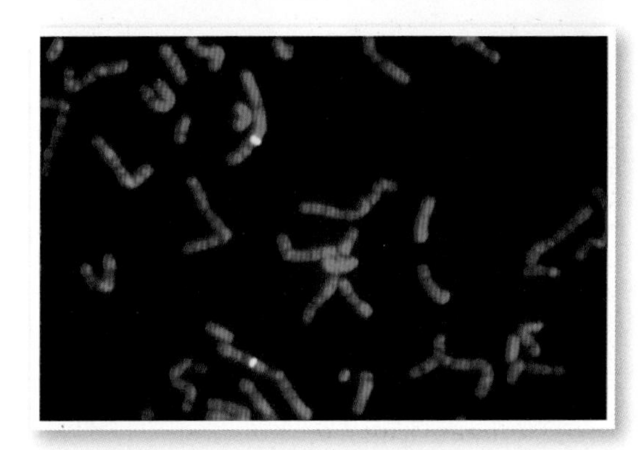

Chromosomes. Some of the genes on some chromosomes have been tagged with a fluorescent chemical and show up as yellow on the red chromosomes.

Key words

gene, chromosome

Exam tip AQA

✔ Remember that we have 23 pairs of chromosomes in our body cells. Our eggs and sperms have half that number, just 23. Remember also that other living organisms have different numbers of chromosomes from us.

Questions

1 Where, in animal and plant cells, are the chromosomes? (E)

2 How many chromosomes are there in each of your brain cells?

3 What is a gene?

4 Where are genes found? (C)

5 What do genes do?

6 Humans have about 20 000 genes. Why do you think we have so many genes?

7 Explain why differences in intelligence between individuals are not just due to genes. (A*)

24: Reproduction

Learning objectives

After studying this topic, you should be able to:

- ✔ know that there are two forms of reproduction – sexual and asexual
- ✔ recall that new plants can be produced quickly and cheaply by taking genetically identical cuttings from older plants

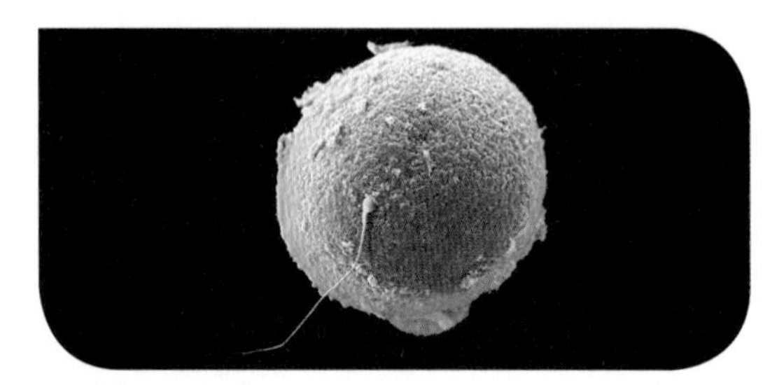

▲ Fertilisation: a human sperm penetrating a human egg (× 500)

▲ This hydra, a small aquatic animal related to jellyfish, can reproduce asexually. The small offspring branching out on the left will eventually break away and become independent. It is a clone of the single parent.

Sexual reproduction

Sexual reproduction involves two different **gametes** (sex cells) joining.

- Each gamete has only half the number of chromosomes of a normal body cell.
- The female gamete is called an egg. It has 23 chromosomes in its nucleus. It is larger than the male gametes. Adult female humans produce one egg each month.
- The male gametes are called sperms. Adult human males make hundreds of millions of sperms each day. Each sperm has a nucleus containing 23 chromosomes.
- One sperm will **fertilise** one egg.
- When this happens, the two nuclei fuse and the chromosomes from the father can pair up with chromosomes from the mother.
- The resulting new individual has chromosomes from two parents. They have a mixture of genetic information. This is why the offspring have genetic variation.

A What are gametes?

B How are male gametes different from female gametes?

C Explain why offspring are genetically different from each other and from their parents.

Asexual reproduction

Some organisms can reproduce **asexually**. Only one parent is needed, and

- there are no gametes
- there is no mixing of genetic information from two parents
- there is no genetic variation among the offspring
- the offspring are genetically identical to each other and to the parent.

Bacteria and some other single-celled organisms can reproduce asexually. Some more complex organisms, like the hydra shown on the left, can also reproduce asexually.

Cuttings

Some plants can reproduce asexually as well as sexually. We can take **cuttings** from plants. New plants produced by cuttings are all genetically identical to their single parent plant.

The advantages of taking cuttings are:

- The new plants are made quickly and cheaply.
- The new plants are genetically identical to the parent plant, and so will have the desirable characteristics of the parent plant.

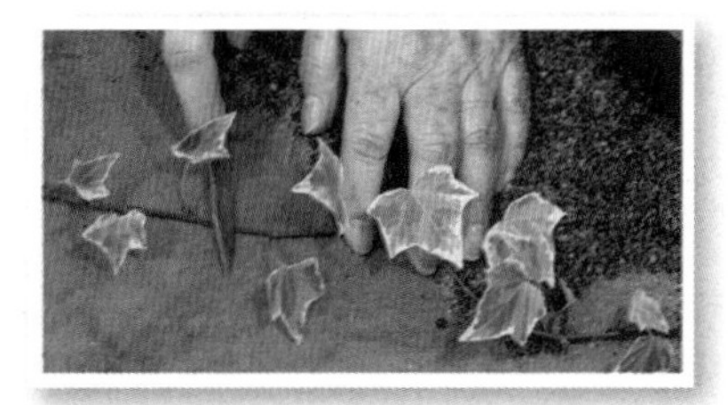

Cuttings can be taken from a piece of ivy. These are then planted to grow into new plants.

Taking cuttings has been done for many years. The plants produced in this way are **clones**, because they are genetically identical to each other and to the parent.

Questions

1 State one way in which male gametes are the same as female gametes.

2 When a sperm fertilises an egg, what happens to the two nuclei?

3 Why are cuttings taken from a plant all genetically identical to the parent and to each other?

E

4 State two advantages of producing plants by taking cuttings.

5 Sometimes the new cutting is placed in a pot and a clear plastic bag placed over the whole thing. Why do you think this is done?

C

6 Plants reproduce sexually by making seeds. A plant grower has a variety of geranium plant that is particularly attractive and sells well. Should he use cuttings or seeds to grow lots of them? Explain your answer.

A*

Key words

sexual reproduction, gametes, fertilisation, asexual reproduction, cuttings, clones

Did you know...?

Identical twins are clones of each other.

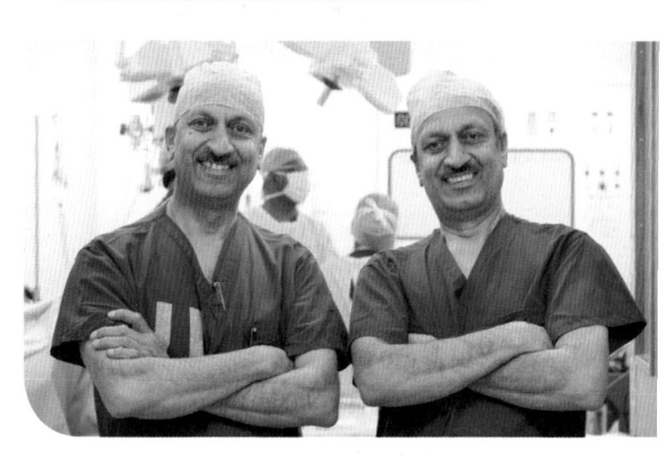

Identical twin brothers. Not only do they look exactly alike, they are both senior operating assistants in a hospital.

Exam tip AQA

✔ When you are describing offspring produced by asexual reproduction, don't just say they are identical, say they are genetically identical to each other and to the parent.

Geranium flowers

Learning objectives

After studying this topic, you should be able to:

- know that modern cloning techniques include tissue culture, embryo transplants, and adult cloning
- interpret information about cloning techniques
- make informed judgements about issues concerning cloning

Exam tip AQA

- Try not to be emotional if you are asked about uses of cloning. Consider the social and ethical issues and try to be objective.

A Why does the liquid or jelly used for tissue culture have to be very clean?

B Why are the calves produced by embryo transplant genetically identical to each other?

Did you know...?

Dolly the sheep was a celebrity and could not be allowed to run in the fields, as she was valuable. However, she gave birth to six lambs, all healthy and born naturally. She died, aged seven years old, of a type of lung cancer caused by a virus. The other (non-cloned) sheep kept in the same barn also died of it.

Tissue culture

Tissue culture can be used to make new plants by asexual reproduction. Scientists and technicians take small groups of cells from part of a plant and put them into a special liquid or jelly. The liquid or jelly has to be very clean so that there are no bacteria or moulds. It may also have some special chemicals to promote the development of these cells into root cells, stem cells, and leaf cells, so that new plants grow.

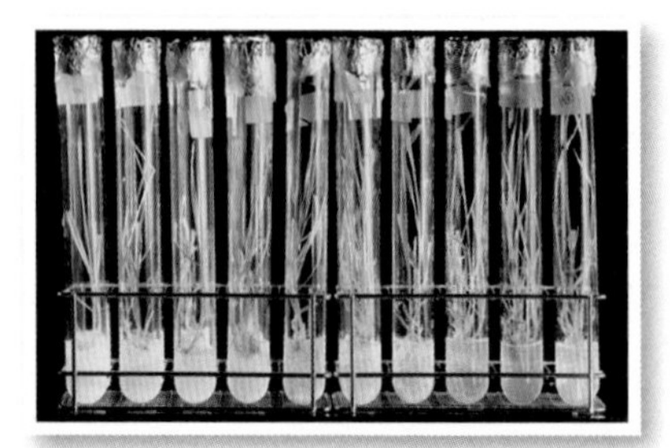

Cereal plants being grown by tissue culture. All of these plants are genetically identical to the parent plant. They will all have the parent plant's desirable characteristics.

Embryo transplants

Embryo transplants may be used to produce cattle with desirable characteristics.

- Eggs are obtained from prized female cows and fertilised with a selected bull's sperm, in a dish.
- Each resulting embryo is allowed to develop in the dish to the eight-cell stage.
- Then it can be split into four two-cell embryos.
- These new embryos then each develop to eight cells.
- Each can then be put into the uterus (womb) of a less-valuable cow.
- These cows are host mothers (surrogates). The embryos develop inside them and the calves are born.
- The calves are all genetically identical to each other.

In this way, one valuable cow with good characteristics can have many offspring within a short space of time.

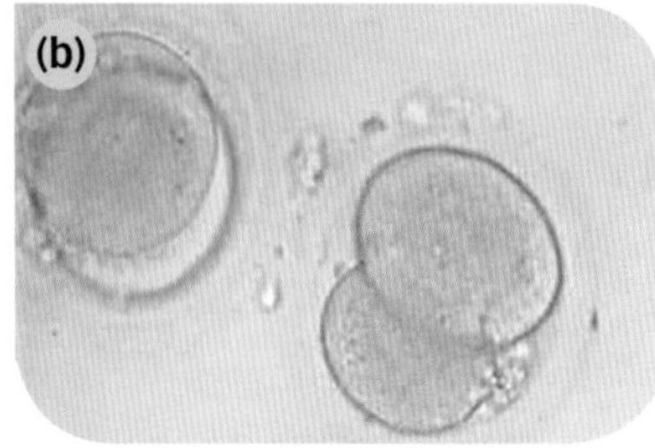

(a) Cow embryos being removed from storage in liquid nitrogen
(b) Embryos at the two-cell stage

Adult cell cloning

Adult cell cloning was first carried out in 1996 in Scotland by Professor Ian Wilmut and his team.

- An unfertilised egg was taken from a Scottish Blackface ewe (female sheep).
- Its nucleus was destroyed.
- A cell was taken from the mammary gland of a six-year-old Finn Dorset ewe.
- Its nucleus was implanted into the empty Scottish Blackface egg.
- An electric shock was given to the resulting egg cell to make it divide.
- It developed into an embryo that was put into a host mother sheep.

The resulting cloned sheep was Dolly, a Finn Dorset ewe.

Professor Ian Wilmut and Dolly

Dolly was produced because scientists had created genetically engineered sheep that can make useful medicines for humans in their milk. These sheep could breed, but their offspring may not inherit the human gene that made their milk useful. Also, half of their offspring would be male and would not make milk. If these sheep could be cloned instead, then many sheep able to make the valuable medicine could be created.

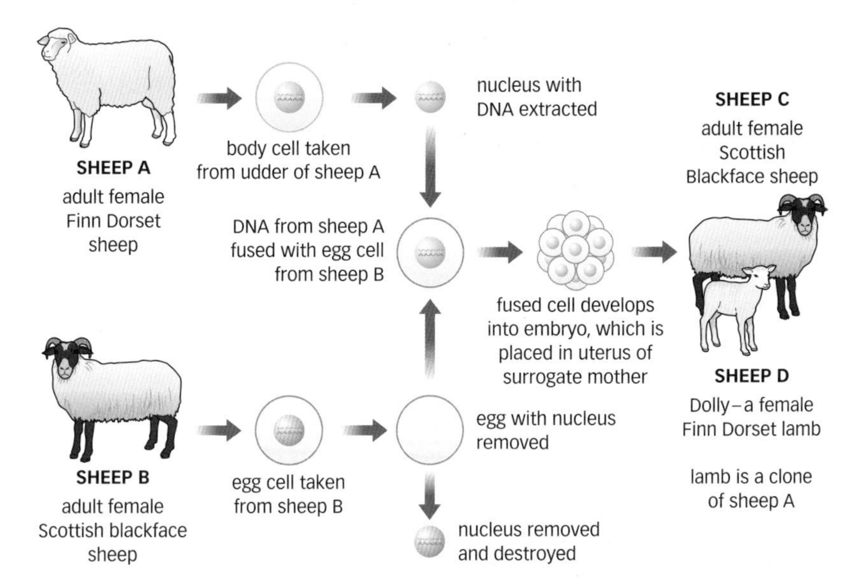

Flow diagram showing adult cell cloning

Key words

tissue culture, embryo transplant, adult cell cloning

Questions

1. What is adult cell cloning?
2. Why are the calves produced by embryo transplant not genetically identical to
   (a) their true mother?
   (b) their surrogate mother?
3. Explain how Dolly, a Finn Dorset ewe, was born to a Scottish Blackface ewe.
4. Dolly's birth was kept secret for six months whilst the researchers wrote their paper and had it peer-reviewed before publication. When the news broke in 1997, some of the headlines included: *'Golly Dolly! It's the abolition of Man.' 'Terrified researcher tells of how Dolly kills and eats a lamb.' 'The clone rangers need to be stopped.' 'Human cloning not far away.'* Comment on the use of such sensationalism.
5. Human cloning is illegal and is not carried out (except in nature in the form of identical twins!). Discuss the social and ethical issues around animal cloning.
6. The press reported that Dolly died young because she was cloned. Does the information about her death on page 68 support their view?

E C A*

26: Genetic engineering

Learning objectives

After studying this topic, you should be able to:

- state that in genetic engineering genes can be transferred from one cell to another cell
- state that genes can also be transferred to the cells of animals or plants at an early stage in their development, so that they develop the desired characteristics

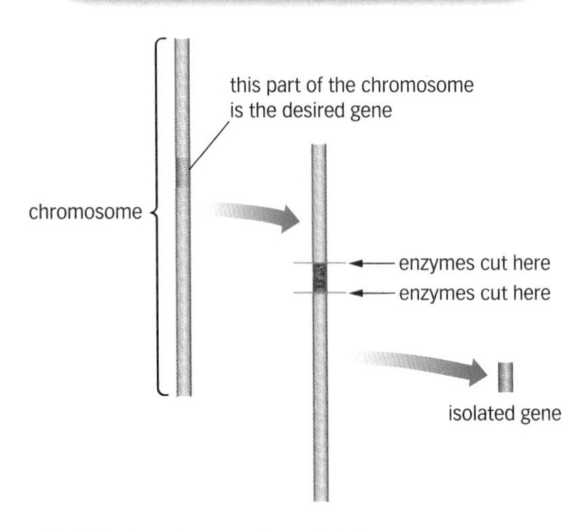

▲ Using an enzyme to cut out a gene

A How can scientists cut out genes from human chromosomes?

B What are the advantages of using genetically modified bacteria to make human insulin?

What is genetic engineering?

Genetic engineering means changing an organism's genes.

In the past, selective breeding programmes have been used to introduce genes from one variety of plant or animal into another. This takes a long time – around 20 years of careful breeding.

Now scientists can make use of various **enzymes** to manipulate genes, and to speed up the process of transferring genes from one organism to another.

Cutting out genes

Chromosomes and genes are made of **DNA**. Scientists have obtained special enzymes from bacteria, and these enzymes can cut DNA at particular places. There are many different types of these enzymes, and different ones can be used to cut DNA. If a gene is cut at either end, it can be cut out from its chromosome.

Making insulin: genetically modified bacteria

Genetic engineering is used to produce human insulin. The gene for human insulin is put into bacteria. The bacteria then multiply and produce insulin. The insulin is harvested from the bacteria and used to treat people with diabetes. This method is better than the old method of obtaining insulin from pigs' pancreases because

- The bacteria can be easily grown in large amounts.
- Scientists can now make enough insulin to treat all the people with diabetes.
- The insulin the bacteria make can be easily purified.
- It is suitable for vegetarians or anyone who objects to obtaining products from pigs.
- There is no risk of transmitting a disease.

Treating emphysema: genetically modified sheep

Transferring genes to early embryos

Human genes that have been 'cut out' can be put into cells of another organism, such as a mammal or a plant. The genes need to be transferred into an early embryo. Then, as the embryo divides, all the cells of the resulting individual will have the new gene.

Genetically engineered female sheep have been produced that have a gene for a human protein. This protein is too large to be made in bacterial cells. The genetically modified sheep make the human protein in their milk. The sheep can be milked and the human protein is extracted from the milk. It is used to treat people with hereditary emphysema.

The genetically engineered sheep are produced in this way:

The human gene for the protein is 'cut out' from a human chromosome.

↓

Several copies of the gene are obtained.

↓

They are put into fertilised sheep eggs, in a glass dish.

↓

The fertilised eggs are allowed to divide into a ball of cells.

↓

One cell is taken from each ball of cells, and the chromosomes are checked to find out which ones are female.

↓

The balls of cells that will develop into female sheep are put back into the female sheep wombs.

↓

The embryos develop and lambs are born.

The advantages of producing the human protein in this way are

- The sheep are not harmed in any way.
- The sheep are kept on farms in good conditions.
- The protein is only made in their milk.
- The protein can be easily collected without hurting the sheep – just as we milk cows.
- These sheep are valuable and so will have a longer life than many other sheep that are killed for meat.
- The protein can be easily separated from the milk, and is pure and uncontaminated.
- A lot of the protein can be made in order to treat people who are ill.

One disadvantage is that this can only be done in parts of the world where sheep can live. However, the medicine produced can be exported all over the world.

Key words

genetic engineering, enzyme, DNA

Exam tip

✔ Always remain objective when discussing ethical issues. Try to give a balanced argument.

Questions

1 Why are genetically modified sheep used to make a human protein to treat hereditary emphysema?

2 Why is the human gene put into an early embryo sheep and not into a developed lamb?

3 Why do you think only female embryos are used?

C

4 Discuss the ethical issues of making the human protein in this way.

Learning objectives

After studying this topic, you should be able to:

- understand that new genes can be transferred to crop plants, which are then called genetically modified crops
- know that some GM crops are resistant to insects or to **herbicides**
- know that GM crops may show increased yields or improved nutritional content

Key words

genetic modification, **herbicide**

Genetically modified (GM) plants

Humans have been selectively breeding plants for about 10 000 years. They have always tried to increase the yield or make a crop grow in an environment where it does not normally grow. **Genetic modification** (GM) speeds up the process.

The human population has increased a lot and is still growing. By 2030 it will be about nine billion. More people, and their roads and buildings, take up more space. So we have to grow more food on less land. It is very unlikely that we can do that without using GM plants.

Using GM crops will not be the only way to improve food production, but it is likely to be an important one. We cannot just keep on using more fertiliser and pesticides, as they have adverse effects on the soil and environment.

Examples of GM crops

GM crop plant	Use
Soya beans	Resistant to weedkiller, so that weeds can be sprayed and killed without killing the soya. Some 77% of all soya grown in the world is genetically modified. Most is grown in the US and Brazil.
Corn (maize)	Resistant to a pest, the corn borer. This is a tiny insect that eats into the corn stalk so that the plant falls over and does not produce seeds for humans or livestock to eat. The holes also let fungi get in, and these produce toxins. GM maize has been grown in the US and Canada since 1997. Some 80% of the maize grown in the US is GM. GM maize is now also grown in Spain, Portugal, the Czech Republic, and Germany.
Golden rice	Contains vitamin A in its grains. White rice does not contain vitamin A. Rice is the major part of the diet in many developing countries. Each year 600 000 children worldwide go blind due to lack of vitamin A. Many of these children die within a year of going blind, as vitamin A is also needed for growth and protection from infections. Golden rice contains vitamin A and is a good way of providing enough vitamin A to children in developing countries at no extra cost.

Did you know...?

If potatoes and tomatoes (both of which originated in South America) were introduced to us today as novel foods, they might well not get past safety tests. People would probably object to their use as foods, as parts of them contain toxins. They are both in the deadly nightshade family.

Cotton	Resistant to pests. Cotton fibres are used for textiles, and the seeds provide oil and protein for animal feed or oil for margarine. The cotton seed capsule (boll) is attacked by caterpillars. The plants used to be sprayed with chemical pesticides but often these did not kill the caterpillars. The chemicals stayed on the outside of the boll and the caterpillars are inside. GM cotton has a gene from a bacterium. The gene codes for a toxin that kills the caterpillars. This Bt toxin has been used for decades by extracting it from bacteria. Now with GM cotton, the cotton plants themselves make the toxin and kill the caterpillars even if they are inside the boll. Most GM cotton is grown in India and the US, but some is grown in Argentina, Mexico, South Africa, Australia, and Columbia. 68% of cotton grown in China is GM.
Tomatoes, potatoes, squash, papaya 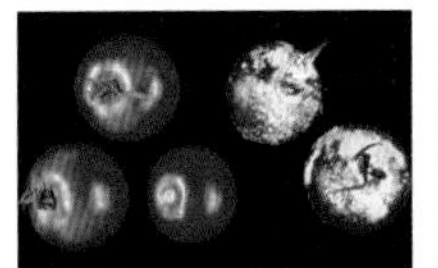	Resistant to pests. If crops are resistant to pests, fewer chemicals have to be used. This reduces the risk of killing useful insects or of the chemicals entering the food chain. Three of the tomatoes on the left are GM. They are resistant to a mould fungus.
Bananas 	Resistant to pests and containing extra nutrients, like zinc. In some African countries bananas form a great part of the diet. Zinc is an important mineral, and where people do not eat much meat, they often do not get enough zinc. Farmers can grow GM bananas that contain more zinc and are resistant to diseases. They do not need to use pesticides. The farmers are not exposed to the harmful chemicals, and neither is the environment.
Corn	Makes 'fish oils'. These oils are normally obtained from oily fish and help human brain development. However, fish do not make these oils. They get them from eating algae. The gene from algae has been put into a variety of corn. This could provide the 'fish oils' humans need, particularly as fish stocks are dwindling.
Tomatoes, melon	Longer shelf life, as ripening is delayed. This means the fruits can be ripened on the plants before being picked. This enables them to have more flavour.

Questions

1. Why do humans need to increase the amount of food grown in the world during the next 20 years and beyond?
2. Humans have increased food production greatly over the last 50 years. They grew new varieties of crop plant that had better yields. How were these varieties produced?

E

3. The 'green revolution' that started in the 1960s involved using much more pesticide and fertiliser chemicals. Why can we not just increase the use of these chemicals over the next 20 years to increase crop yields?
4. Were there any harmful effects from these chemicals?
5. Explain why using GM crops can reduce the amount of fertiliser and pesticide chemicals used.
6. What is the advantage of golden rice, compared to normal white rice?

C

7. Why may GM corn that contains 'fish oils' be important for humans in the future?

A*

28: Concerns about GM crops

Learning objective

After studying this topic, you should be able to:

- ✔ make informed judgements about the economical, social, and ethical issues surrounding GM crops

Key words

hectare, fertiliser, pesticide, yield

▲ A researcher at a research station in the UK compares growth of GM crops with non-GM crops

GM crops are controversial

Trials have to be carried out on new GM crops to see if there are any associated health risks or risks to the environment. Protesters have sometimes destroyed the trials, so they have to be carried out in secret places. Because they are controversial, many GM crops developed nearly 20 years ago are still not being grown, although they are safe and could help many people.

Genetic engineering was used to develop GM crops because it was faster than using a selective breeding programme. However, the 'red tape' of bureaucracy, together with public pressure, has held up the commercial growing of many of these crops. Some, such as golden rice, were developed 20 years ago and are still not being grown. Had this rice been developed using selective breeding it would also have taken 20 years and would have been introduced by now without protest. It would have saved many lives.

What are the objections?

When Sainsbury's first introduced tomato purée made from GM tomatoes in 1991, it was clearly labelled and people had no objections. It tasted good and sold well.

▲ Tomato puree made from GM tomatoes

Monsanto, a large US-based company, has developed strains of GM corn and soya. At first they did not label their GM produce, and this upset many people, as it took away their ability to choose. One supermarket chain reacted by saying they would not sell any GM produce. Other supermarkets followed suit in order not to lose trade, and the press joined in with sensationalised articles about 'Frankenfoods'.

We need to look objectively at some of the arguments.

- Many people oppose the idea of eating GM crops. Some think it is 'interfering with nature'. However, this is what humans regularly do – vaccination, surgery, curing illnesses, and growing more food per **hectare** (by using **fertilisers** and **pesticides**) to feed a growing population are all interfering with nature. You could argue that farming is interfering with nature. However, humans are also part of nature, and we harness it to meet our needs.
- Some trials show that GM crops do not produce a higher **yield**. Other trials show that they do. More trials need to be carried out. In some cases the purpose of genetic modification is to improve the nutrition or to make the plants resistant to disease. So individual plants will not produce more yield, but each hectare will if pests are not destroying crops.
- Some GM crops are resistant to pests, so they need fewer pesticides. This is good for the environment, as the pesticides often kill useful insects as well as harmful ones. A particular pesticide has recently been banned in Germany and France because many dead bees have been found to have residues of it in their tissues. However, this pesticide has not yet been banned in the UK.
- In 2003, Zambia banned the import of GM crops and this led to widespread famine. In 2005, they allowed GM maize in.
- Most people who oppose GM crops live in developed countries and are not at risk of starvation or of becoming blind due to lack of vitamin A.
- GM foods have been eaten in the US for well over 10 years and have not caused health problems. This is a good natural experiment – the US population compared to the control group in Europe.
- Golden rice had to be trialled to see if it caused allergies now that it contained vitamin A, even though humans have been eating carrots (a good natural source of vitamin A) for thousands of years. Much of the non-GM food we eat has never been tested.
- Many non-GM foods (such as processed foods) do cause health problems because they contain a lot of saturated fat and salt. Both of these are clearly linked to heart disease and to cancer (35% of all cancers are diet-related).
- Studies of areas where GM crops are grown, compared with areas where they are not grown have shown that GM crops do not appear to affect the local wildlife, neither plants nor animals. However, many other human activities do affect wildlife, such as extracting and burning oil.

Did you know...?

At the time of writing there are no commercially-grown GM crops in the UK. A few are grown in Spain, Portugal, and Germany. Most are grown in the US, Canada, Brazil, China, India, and Australia.

Questions

1 Why was genetic engineering used to produce new varieties of crops?

E

2 Make a table. On one side list arguments in favour of GM crops. On the other list arguments against GM crops. Share your ideas with the rest of the class. Make a poster summarising the class's pros and cons.

C

3 Use the Internet to find out more about:

(a) golden rice
(b) Flavr Savr tomatoes
(c) GM soya
(d) GM potatoes
(e) GM maize.

A*

You might like to work in small groups, each group researching one of the topics. Then prepare a presentation to share your information with the rest of the class.

Learning objectives

After studying this topic, you should be able to:

- ✔ know that in classification, organisms are grouped based on similarities and differences
- ✔ know the characteristics of some of the major groups

Key words

classification, kingdom

Similarities and differences

The world is full of millions of different types of living things. Biologists put living things into groups, which makes them easier to study. This grouping process is called **classification**.

To classify living things biologists observe their features (characteristics). You can observe both similarities and differences in these two types of daffodil.

Similarities:
- same-shaped leaves
- same number of petals

Differences:
- different colour
- different height

Living things that share lots of similar characteristics are grouped together. If organisms have lots of differences in their characteristics then they are classified in different groups.

Kingdoms

All living things are placed into major groups called **kingdoms**. Each kingdom has different characteristics.

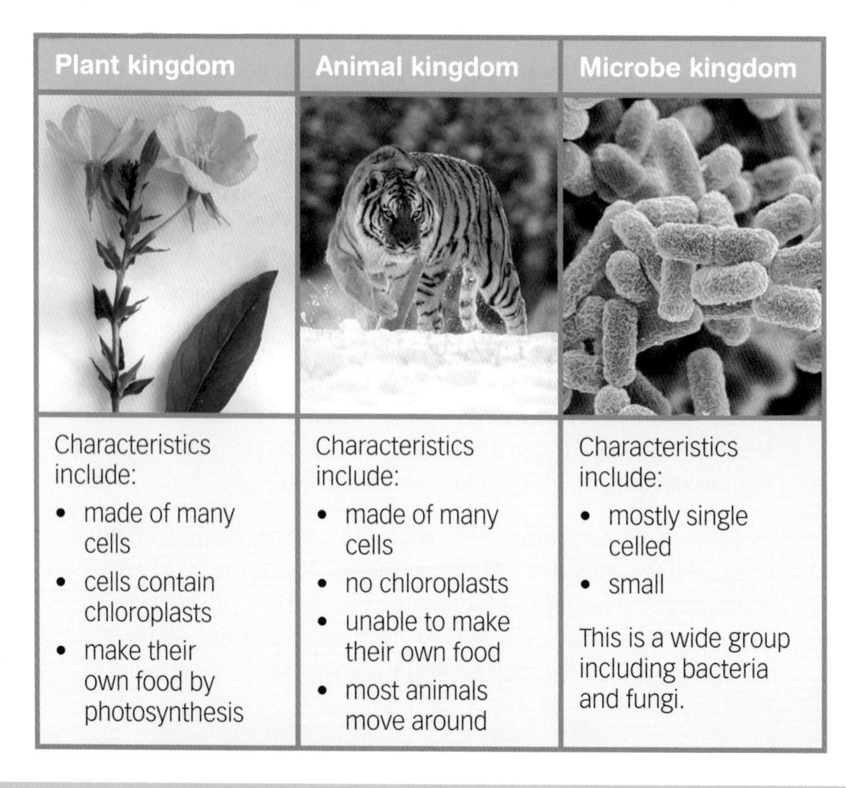

Plant kingdom	Animal kingdom	Microbe kingdom
Characteristics include: • made of many cells • cells contain chloroplasts • make their own food by photosynthesis	Characteristics include: • made of many cells • no chloroplasts • unable to make their own food • most animals move around	Characteristics include: • mostly single celled • small This is a wide group including bacteria and fungi.

A List features that are common to both cats and dogs, ie characteristics used to place them in closely related classification groups.

B List features of goldfish and crabs that they don't share, ie characteristics used to place them in different classification groups.

C Why do biologists classify living things?

D New organisms are discovered every day. How would you go about deciding which group a new discovery should be placed in?

Evolutionary links between species

If two species are closely related, then they tend to share more characteristics. This is because they both had the same close ancestor. Chimpanzees and humans share a number of features. They both

- have a backbone
- have skin covered in hair
- feed their young with milk
- have grasping hands.

This is because they both had the same common ancestor. If you compare these two species with a lizard, there are far fewer similarities, they all have a backbone, for example.

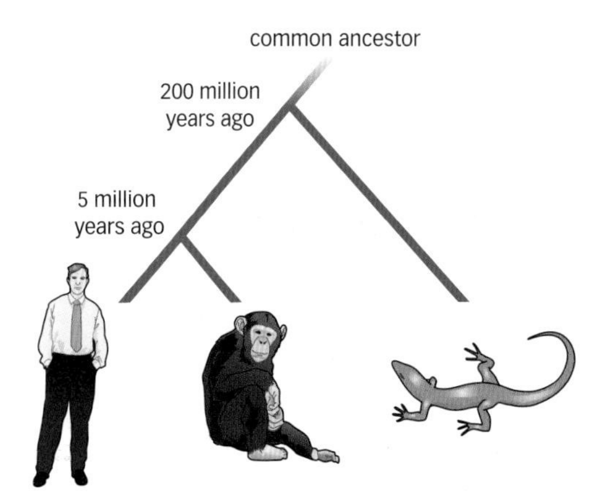

This family tree shows the relationship between three species. The human and chimp are far more closely related to each other than they are to the lizard.

Variations to suit a habitat

Closely related species may look very different if they live in different habitats. The Arctic fox lives in an arctic habitat and the fennec fox lives in a desert habitat. They still share characteristics.

On the other hand, organisms that are not closely related may share several features if they live in the same habitat. For example, the shark and the dolphin both have an ocean habitat. They show some similarities but they are not closely related.

Arctic fox

Fennec fox

Did you know...?

The ancient Greeks were the first to try to organise and describe living things. Aristotle attempted to group animals, whilst his student and friend Theophrastus did the same for plants.

Exam tip AQA

✔ When classifying organisms, list their similarities and differences.

Questions

1. State the main features of a plant. (E)
2. Explain why humans are classified as animals. (C)
3. Sketch a possible family tree for the following four organisms, based on their similarities: zebra, horse, lizard, and goldfish. (A*)

30: Surviving change

Learning objectives

After studying this topic, you should be able to:

- know the causes and effects of evolution

Key words

evolution, natural selection, mutation

A Define evolution.

B What is an ancestor in the theory of evolution?

Evolution in action

Evolution

The world is full of millions of different species. Where did they all come from? This has puzzled biologists for many years. One big question is that of how life first began. Biologists now believe that simple life forms evolved in primitive oceans from large complex molecules. From these simple beginnings, all other organisms have evolved. This has taken over three billion years.

Other questions about where all the different organisms came from include:

- Why are there so many different species?
- How do species change or adapt over time?
- How did all the different species form?
- Why are some species closely related to each other?

Most biologists believe that an idea known as evolution best answers all four questions.

There are many different habitats in the world. In each habitat there are organisms that are well adapted to survive. This results in lots of species. But the habitats of the world are constantly changing. The organisms must also change to survive.

Evolution is the gradual change of an organism over time. This idea suggests that gradually one type of organism, called the ancestor, might change over many generations into one or more different species. This generates lots of different species over time. But how do organisms change?

Natural selection

The biologist Charles Darwin suggested an idea to explain how one species can change into another. This is now widely accepted by most biologists and is called **natural selection**. This theory can be used to explain how the giraffe's long neck evolved. There are four major steps in the theory.

1. Large populations showing variation

- Most species produce lots of offspring. This should cause a massive population growth for every species. One original pair of giraffes would produce millions of giraffes over a few hundred years.
- Individuals in a population will show a wide range of variation, because of differences in their genes.
- This variation is caused by chance **mutations** in genes.

2. Survival

- Not all of the organisms in a population survive and reproduce.
- Their survival is affected by changes in the environment. Some die from disease. Some starve. Some are eaten by predators. Some cannot find a mate.

3. The fittest

- Some of the variations will be an advantage. For example, some giraffes have longer necks than others.
- Longer necks allow those giraffes to reach leaves higher on the tree.
- When food lower down is scarce, those giraffes without the advantage of a long neck will die.

4. Passing on the advantage

- The surviving giraffes are the only ones that reproduce.
- Their offspring inherit the advantage. The gene for long necks has been passed on.
- Over many generations the number of giraffes with long necks increases.
- The result is that the giraffe species has evolved a long neck.

Questions

1 Darwin used the term 'survival of the fittest'. Explain what this means.

2 Pet shops sell white and brown rabbits. White rabbits are easily seen by foxes. Use Darwin's theory of evolution to explain why white rabbits are rare in the wild.

3 Explain why evolution does not often happen over short time periods.

Learning objectives

After studying this topic you should be able to:

- know that evolution can result in the formation of new species
- give several examples of evolution in action

A Explain why Darwin could not suggest how a new species forms.

B Explain why mutations are central to the formation of new species.

Forming new species

Sometimes two separate groups of the same **species** evolve differently. Each group gradually changes over time, becoming more different from each other. This can result in two new species.

Darwin did not suggest a mechanism for the formation of two new species, because at the time people did not understand inheritance and variation. Modern biologists have explained the process of forming new species:

- A single population becomes separated into two groups. The environment may be different for the two groups.
- Over time, each separated group of the population evolves differently.
- The longer they are separated, the more different they become.
- Eventually the two sub-populations have changed so much that they can no longer interbreed.
- They have formed separate but closely related species.

▲ The African elephant (left) and the Asian elephant (above) are closely related species, which both evolved from a common ancestor many years ago when they were separated by great distances.

Variation is the key to the development of new species. Mutations of genes lead to inherited variation in characteristics. Where useful mutations occur, they may cause a rapid change in the species.

Evidence for evolution

Biologists have looked for evidence that evolution happens by natural selection and survival of the fittest. They have found scientific observations that support the idea of evolution.

The peppered moth (*Biston betularia*)

Perhaps the best known example of evolution in action involves a moth called the peppered moth. This moth is pale in colour, and looks as though it has been sprinkled with pepper.

The pale moth was well camouflaged against the light bark of trees. During the 1800s, trees in industrial areas became covered in soot particles, and the bark became much darker. The moths were no longer camouflaged, and they were eaten by birds.

A mutation occurred in some moths, making them much darker in colour. The darker moths now had the advantage of camouflage.

▲ Light and dark varieties of peppered moth (*Biston betularia*)

Over the next 50 years, the dark variety became more common.

Today, in cleaner areas the light form of the moth is more common again. In industrial areas the dark form is still the more common form.

Key words

species

Questions

1 Use Darwin's theory of natural selection to explain the steps involved as the moths changed to the darker form.

E

2 What happens to the dark-coloured moths in the cleaner areas of Britain?

3 Lions and tigers are two closely related species which evolved from a common ancestor. Explain how moving to different habitats has resulted in the formation of these two different species.

4 When explorers discovered Australia they were amazed by how different the animal species were from those back home. Explain why Australian animals are so different from European ones.

Learning objectives

After studying this topic, you should be able to:

- ✔ know how Darwin collected his evidence
- ✔ evaluate other theories of evolution

▲ Charles Darwin aged 40

▲ HMS *Beagle*

How did Darwin gather his evidence?

Charles Darwin took years to come up with his theory of evolution. He had studied living things at university, and became the ship's naturalist on the HMS *Beagle*. During his voyage he collected evidence which helped him develop his ideas on evolution.

The ship visited the Galapagos Islands off the coast of South America. Darwin visited each of the islands and made notes on the different types of plants and animals. Darwin was struck by the variety of types of finches he found on the islands, which he thought were unrelated species. But after studying them, he realised that they were closely related species.

Darwin developed his ideas of evolution to explain how the different species of finch might have arisen.

- An ancestral finch species arrived on all the islands.
- Gradually over time the finches on separate islands changed independently.
- Some birds became adapted to feed on insects.
- Other birds adapted to eat seeds or fruit.
- This gave rise to lots of different finch species, that were different on the different islands.

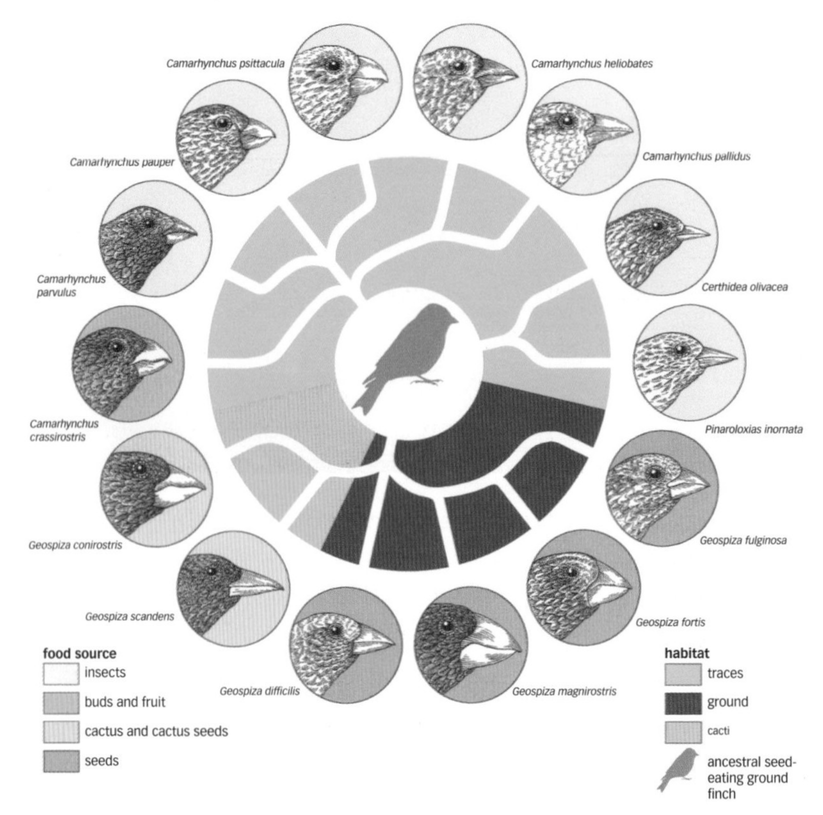

▲ These Galapagos finches all evolved from a common ancestor

How was Darwin's theory accepted?

Charles Darwin published his ideas in a book called *On the Origin of Species by Means of Natural Selection* in 1859. Many people were horrified by his ideas. The main objections were:

- Most religions have an account of how life started – they describe humans and other species being created. Darwin's theory seemed to disagree with the religious viewpoint. People particularly did not like the idea that humans could have evolved from apes.
- There was not much evidence at that time to convince other scientists that species evolve from each other.
- Scientists then did not know about genes, and so could not explain how characteristics could be passed on to allow inherited variation and evolution.

A Scientists work by making observations. What sort of observations would Darwin have made of the finches to be able to explain their evolution?

An alternative idea: the work of Lamarck

Jean Lamarck was a French biologist who worked before Darwin. He suggested a different way to explain how species are formed in evolution. He would have explained the evolution of the giraffe as follows:

In a giraffe population, some giraffes want to feed off the leaves high on a tree. They have an 'inner need' to stretch their necks.

- The giraffes stretch their necks, becoming more successful.
- They pass their longer necks on to their offspring.

The ideas put forward by Lamarck are not accepted today. Living things cannot decide to alter their body because of an 'inner need'. Their bodies might change during their lifetime – for example, someone who works out in the gym will develop bigger muscles. However, this characteristic will not be passed on to their children. The only changes in organisms that can be passed on to offspring are changes in genes, not changes acquired during the organism's lifetime.

Exam tip AQA

✔ When comparing the ideas of two different scientists, list the key points of each one and say what is similar and what is different.

B Explain why it was important that Darwin wrote his book and spoke at scientific lectures to tell people about his ideas.

▲ Jean Baptiste Lamarck

Questions

1. Give three clear reasons why Darwin's theory was not immediately accepted. (E)
2. Explain why modern biologists think that Lamarck's ideas are not correct. (C)
3. The ideas of both Darwin and Lamarck can be used to explain the evolution of the giraffe's long neck. Explain how each of these scientists would have accounted for the giraffe's long neck. (A*)

Course catch-up

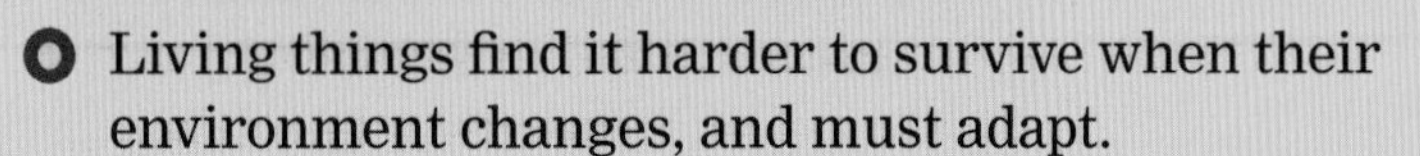

Revision checklist

- Living things find it harder to survive when their environment changes, and must adapt.
- Indicator species show how polluted air or water is. Sulfur dioxide levels affect lichen distribution, and dissolved oxygen concentrations control invertebrate distributions.
- The mass of living material (biomass) gets less as you move along food chains or up pyramids of biomass.
- Energy flows from the plants that capture it to the animals that make up the rest of the food chain, but some is lost at each step as heat and in waste materials.
- The elements in living things are recycled by detritivores, like worms, and decomposers, like fungi and bacteria.
- Photosynthesis traps carbon from the atmosphere in biomass. It passes along food chains and returns to the atmosphere when living things respire or decompose.
- Our characteristics are controlled by the genes we inherit, and by environmental factors like our nutrient intakes.
- Sexual reproduction fuses sex cells to produce offspring with genetic variation.
- Asexual reproduction gives identical clones from one parent.
- Clones are produced from plant cells, separated embryonic stem cells, or egg cells controlled by body cell nuclei.
- Genetic engineering transfers genes to other organisms. If early embryos are used, they develop the characteristics the genes code for.
- GM crops can resist insects or herbicides and produce increased yields or more nutritious crops.
- Some people worry that GM crops could harm human health or spread their genes to wild flowers and insects.
- To show the relationships between living things, we use their similarities and differences to classify them into groups.
- Darwin's theory of evolution by natural selection is the most widely accepted explanation of evolution.
- New species can form when populations are divided between different environments.
- Darwin's theory took years to be accepted because it conflicted with religious accounts, there was limited evidence, and no-one could explain inheritance.

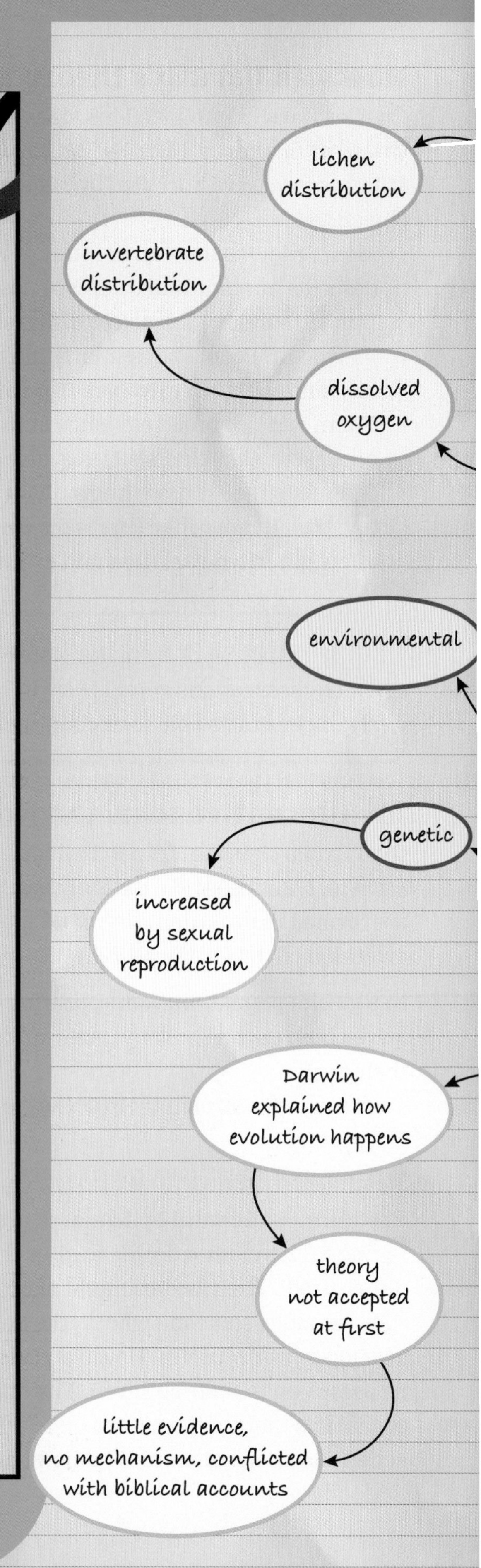

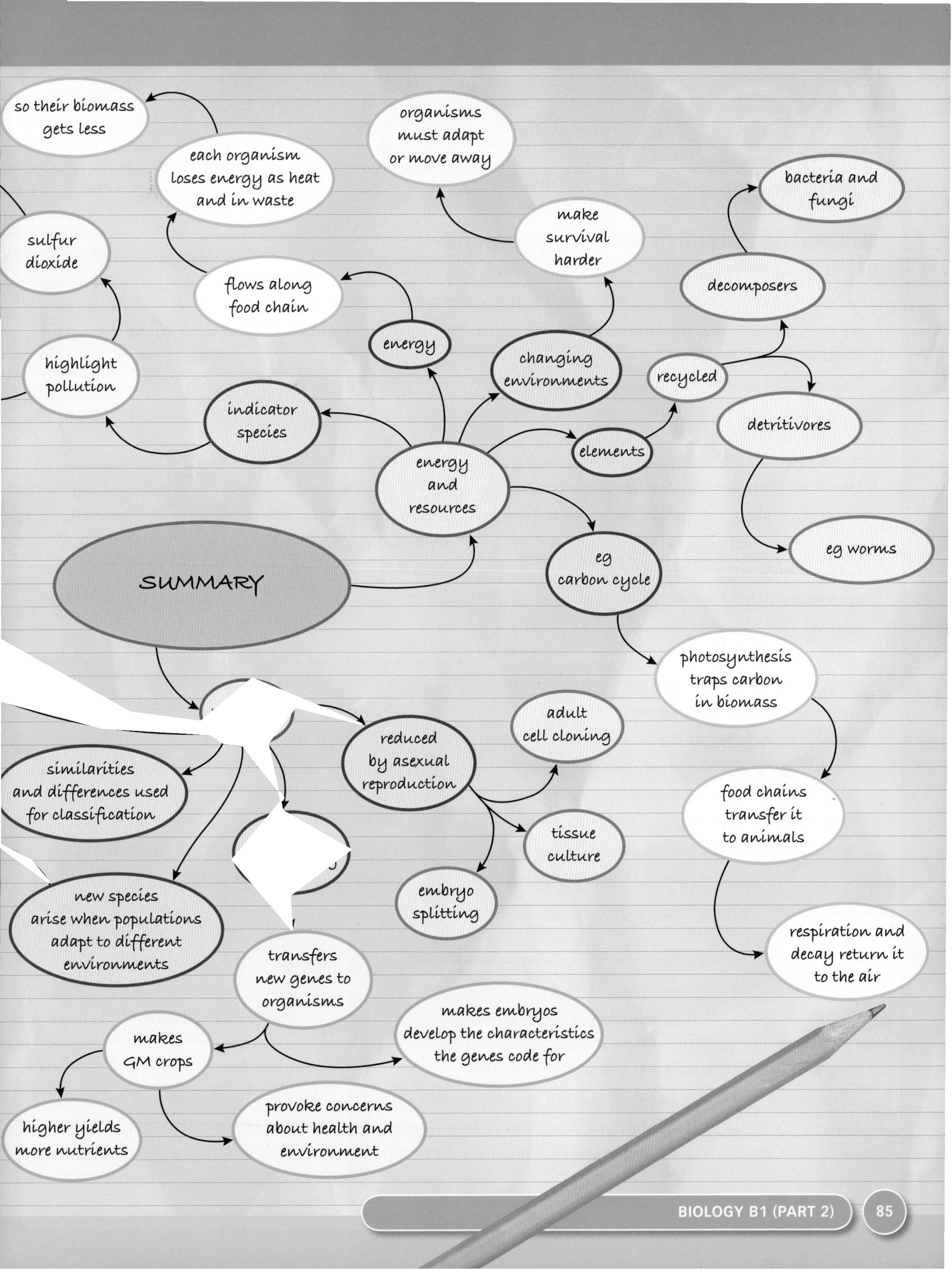
so their biomass gets less
each organism loses energy as heat and in waste
organisms must adapt or move away
bacteria and fungi
make survival harder
sulfur dioxide
flows along food chain
decomposers
energy
changing environments
recycled
highlight pollution
indicator species
detritivores
elements
energy and resources
eg worms
SUMMARY
eg carbon cycle
photosynthesis traps carbon in biomass
adult cell cloning
reduced by asexual reproduction
similarities and differences used for classification
food chains transfer it to animals
tissue culture
embryo splitting
new species arise when populations adapt to different environments
respiration and decay return it to the air
transfers new genes to organisms
makes embryos develop the characteristics the genes code for
makes GM crops
provoke concerns about health and environment
higher yields more nutrients

Answering Extended Writing questions

QUESTION

Genetically modified organisms (GMOs) can be created that produce medicines for humans.

What are the benefits and risks of producing medicines in this way? Explain why some people are concerned about the use of GMOs.

The quality of written communication will be assessed in your answer to this question.

G–E

Some animals different geans. People say this is against nature. However people used to say that about IVF but the man who did it got the No Bell prize and millions have now got kids.
If I had an illness Id want to be cured and woudnt care where the medicine came from.

Examiner: This candidate knows what genetic engineering is. No specific examples are mentioned, such as bacteria making insulin or sheep making medicine to treat emphysema. The candidate mentions one concern that some people have and tries to give a counterargument, but goes off the point. Quite a lot of grammatical errors.

D–C

Some sheep have been engineered to make medicine. Some people think this is playing god and interfering with nature. But if they had emphysemia they would want to be cured. The medicine is in the sheeps milk so they are not harmed, just milked and wont be killed to eat. Genetic engineering means changing your genes.

Examiner: This starts well with a good example. It gives some advantages and describes the concerns of some people. It explains what genetic engineering is, but the answer is not very well organised. There are a few grammatical and spelling errors. This answer is in the 3–4 mark band.

B–A*

Genetic engineering means changing an organism's genes. Bacteria can be engineered to make insulin for diabetics. This is better than getting it from pigs' pancreases as vegetarians won't object. Lots of bacteria can be grown in labs anywhere to make lots of insulin.
Some people think that scientists should not change an organism's genes as this interferes with nature. They may worry that the bacteria could escape from labs and make people ill.

Examiner: This is a well expressed answer. It is well organised and explains what genetic engineering is, covering some benefits and possible risks of genetic engineering. The spelling and punctuation are good. Some of the concerns that some people have are described.

Exam-style questions

1 A02 Match these ways of producing offspring to their uses.

Reproduction method	Use
sexual reproduction	growing identical plants
tissue culture	making a cat with the same genes as its mother
embryo splitting	growing plants with genetic variation
adult cell cloning	producing identical farm animals

2 Jen measured oxygen concentrations in a stream that runs past a pig farm.

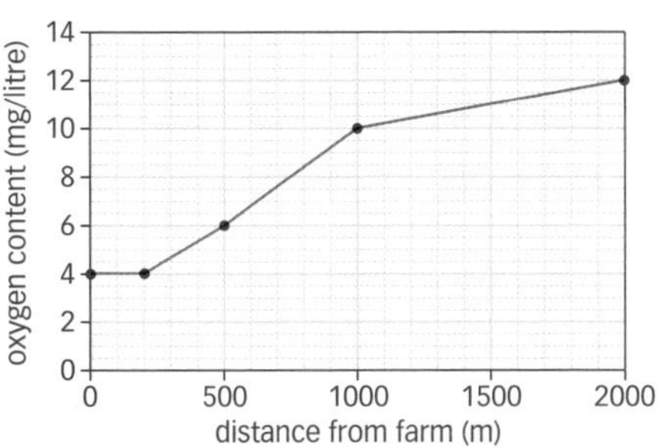

A03 **a** Where was the oxygen content of the water lowest?

A03 **b** How much did it rise between 500 m and 1000 m from the farm?

Jen took water samples and counted the number of mayfly nymphs in them.

Distance from farm (m)	Mayfly nymphs
0	0
200	0
500	6
1000	34
2000	35

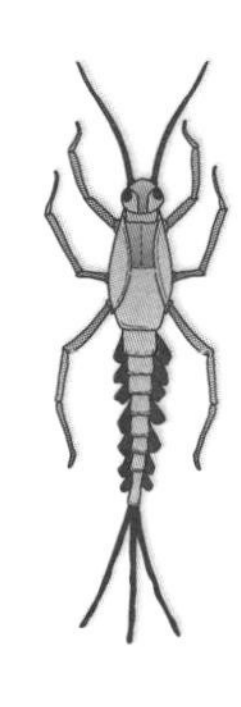

A03 **c** What can Jen conclude about mayfly nymphs?

A02 **d** Jen has a hypothesis. She thinks that the pig farm is polluting the stream. Suggest other measurements she could take to strengthen her evidence.

G–E D–C

3 The UK public has a history of opposition to GM crops. As at 2010, no GM crops had been grown in the UK for sale.

A02 **a** Suggest a useful characteristic that a GM crop could have.

A02 **b** What concerns might people have about growing GM crops in this country?

A02 **c** What concerns might people have about eating GM foods?

D–C B–A*

Extended Writing

4 A02 Kasia is a vegetarian. She says if everyone ate cereals instead of meat we could feed more people. Explain why she is right.

G–E

5 A01 The genes that code for an important human protein have been identified. Now scientists want to genetically engineer goats to produce it in their milk. List the steps the scientists need to take to make large quantities of the protein.

D–C

6 A03 Lamarck and Darwin both proposed theories that explained evolution. Using giraffes as an example, explain the similarities and differences between their theories.

B–A*

A01 Recall the science

A02 Apply your knowledge

A03 Evaluate and analyse the evidence

Cells and the growing plant

Why study this unit?

Photosynthesis is one of the most important biological processes. It is through photosynthesis that energy is trapped into the living world. Once trapped, this energy is used to power the entire living world in all its glory.

In this unit you will study photosynthesis as a process, and look at where the process occurs. You will also look at processes for sampling the distribution of living organisms in the environment.

The cell is often considered to be the basic building block of living things. This unit looks at the structure of plant and animal cells. You will observe how cells develop special structures to perform special functions. The principles of how cells work together to form organs, and how organs work together in organ systems, will form part of your study of cells. Finally, the problem of how molecules can get into and out of cells will be explored.

You should remember

1. You are made of cells that are organised into tissues, organs, and systems – such as the reproductive system.
2. Plant and animal cells both have a membrane, cytoplasm, nucleus, mitochondria, and ribosomes, but plant cells also have a cell wall and a large vacuole.
3. Cell structure and function.
4. The environment can be studied by sampling the distribution of organisms.
5. Plants make food by the process of photosynthesis.
6. Photosynthesis occurs in the leaves.
7. Diffusion is the movement of particles from a high concentration to a low concentration.

Have you ever considered how garden centres and nurseries get plants ready for sale at just the right time? Just how do poinsettia plants reach their stunning best in time for Christmas? It's all down to controlling plant growth.

The poinsettia has become a symbol of Christmas. In 2009, over 4 million British-grown plants were sold, plus many more imported plants. The production process for these plants starts in late August. Large greenhouses are used to grow them in. Shoots are pinched out to keep the plants short and full. To achieve the characteristic red colour, the plants are exposed to shortened days of less than 12 hours' light, with no light during the night. To produce strong, healthy plants, the daylight must be very bright, and the temperature must be kept above 10 °C, to promote the maximum rate of photosynthesis. The result will be strong, bushy plants, with glorious red colour.

Learning objectives

After studying this topic, you should be able to:

- know that plants and animals are built out of cells
- know the parts of cells and their functions

Key words

cell, microscope, nucleus, cytoplasm, cell membrane, cell wall, chloroplast, permanent vacuole

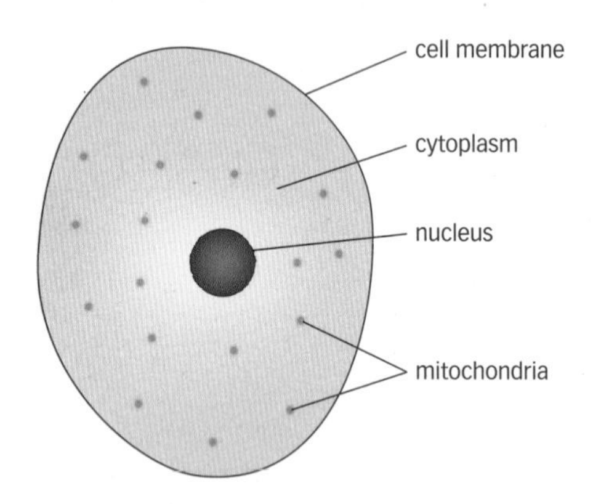

▲ A typical animal cell

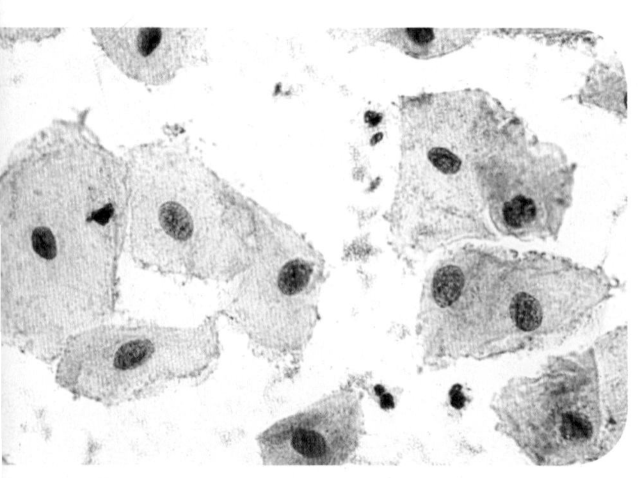

▲ Human cheek cells as seen through a microscope (×2600)

Building blocks

All living things are made of one or more **cells**. Cells are the building blocks of living things. The larger the organism, the more cells it will contain. Usually, cells are very small and can only be seen using a **microscope**. Cells were first seen by Robert Hooke in 1665.

▲ The type of microscope used by Robert Hooke

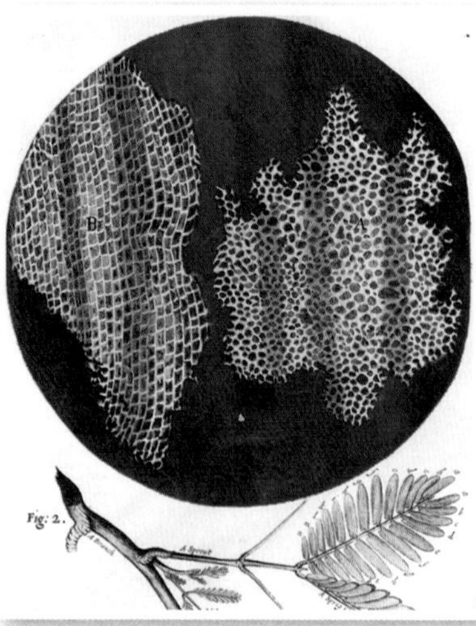

▲ Coloured enhanced image of the cells of bark seen by Robert Hooke

Cell structure

Typical animal cells

Animal cells come in many different types, but they have certain features in common.

Cell part	Description and function
Nucleus	A large structure inside the cell. It contains chromosomes made of DNA. The nucleus controls the activities of the cell, and how it develops.
Cytoplasm	A jelly-like substance containing many chemicals. Most of the chemical reactions of the cell occur here.
Cell membrane	A thin layer around the cell. It controls the movement of substances into and out of the cell.
Mitochondria	Small rod-shaped structures that release energy from sugar during aerobic respiration.
Ribosomes	Small ball-shaped structures in the cytoplasm, where proteins are made.

A Explain why cells could not be seen until the development of the microscope.

B Which part of a cell controls the activities of the cell?

C As you grow, what happens to the number of cells in your body?

Did you know...?

The cellulose in a plant cell wall helps support the plant, but it is also very useful to us. It makes the paper you are holding, and it is the fibre in our diet.

Typical plant cells

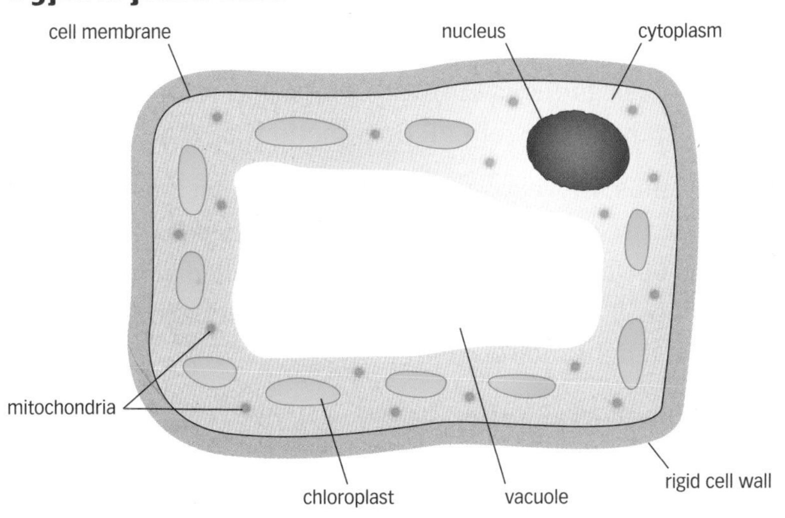

▲ A typical plant cell

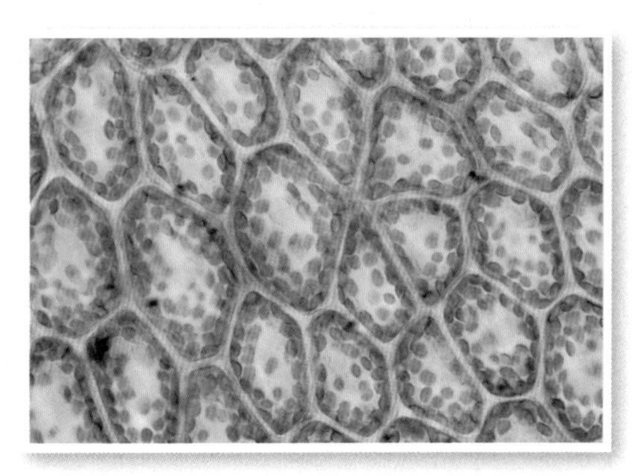

▲ Plant leaf cells seen through a microscope (× 500)

Plant cells have all of the structures seen in animal cells. But they also have one or two extra parts.

Cell part	Function
Cell wall	A layer outside the cell membrane. It is made of cellulose, which is strong and supports the cell.
Chloroplasts	Small discs found in the cytoplasm. They contain the green pigment chlorophyll. Chlorophyll traps light energy for photosynthesis.
Permanent vacuole	A fluid-filled cavity. The liquid inside is called cell sap. The sap helps support the cell.

A group of organisms called algae are closely related to plants. They include seaweeds. Algal cells have a cell structure exactly like that of plant cells.

Exam tip 

✔ Remember the difference between the cell membrane and the cell wall. Students often confuse these.

Questions

1 What is the function of the cell membrane? (E)

2 Plants are not as flexible as animals. Can you suggest a reason why? (C)

3 Not all plant cells contain chloroplasts. Suggest why root cells do not contain chloroplasts. (A*)

2: Bacterial and fungal cells

Learning objectives

After studying this topic, you should be able to:

- know the structure of bacterial cells
- know the structure of fungal cells

Key words

bacteria, fungi

Other types of cell

As well as animals and plants, two other important groups of organism are bacteria and fungi. These organisms are built out of cells that have a slightly different structure from animal or plant cells.

Bacterial cells

Bacteria are a large group of organisms. Some are useful to us, for example, for breaking down waste or making food, whilst others cause problems such as diseases. Bacteria are single-celled organisms. At first sight bacterial cells look simple, but they carry out all the functions of other cells.

Bacterial cells are very small and can only just be seen using a light microscope. To see the detail of bacterial cells, biologists use high-powered microscopes called electron microscopes. These microscopes magnify thousands of times more than a light microscope.

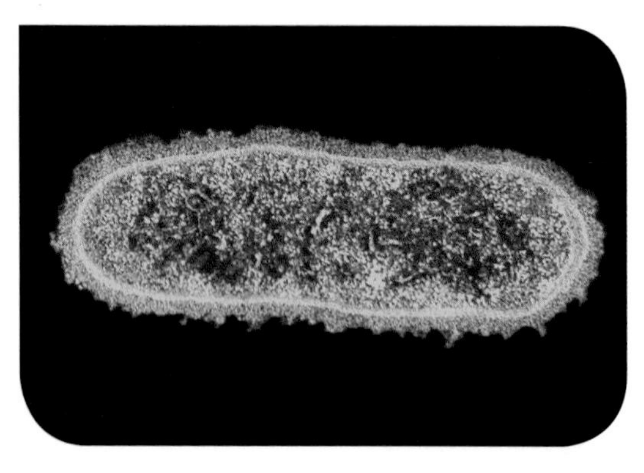

The bacterium *E. coli* seen under an electron microscope (× 15 000)

Typical bacterial cells

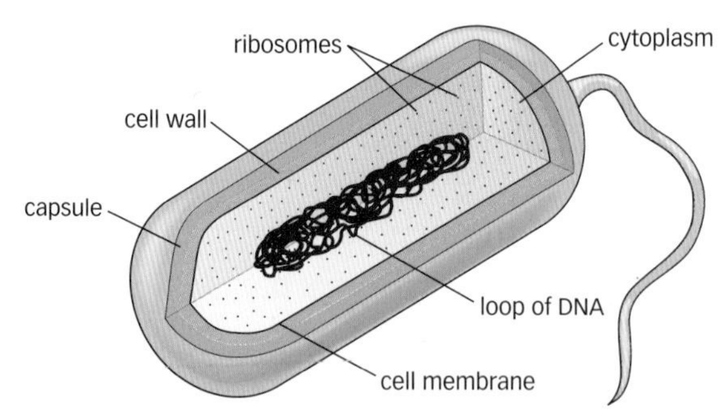

A typical bacterial cell

Bacterial cells all share some features in common. Some parts are similar to those of plant and animal cells.

Cell part	Function
Cytoplasm	A jelly-like substance where most of the cell's reactions occur.
Cell membrane	Controls the movement of molecules into and out of the cell.
Cell wall	Having the same function as in a plant cell of maintaining the shape of the cell, but made of a different chemical instead of cellulose.
Ribosomes	Make proteins.

Bacterial cells also have parts not found in plant and animal cells.

Cell part	Function
Loop of DNA	DNA which controls the cell, as bacterial cells do not have a nucleus.
Capsule	Some bacteria have a slimy capsule around the outside of the cell wall, which protects them against antibiotics, for example.

A Explain why we need to use powerful electron microscopes to see bacteria.

B Which part of a bacterial cell performs the function of a nucleus?

C Name two parts of a bacterial cell that have the same structure and function as they do in an animal cell.

Fungal cells

Fungi are another important group of organisms. They include mushrooms, moulds, and yeasts. Yeasts are commercially useful to us in the making of bread and beers. Yeasts are single-celled, but larger than a bacterial cell. They can be clearly seen using a light microscope, but the detail can be seen more clearly using an electron microscope.

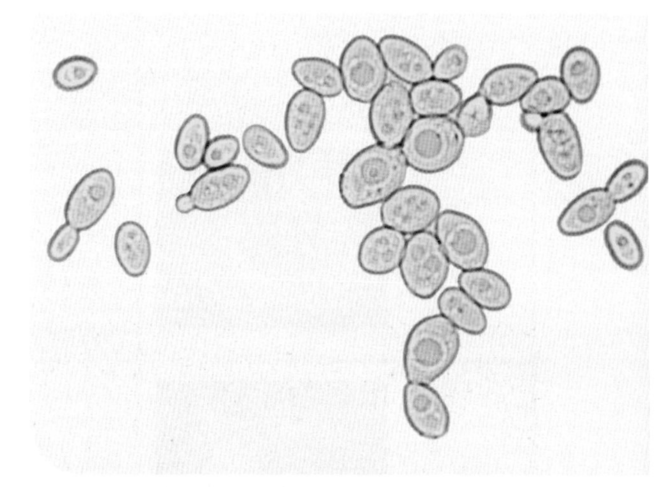

Baker's yeast seen under a powerful light microscope (×750)

Typical fungal cells

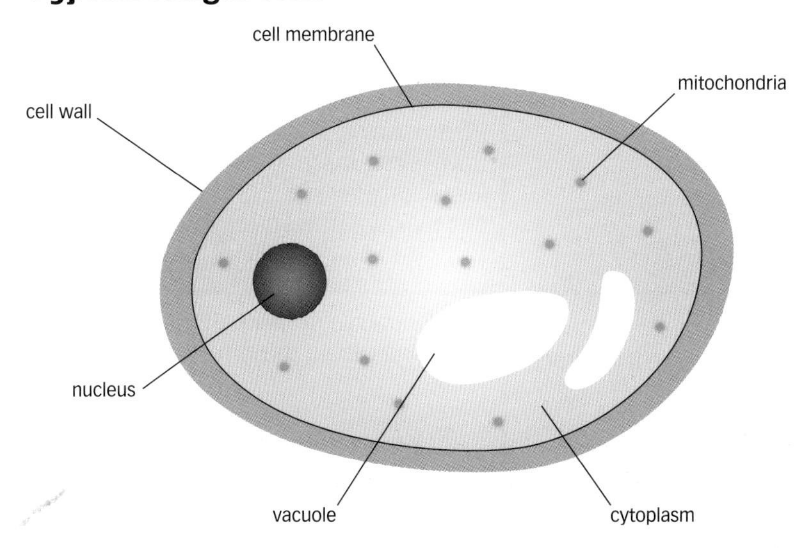

A typical fungal cell

Fungal cells have many parts in common with other cells. The fungal cell has a membrane, cytoplasm and a nucleus, which function as they do in plant cells. The fungal cell wall is similar to that of a plant or bacterial cell. It has the same function, but is made of a third chemical called chitin.

Questions

1 State one way in which a fungal cell and a bacterial cell are (a) similar and (b) different. **E**

2 How could you tell a bacterial cell and a plant cell apart? **C**

3 Explain why biologists discovered bacterial cells much later than plant cells. **A***

Learning objectives

After studying this topic, you should be able to:

- ✔ know that cells can become specialised to perform different functions

Differentiation

Cells all have the same basic structure. However, not all cells end up having the same function. They become **specialised** to carry out their particular job. This is called **differentiation**. In becoming specialised, the cell may develop particular structures or become a specific shape. These new structures or shapes help the cell perform its function efficiently.

Specialised animal cells

Here are some examples of how animal cells may become specialised. These types of cell are all found in humans.

Cell type	Specialised structure	Function of cell
Red blood cell	Lacks a nucleus. Large surface area. Cell is small so fits into narrowest blood vessels. Contains haemoglobin which binds reversibly to oxygen.	Haemoglobin binds to oxygen and transports it around the body. The red blood cell gives up the oxygen to other body cells that need it.
Nerve cell	Many short extensions at the ends of the nerve. One long nerve fibre extension. Nerve fibre insulated with fatty sheath.	Receives impulses from other nerve cells via its many extensions. The impulses travel along the long nerve fibre. The insulation prevents loss of the impulse and makes it travel quickly.
Muscle cell	Cell is long and thin. Full of proteins that can make it contract.	The contractile proteins shorten the cell. This brings about movement.
Sperm cell	Cell has a head containing a nucleus, and a long tail.	The tail helps the cell to swim to the egg. The nucleus contains DNA which combines with the DNA of the egg cell.
Ciliated epithelial cell	Tall column-shaped cells. Cells can pack tightly together. Each cell covered at the top with fine hairs called cilia.	Tightly packed cells form a covering layer of cells. The cilia beat to create a current which can move particles such as bacteria up and out of the windpipe.

A Why are there many different types of cell in the human body?

B Explain how the following cells use their specialised structure to carry out their function:
(a) red blood cells (b) nerve cells

C Explain why a red blood cell cannot carry out the function of a muscle cell.

Key words

specialised, differentiation

Specialised plant cells

Cell type	Specialised structure	Function of cell
Palisade mesophyll cell	Found in the upper part of the leaf. Column-shaped cells with many chloroplasts.	The shape means that many cells can pack side by side. The chloroplasts contain chlorophyll for trapping light.
Root hair cell	Found in the young root. Long extension that protrudes out into the soil.	The extension increases the surface area of the cell, which improves its ability to absorb water and minerals from the soil.
Xylem	Found in roots, stems, and leaves. Hardened cell wall. Hollow inside with no living contents.	The hard cell wall gives strength, which helps support the plant. Being hollow allows the xylem to transport water.
Phloem	Found in roots, stems, and leaves. End walls of cells perforated. Cells largely hollow inside with small living cells next to them.	The hollow cavity and perforated end walls allow sugars to move through the plant. The living neighbouring cells supply energy for the transport of sugars.

Questions

1 Explain how the palisade mesophyll cell is adapted to its function of photosynthesis.

2 Why are xylem cells needed in the root, stem, and leaves of a plant?

E

3 Explain how the xylem cell is specialised to carry out its functions.

C

4 Explain how a large surface area helps the root hair cell to absorb water and minerals.

A*

4: Diffusion

Learning objectives

After studying this topic, you should be able to:

- understand the process of diffusion
- know how diffusion allows particles to enter and leave cells

Potassium permanganate crystals have been placed in a beaker of water, and after two hours the particles have diffused throughout the water

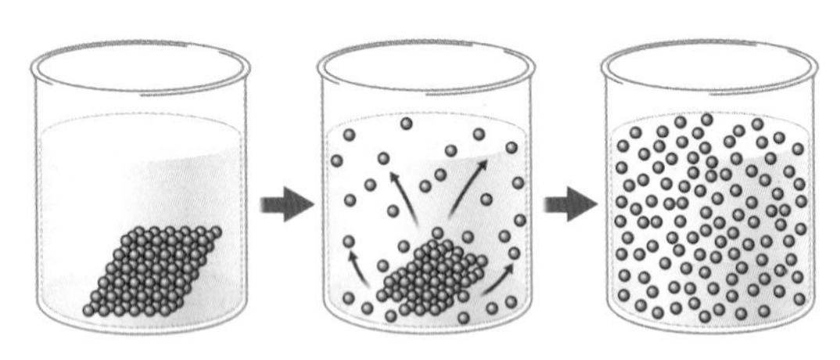

Diffusion is the movement of particles along a concentration gradient

Getting in and out

Cells carry out many reactions. They need a constant supply of some substances, and need to get rid of others. So dissolved particles (molecules and ions) need to get into and out of cells. One important way that particles can move into or out of a cell is by **diffusion**.

Diffusion

Particles in a gas or in solution constantly move around. Particles tend to move from an area where they are in high concentration to an area where they are in lower concentration, along a **concentration gradient**. The particles move until they are evenly spread. This random movement of particles along a concentration gradient is called diffusion.

Diffusion in cells

Many dissolved substances enter and leave cells by diffusion, including important molecules like oxygen, which is needed for respiration in plant and animal cells. Carbon dioxide also gets into and out of cells by diffusion. Substances can diffuse as gases, or as dissolved particles in solution.

To get into a cell, particles pass through the cell membrane. The membrane will only allow small molecules through. This is fine for oxygen and carbon dioxide, as they are both small molecules. The process of diffusion does not use energy, because the molecules move spontaneously from regions of high concentration to regions of low concentration.

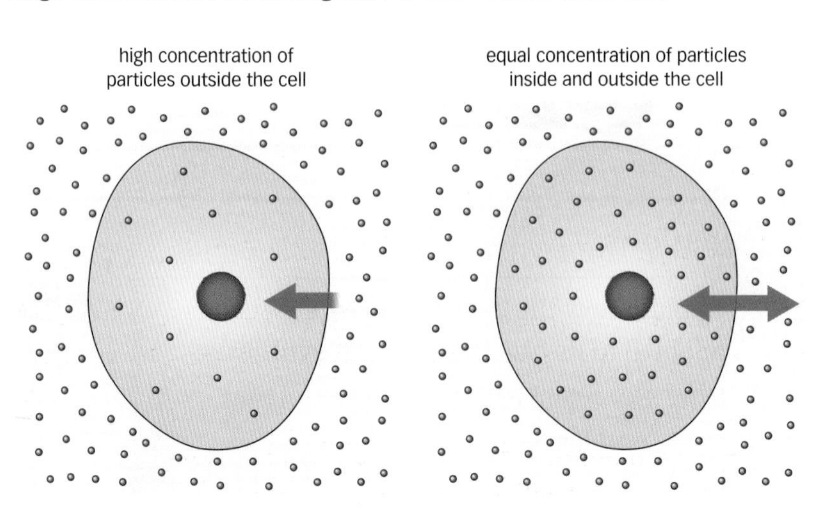

Particles moving into cells by diffusion

A Define diffusion.

B List some important molecules that diffuse into and out of cells by diffusion.

C Explain how the cell membrane can control which substances enter or leave the cell.

Factors that affect the rate of diffusion

Diffusion happens because particles in solution constantly move. They can move in any direction, but far more particles tend to move from high to low concentration than the other way. This gives a net movement of particles from high concentration to low, along the concentration gradient. However, the rate of diffusion can vary. For example, increasing the temperature will give molecules more energy and the rate of diffusion is faster. There are several factors that can affect the rate of diffusion in cells.

Distance

The shorter the distance the particles have to move, the quicker the rate of diffusion will be. For example, if carbon dioxide has to reach cells in the centre of the leaf, then the thinner the leaf, the shorter the distance the gas has to travel and the quicker it will reach the cells.

Concentration gradient

The greater the difference in concentration between two regions, the faster the rate of diffusion. For example, leaf cells produce oxygen as a waste gas during photosynthesis. There is a build-up of oxygen in the leaf, giving a steep concentration gradient of oxygen between the inside and outside of the leaf. This leads to rapid diffusion of oxygen out of the leaf.

Surface area

The greater the surface area that the particles have to diffuse across, the quicker the rate of diffusion. For example, the lungs of animals and the internal structures of a leaf have a large surface area. This allows gases to diffuse rapidly into and out of cells.

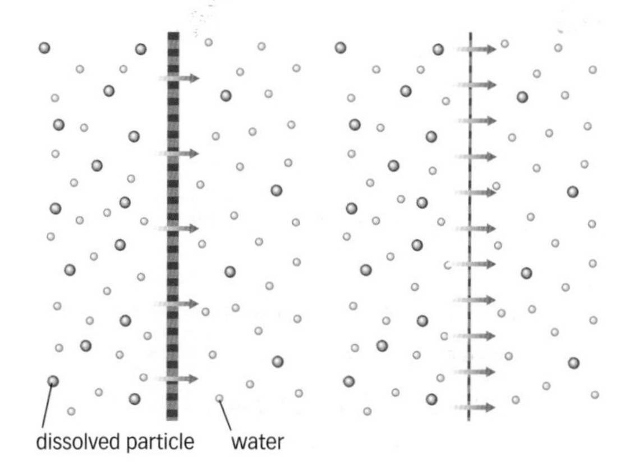

▲ The rate of diffusion depends on the distance the dissolved particles have to travel

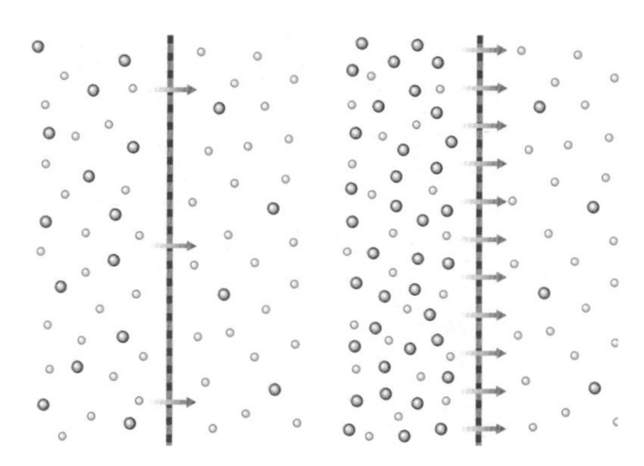

▲ The rate of diffusion depends on the concentration gradient

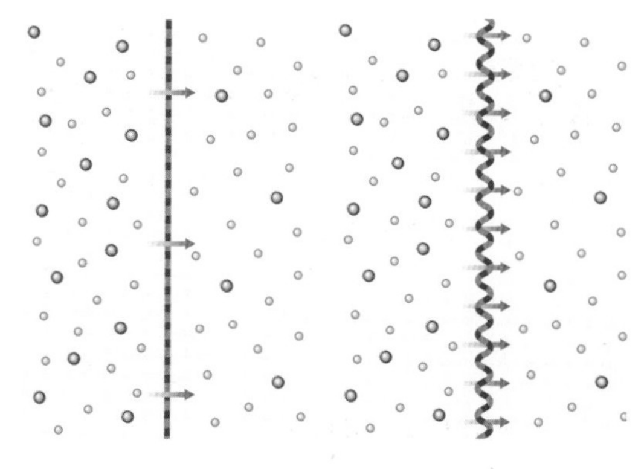

▲ The rate of diffusion depends on the surface area

Questions

1 State three factors that affect the rate of diffusion. (E)

2 Oxygen diffuses across the gills of a fish. Do you expect the cells lining the gills to be thick or thin? Explain why. (C)

3 Oxygen is constantly being used up in cells during respiration. Explain how this helps maintain the diffusion of oxygen into cells. (A*)

Key words

diffusion, concentration gradient

Learning objectives

After studying this topic, you should be able to:

- know that cells work together as tissues
- know that tissues work together as organs
- appreciate the working of some animal tissues and organs

Key words

multicellular, tissue, organ

Working together

Animals such as humans are built out of many cells – they are **multicellular**. Their cells do not work alone. It is more organised and efficient for the cells to work together.

You have seen that animal cells become specialised for a particular function. In the body of a multicellular animal, similar cells are organised together as a **tissue**. A tissue is a group of cells with a similar structure and function, working together. Organising cells into tissues allows life functions to be carried out more efficiently.

Animal tissues

Animal tissues include:

Muscle tissue

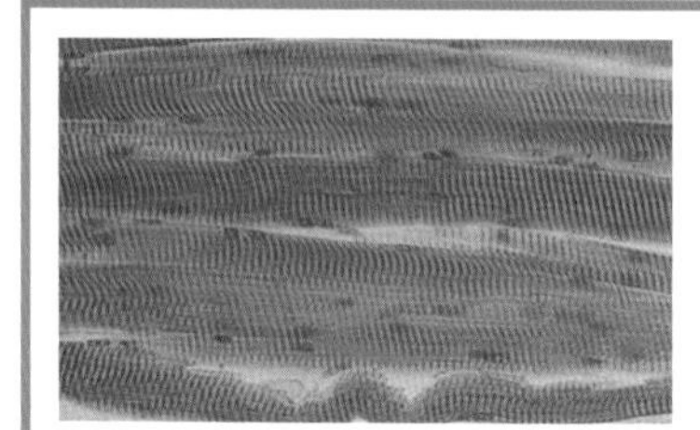

Skeletal muscle

A group of muscle cells. When the cells contract together the entire muscle shortens, bringing about movement.

Glandular tissue

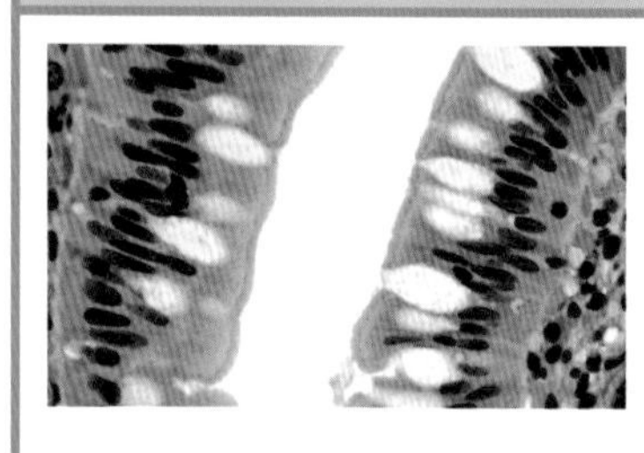

Gut lining with goblet cells releasing liquid

A group of cells that produce a substance and then release it. Glandular tissue produces hormones in glands such as the pancreas, or enzymes in the cells lining the gut.

Epithelial tissue

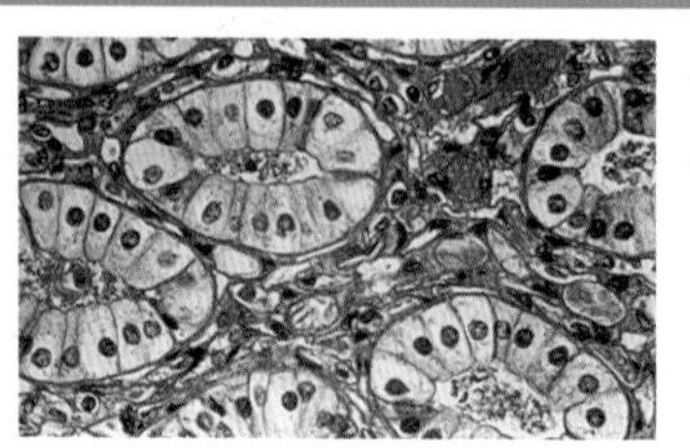

Epithelial tissue lining the tubes of the kidney

A group of cells that form a covering layer for some parts of the body. These cover and protect parts of the body, or act as lining.

A What is a tissue?

B Why do multicellular animals have cells organised into tissues?

C What substances can glandular tissue produce?

Organs

Tissues are grouped together to form **organs**. Organs are usually made of several different types of tissues. The tissues of an organ work together to perform one major task.

Cell, eg muscle cell → Tissue, eg heart muscle → Organ, eg the heart

▲ Cells are organised into tissues and organs

Exam tip AQA

✔ Understand the idea of scale here. Cells are small. Tissues are made of many cells, so they are larger, while organs are larger still.

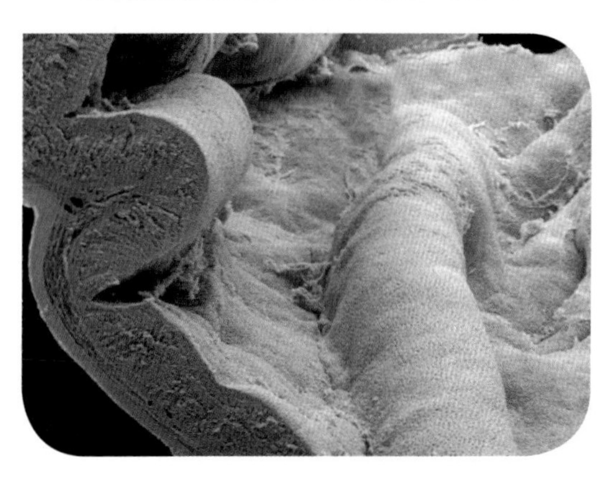

▲ A scanning electron micrograph of the stomach surface and a cross-section of the stomach wall (× 30)

The stomach

The stomach is an example of a human organ. It is made of several different tissues including muscle, glandular, and epithelial tissue.

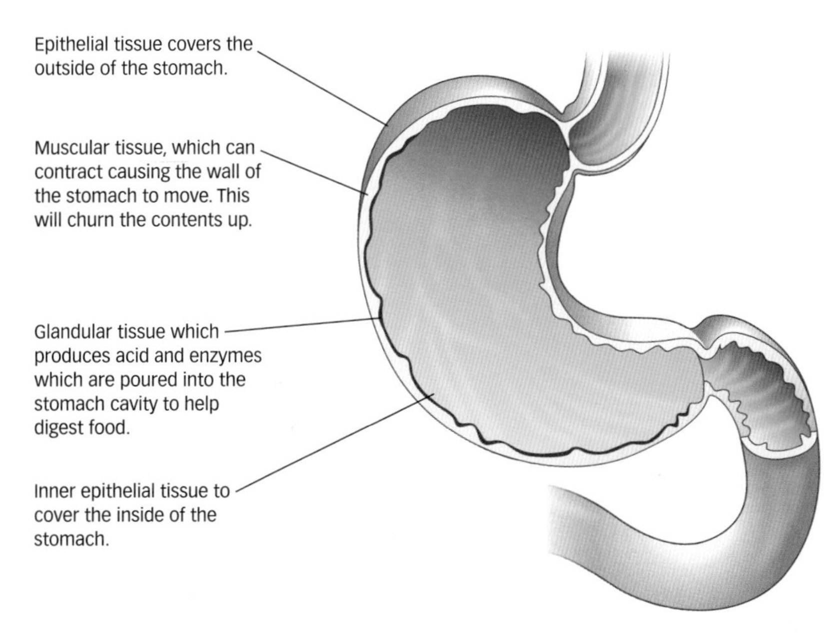

▲ The stomach is an organ that has several different types of tissue working together

Questions

1. What is the difference between a tissue and an organ?
2. Name three human organs other than the stomach.
3. Explain why tissues are larger than cells.
4. Explain how muscular tissue and glandular tissue work together to help bring about digestion of food in the stomach.

E C A*

Learning objectives

After studying this topic, you should be able to:

- know that organs work together in organ systems
- know the major organs of the digestive system

Key words

organ system, digestive system

Putting systems in place

Specialised cells are combined in tissues, and tissues work together in organs. But even organs do not work alone – several organs may be organised to work together to achieve the life processes of the organism. A group of organs working together is called an **organ system**. There are many organ systems in the human body, but perhaps the best known is the **digestive system**.

The organs of the digestive system

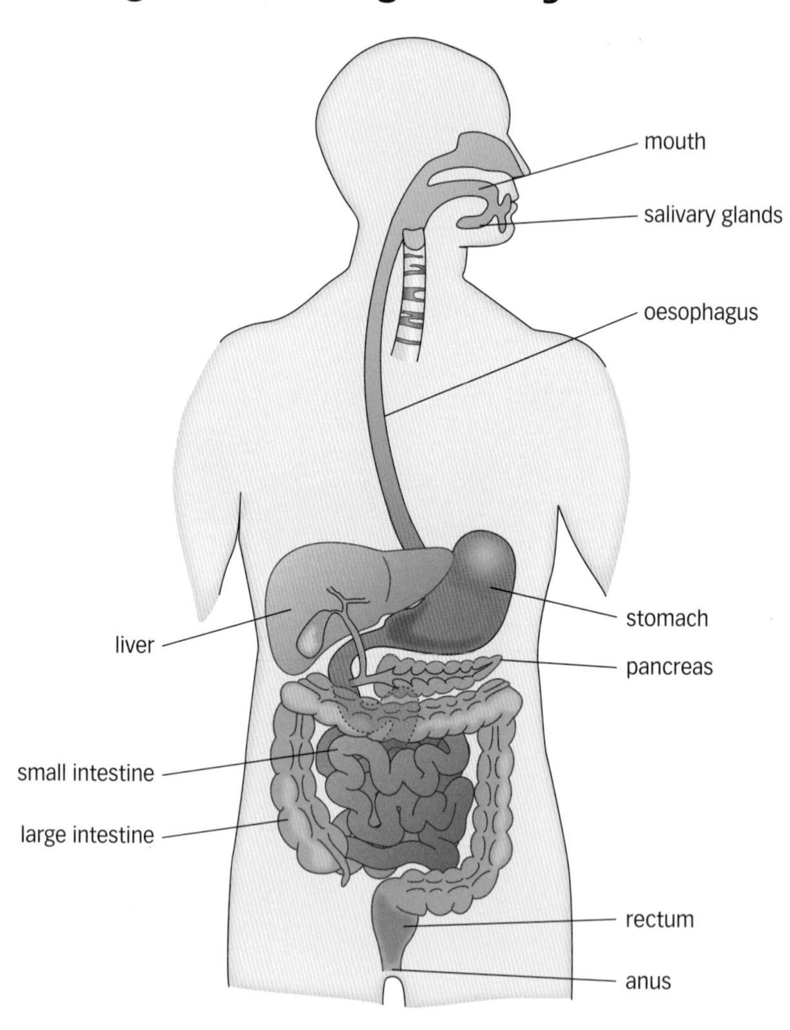

▲ In the digestive system several organs work together to bring about digestion

Exam tip AQA

- Make sure you can label the organs of the digestive system on a diagram.

A What is an organ system?

B Apart from the digestive system, name another organ system in the human body.

C Explain why the digestive system needs several different organs to carry out its function.

Division of labour

The function of the digestive system is the digestion and absorption of food. The different organs in the digestive system each have different functions in this overall task.

The digestive system allows the exchange of substances between the body and its environment. The body takes in food into the gut, and releases digestive juices to mix with the food. The resulting digested food particles and water pass through the walls of the digestive system. They are taken into the blood to be transported around the body.

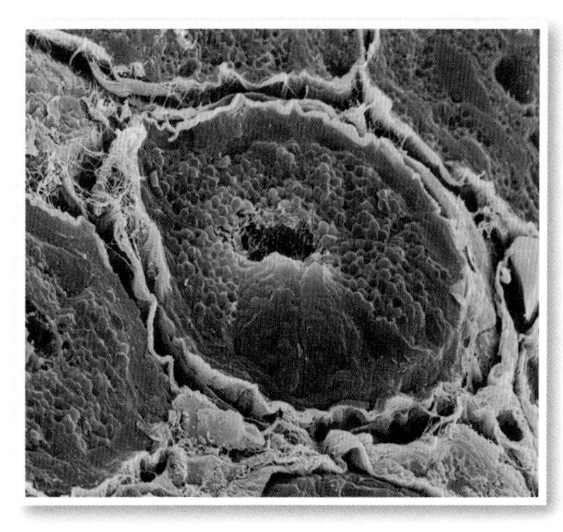

▲ A scanning electron micrograph of part of a salivary gland, where digestive juices are produced (× 1000)

Organs for releasing digestive juices

Some organs in the digestive system are involved in the production and release of digestive juices into the gut. These organs are glands. The pancreas and salivary glands are two major glands that carry out this function.

Digestive juices can have two functions:

- to lubricate the food
- to carry enzymes to aid digestion.

In addition, the juice released by the pancreas together with the bile from the liver also helps to neutralise acid formed in the stomach.

Organs for digestion

Digestion is the breakdown of large food molecules into smaller particles that can pass through the gut wall and be absorbed into the blood. This breakdown occurs inside the mouth, stomach, and small intestine. Each of these organs has glands that produce enzymes, and these are mixed with the food. The mixing is brought about by either the teeth and tongue in the mouth, or muscles in the wall of the stomach and small intestine. The digestive enzymes bring about the digestion of the food.

Organs for absorption

Absorption is the process by which the smaller digested food particles are taken into the blood. The particles pass through the gut wall and are taken into the blood from the gut environment.

Digested food particles are mainly absorbed in the small intestine. The lining of the small intestine has a very large surface area, making it efficient at absorption. The water in the undigested food is absorbed in the large intestine. At the end of the digestive process, the material that is left is faeces, which leave the body through the anus.

Questions

1 Name two organs that produce digestive juices.

2 What substances are taken into the body from the external environment via the digestive system?

E

3 Explain why the digestive system is regarded as an organ system.

4 Name one other organ system in the human body involved in exchange, and state what it exchanges.

C

5 Explain how the small intestine is adapted for absorption.

A*

Learning objectives

After studying this topic, you should be able to:

- ✔ know that animal and plant cells are organised into tissues and organs
- ✔ know the main organs of the plant
- ✔ understand the distribution of tissues inside the plant

Key words

epidermal tissue, xylem, phloem, palisade mesophyll cells

Organising an organism

As with animals, plant cells are organised in a specific way within the plant:

- Groups of similar cells work together as a tissue.
- Groups of different tissues work together as an organ.
- All the organs build the whole organism.

In plants there are a number of different organs, each with a different function.

Plant organs

Organ	Function
Stem	Supports the plant. Transports substances through the plant.
Leaf	Produces food by photosynthesis.
Root	Anchors the plant. Takes up water and minerals from the soil.
Flower (this is an organ system consisting of three organs: the petal, the stamen, and the carpel)	Reproduction.

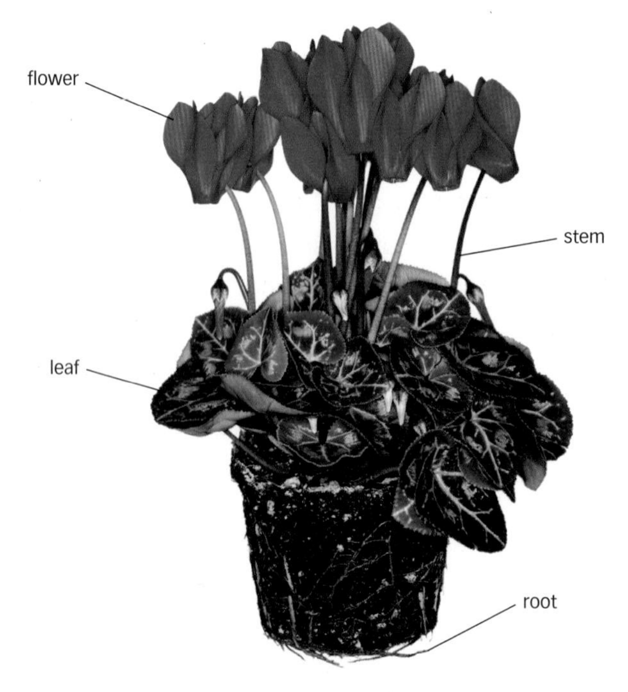

▲ The stem, root, and leaf are organs. The flower is an organ system.

A Name three plant organs.

B In an organ system, different organs work together. Why is the flower classed as an organ system?

Inside a plant

Inside a plant organ are tissues made up of similar cells working together:

- The outside of plant roots, stems, and leaves are covered in **epidermal tissue**. This protects the organs, although the root epidermis may get damaged by soil.
- The bulk of the stem and root is composed of packing cells. These cells are filled with watery fluid, which makes them firm so they can help support the plant.
- Two major tissues inside roots, stems, and leaves are **xylem** and **phloem**. These tissues are involved in transport, and they are found in the tube-like vascular bundles that run up through the roots, leaves, and stems. The cells of the xylem have thickened cell walls. These cells are strong and help support the plant.
- The leaf has many cells specialised for photosynthesis. These are the **palisade mesophyll cells**, located on the upper surface of the leaf. They contain lots of chloroplasts, and so can absorb sunlight energy for photosynthesis.

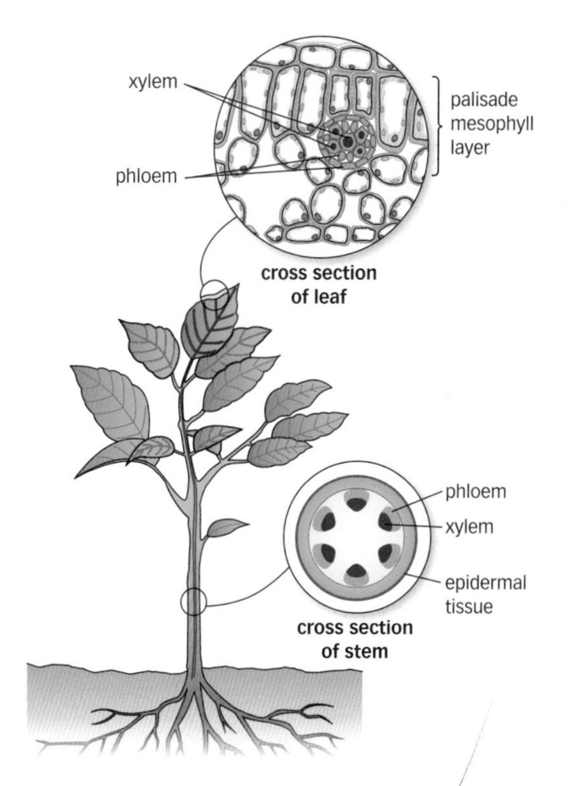

▲ A section through a leaf, showing the different tissues

Transport in the vascular bundles

The vascular bundles form a continuous transport system from the roots, through the stem, and into the leaves.

There are two tissues inside the vascular bundles. Both are involved in the transport of water and dissolved substances through the plant:

- Xylem: these cells are dead and stacked on top of one another to form long hollow tube-like vessels. Xylem cells are involved in the transport of water and dissolved minerals from the roots to the shoots and leaves.
- Phloem: these cells are living and are also stacked on top of one another in tubes. They transport the food substances made in the leaf to all other parts of the plant.

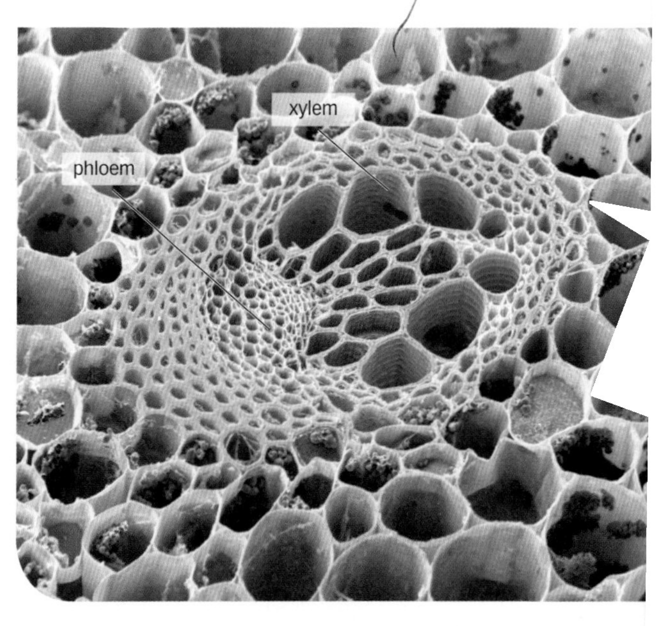

▲ A section through a buttercup stem to show the vascular bundles (×200)

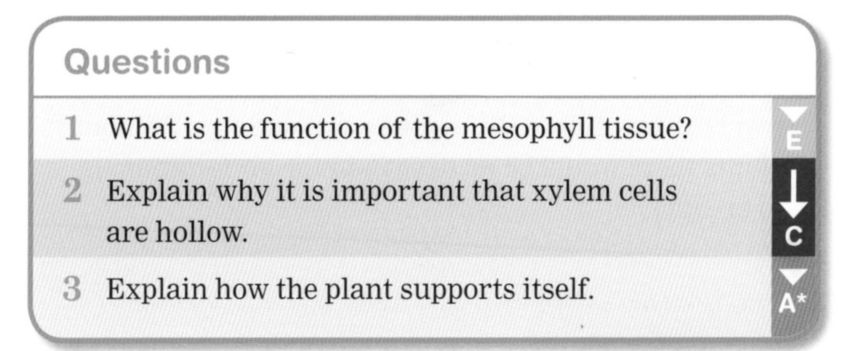

Questions

1 What is the function of the mesophyll tissue?

2 Explain why it is important that xylem cells are hollow.

3 Explain how the plant supports itself.

Learning objectives

After studying this topic, you should be able to:

- know that photosynthesis is the process by which plants make their own food
- appreciate the source of the raw materials for photosynthesis
- understand the fate of the products of photosynthesis

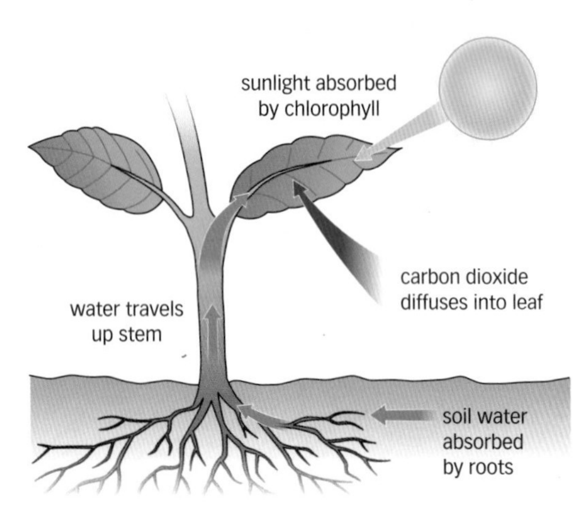

▲ In photosynthesis the plant uses sunlight energy to convert water and carbon dioxide into carbohydrates

Exam tip 

- If you are asked to write out an equation, make sure you know the difference between a word equation and a chemical equation. Learn the equations; it will gain you marks.

Feeding in plants

Plants do not take in ready-made food like animals do. They have to make their own food. To do this plants take in:

- carbon dioxide from the air
- water from the soil.

Some plants, algae, and seaweeds trap the Sun's energy in **chlorophyll**, in the chloroplasts in their cells. They use this energy to build up the carbon dioxide and water into carbohydrates and oxygen. This process is called **photosynthesis**.

Word equation for photosynthesis

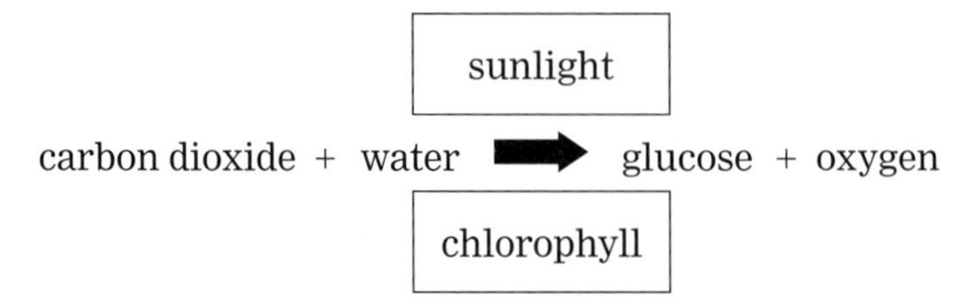

$$\text{carbon dioxide} + \text{water} \xrightarrow[\text{chlorophyll}]{\text{sunlight}} \text{glucose} + \text{oxygen}$$

Chemical equation for photosynthesis

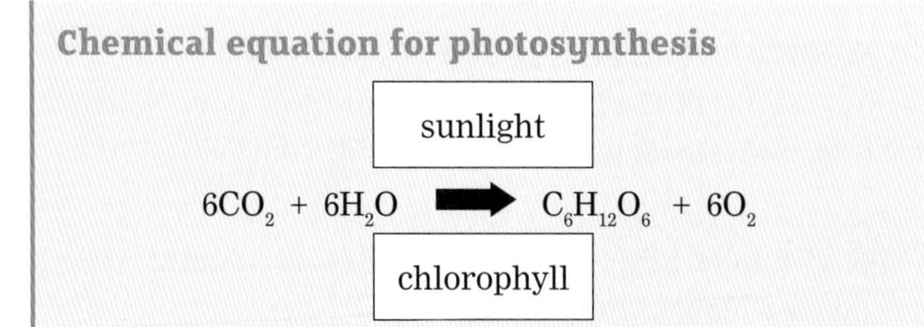

$$6CO_2 + 6H_2O \xrightarrow[\text{chlorophyll}]{\text{sunlight}} C_6H_{12}O_6 + 6O_2$$

What does the plant make in photosynthesis?

You can see from the equations that there are two products of photosynthesis:

1. Glucose: this is food for plants, it is a carbohydrate. Some is used for respiration in the plant's cells. The rest can be stored in the plant.
2. Oxygen: this is a waste gas produced in photosynthesis. Some is used for respiration in the plant's cells. The rest is given off into the plant's surroundings. Without plants there would be no oxygen in the air for animals to breathe.

Converting glucose to other substances

The glucose produced in photosynthesis by plants and algae can be converted to other substances that the organisms need. For example, it may be used to make the sugar sucrose, found in sugar cane.

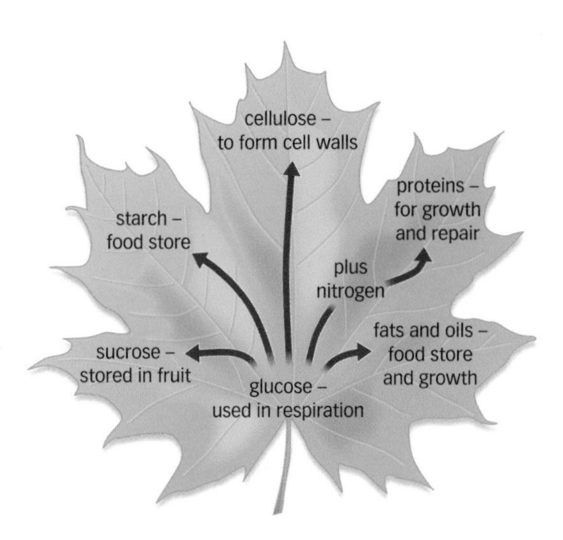

▲ Glucose from photosynthesis is converted to all the substances that a plant needs

If it is not used, the glucose can be changed into insoluble starch and stored until it is needed. Stored glucose can be used for respiration at night, when there is no sunlight and the plant is not making glucose by photosynthesis.

The glucose made in photosynthesis is converted to sucrose to be transported around the plant to parts that need it. Sucrose is good for transport because it dissolves in water and flows easily.

Plants are not made of sugars alone. The plant converts sugars to other substances such as cellulose, proteins, fats, and oils which it needs to grow and function. When producing proteins, plants do not just use sugars. They also use nitrite ions from the soil.

Storing glucose

Glucose is stored in the plant as starch. This has three advantages:

1. Starch can be converted back into glucose for respiration in plant cells.
2. Starch is insoluble and so will not dissolve in water and flow out of the cells where it is stored.
3. Starch does not affect the water concentration inside cells.

Key words

chlorophyll, photosynthesis

A What are the two raw materials a plant needs for photosynthesis?

B What else does a plant need in order for it to photosynthesise?

C Explain why humans could not survive without photosynthesis.

Questions

1 Where does the energy for photosynthesis come from? E

2 (a) What element is added to the glucose produced in photosynthesis to make proteins?
(b) Where does this element come from? C

3 Explain why plant cells do not store carbohydrate as sugars. A*

9: The leaf and photosynthesis

Learning objectives

After studying this topic, you should be able to:

- know that the leaf is the site of photosynthesis
- appreciate the internal and external structure of the leaf
- understand the adaptations of the leaf for photosynthesis

Key words

leaf, palisade layer, stomata

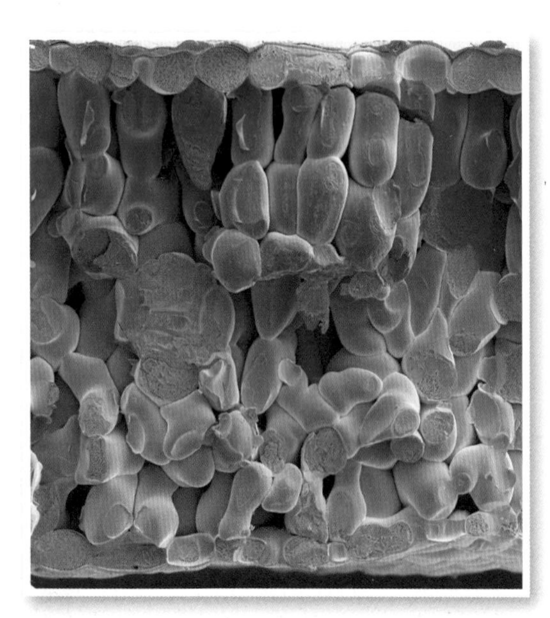

Cross-section of a spinach leaf seen through a powerful electron microscope (×250)

Exam tip AQA

- You need to learn the structure of the leaf.

Leaves

The main plant organs for making food are the leaves.

A leaf

The external structure of a leaf

Inside the leaf

The **leaf** is made up of many specialised cells. Each type of cell has its own function. They work together, making the leaf well-adapted to carry out photosynthesis.

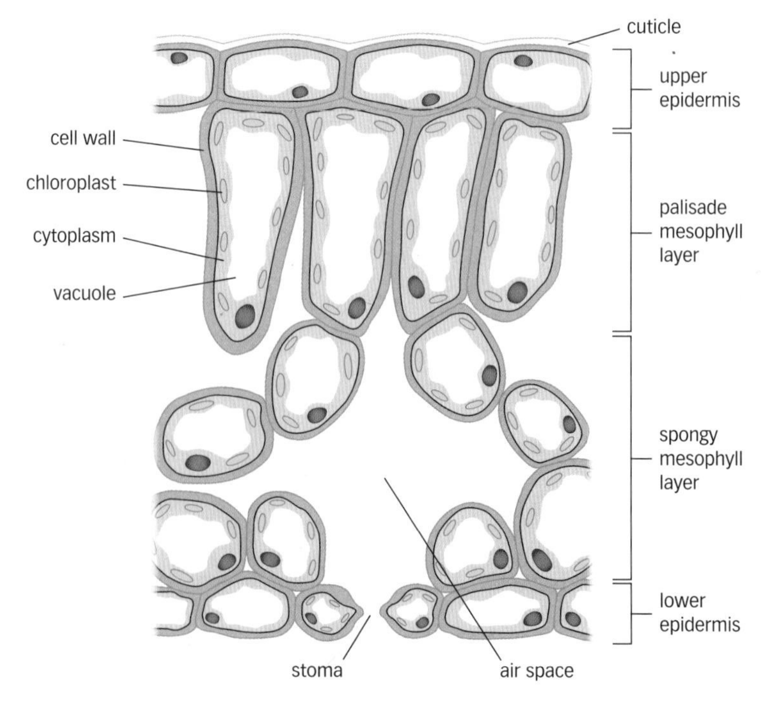

The internal structure of a leaf

Top ten adaptations of the leaf for photosynthesis

- ✓ Many leaves are broad and flat, giving a large surface area to absorb as much light as possible.
- ✓ Leaves are thin, so that carbon dioxide has a short distance to travel to the mesophyll and palisade cells.
- ✓ The leaf cells contain chlorophyll within chloroplasts. This absorbs light energy for photosynthesis.
- ✓ The upper palisade layer, which receives the most light, contains the most chloroplasts.
- ✓ The cells of the **palisade layer** are neatly packed in rows, to fit more cells in.
- ✓ Veins carry water from the roots to the leaf cells, and carry glucose away.
- ✓ Veins support the leaf blade.
- ✓ There are plenty of **stomata**, pores in the lower epidermis, which allow carbon dioxide in and oxygen out.
- ✓ There are air spaces in the spongy mesophyll layer to allow carbon dioxide to diffuse from the stomata to the palisade cells.
- ✓ The air spaces inside the leaf give a large surface area to volume ratio. This allows maximum absorption of gases.

▲ This digital meter is being used to measure chlorophyll and photosynthesis in a cotton leaf

A On a plant, leaves are angled so plenty of sunlight reaches them. Explain why this is important to the plant.

B The leaf epidermis is transparent. Why is this an advantage to the leaf?

C What is the name for the pores in the leaf?

Questions

1. Name the layer in the leaf that carries out most photosynthesis.

E

2. Which adaptations of the leaf allow it to trap as much sunlight as possible?
3. Explain the advantages of the air spaces in the spongy mesophyll layer.

C

4. Leaves of plants that are often in bright sunlight tend to have more stomata. Explain what you think the effect of this will be.

A*

Learning objectives

After studying this topic, you should be able to:

- ✔ know how some factors affect the rate of photosynthesis
- ✔ understand limiting factors

Key words

rate of photosynthesis, limiting factor

▲ The rate of photosynthesis in this greenhouse is increased using artificial lighting

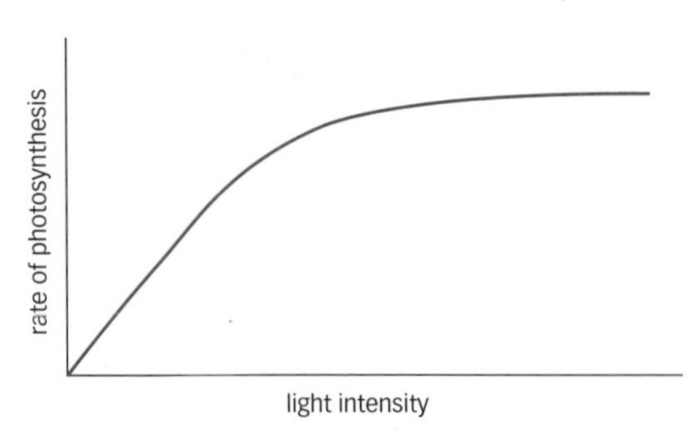

▲ Graph to show how the rate of photosynthesis changes as light intensity increases

The growing season

Plants do not grow at the same rate all year round. Most plants grow best in the spring and summer. This is when the conditions for growth are best. In spring and summer, the weather is usually warmer and there is more sunlight. These conditions are good for photosynthesis and therefore for growth, because the light energy is needed for photosynthesis, and the warmth speeds up the reactions of photosynthesis.

Increasing the rate of photosynthesis

The **rate of photosynthesis**, or how quickly the plant is photosynthesising, depends on several things. The following factors will speed up photosynthesis:

- more carbon dioxide
- more light
- a warm temperature.

People who grow plants commercially in a greenhouse try to make sure their plants have the best conditions. They use lighting systems which increase the hours of daylight available to plants, and they use heaters that burn gas, or other fuels, to add warmth and release carbon dioxide.

A List three things that will increase the rate of photosynthesis.

B Why do you think British woodland flowering plants such as bluebells flower in May?

Factors affecting the rate of photosynthesis

The rate of photosynthesis may be limited by the following factors.

Availability of light

Light provides the energy to drive photosynthesis. The more light there is, the faster the rate of photosynthesis. This is true provided that there is plenty of carbon dioxide, and the temperature is warm enough.

Amount of carbon dioxide

Carbon dioxide is one of the raw materials for photosynthesis. The more carbon dioxide there is available, the faster the rate of photosynthesis. (Again, this is only true if there is plenty of light and a suitable temperature.) Carbon dioxide is often the factor in shortest supply, so it is often the limiting factor for photosynthesis.

A suitable temperature

Temperature affects how quickly enzymes work. Enzymes make the reactions of photosynthesis happen. As the temperature rises, the rate of photosynthesis increases (providing there is plenty of carbon dioxide and light). However, if it becomes too hot, then the enzymes will be destroyed and photosynthesis stops.

Warm, sunny conditions mean light and temperature are not limiting factors for photosynthesis

Limiting factors

When a process is affected by several factors, the one that is at the lowest level will be the factor which limits the rate of reaction. This is called the **limiting factor**.

If the limiting factor is increased, then the rate of photosynthesis will increase until one of the other factors becomes limiting. For example, if photosynthesis is slow because there is not much light, then giving the plant more light will increase the rate of photosynthesis, up to a point. After that point, giving more light will not have any effect on photosynthesis, because light is no longer the limiting factor. The rate may now be limited by the level of carbon dioxide, for example.

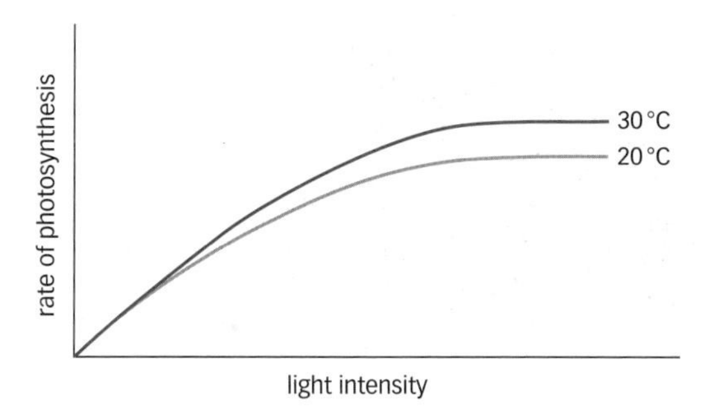

Light levels are limiting initially. The rate of photosynthesis then levels off. It increases at a higher temperature, so at higher light levels, temperature becomes the limiting factor.

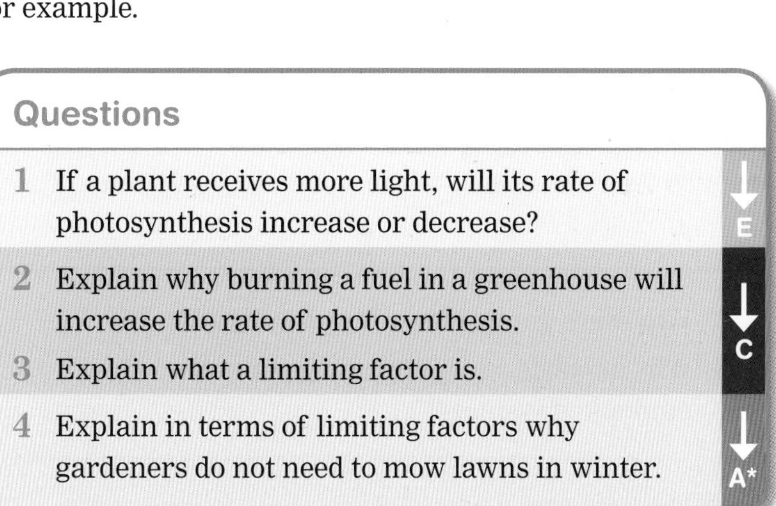

Questions

1 If a plant receives more light, will its rate of photosynthesis increase or decrease? (E)

2 Explain why burning a fuel in a greenhouse will increase the rate of photosynthesis.

3 Explain what a limiting factor is. (C)

4 Explain in terms of limiting factors why gardeners do not need to mow lawns in winter. (A*)

Exam tip AQA

- If you increase a limiting factor, then you will increase the rate. If you decrease the factor, then you decrease the rate. Remember to describe an increase or a decrease, rather than saying 'photosynthesis depends on the factor'.

Learning objectives

After studying this topic, you should be able to:

- understand that the factors needed for photosynthesis can be controlled
- appreciate that there are commercial benefits to controlling photosynthesis in greenhouses

The Victorian Palm greenhouse at Kew Gardens in London

Did you know...?

The greenhouses at Kew, perhaps some of the best known in the world, originally had green glass.

Greenhouses

People have been growing plants in **greenhouses** for many years. The Victorians used greenhouses to grow rare tropical plants collected from their travels around the world. The great greenhouses at Kew Gardens date from that time. Greenhouses have also been used to grow food plants at unseasonal times of the year.

In modern greenhouses, growers use their understanding of photosynthesis to maximise the growth of plants. They artificially control the factors that limit the rate of photosynthesis.

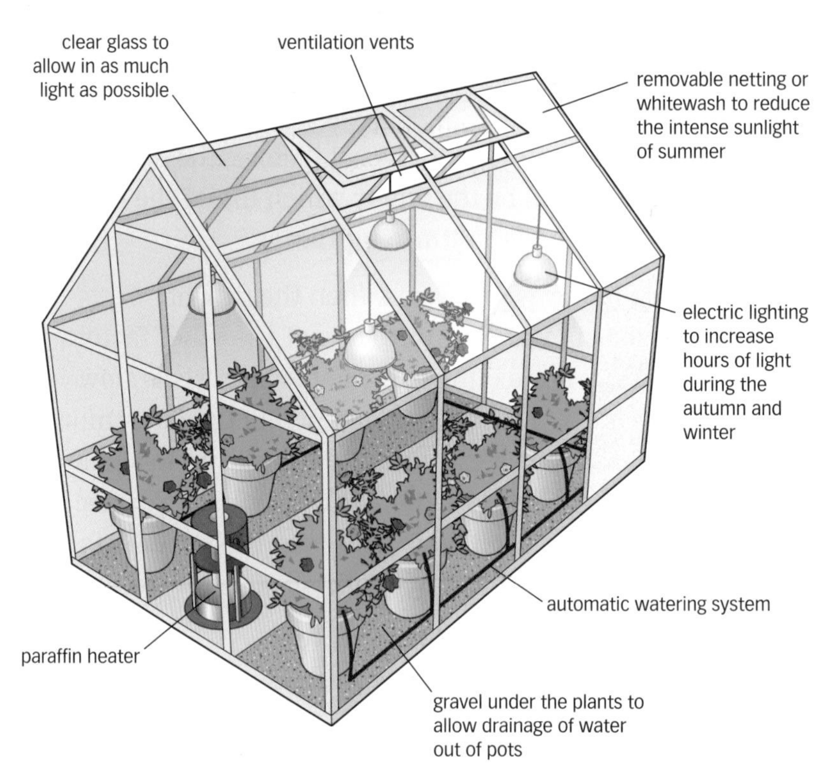

Modern greenhouses use automated systems to give the best conditions for photosynthesis

A Why were greenhouses first developed?

B Why do modern commercial greenhouses use automated systems?

Making the most of plants

Controlling light

Year-round light is provided by the Sun and by electric lighting systems. In the summer the sunlight can be too strong. Netting or whitewashing the windows can reduce the amount of light entering the greenhouse.

- **Advantage:** plants can receive light all year round, so they can be grown throughout the year.
- **Disadvantage:** cost of electric lighting.

Controlling temperature

The glass traps heat from the Sun inside the greenhouse, and shields the plants from the wind. This allows cultivation of plants that need warm temperatures, and allows plants to be grown out of season. If it gets too hot, automatic vents may open to allow hot air out of the greenhouse. In winter, additional heat can be supplied using heaters.

- **Advantage:** plants can be grown out of season, and more tropical plants can be grown in the UK.
- **Disadvantage:** cost of fuel for heaters.

Controlling carbon dioxide

Carbon dioxide is often the limiting factor for plant growth. Additional carbon dioxide can be added by burning fossil fuels such as paraffin in heaters.

- **Advantage:** additional carbon dioxide speeds up the rate of photosynthesis.
- **Disadvantage:** cost of fuel.

Controlling the water supply

Water is needed for photosynthesis. However, too much water can lead to plants rotting. Any watering system used by the gardener has to supply enough water, but not too much.

Automatic watering systems water at set times, or have sensors in the soil to detect how dry the soil might be. Water can be sprayed over the plants, or put directly into the gravel shelves beneath the plants. This will allow water to drain out of the plants so that they are not too wet, but will also act as a reservoir to allow water to soak up into the pots.

- **Advantage:** plants have a constant supply of water so they can photosynthesise.
- **Disadvantage:** cost of setting up the systems and of the electricity and water to run them.

Key words

greenhouse

Questions

1. Name four of the factors limiting photosynthesis that can be controlled in a greenhouse. (E)
2. Explain why out-of-season strawberries are more expensive than those grown outside in the summer.
3. Explain why low-value crops like potatoes are not grown in greenhouses. (C)
4. If a market gardener wanted to grow a crop of strawberries out of season, what factors would they have to consider to make a profit? (A*)

Learning objectives

After studying this topic, you should be able to:

- ✔ know about common sampling techniques
- ✔ understand how to use sampling techniques to collect good quality data

▲ Where do you start studying all the different species in this meadow?

▲ Students count how many organisms of a certain species are inside the quadrat. This gives a sample.

▲ Using a transect and quadrat to study the distribution of organisms across the field

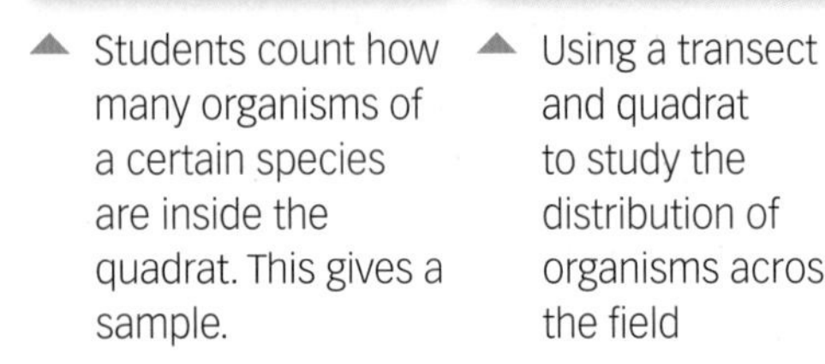

There's a lot out there!

When biologists investigate where organisms live, they meet problems:

- There are very many different organisms.
- They seem to live all over the place.

It is difficult to make sense of the huge amounts of data.

To overcome these problems, biologists have devised a series of techniques to collect information about two things. First, they record the location of organisms of one species; this describes their **distribution**. Second, they record the number of organisms of a particular species in an area; this is the **population**.

Different populations live together in one area, and together they form a **community**. Biologists look for **relationships** between the organisms in a community by studying how their distributions overlap. They also study how factors in the environment affect their distributions.

To collect this information, biologists need techniques to:

- collect organisms
- count the number of organisms in each species
- record where the organisms are found
- collect accurate data
- collect the data fairly
- collect reliable data.

Biologists use a technique called **sampling**. This means counting a small number of the total population and working out the total from the sample.

Sampling techniques

1. Quadrats are square frames of a standard area. They are put on the ground to define an area. The numbers of organisms of particular species in the frame can then be counted.
2. Transect lines are tapes that are laid across an environment. You can count the organisms that touch the tape, such as plants on the ground, in order to study their distribution. Alternatively you can lay quadrats at regular intervals down the tape in order to record the distribution of the organisms inside.

When you have enough readings, it is possible to make estimates of the size of a population from your sample. You can also estimate the distribution of the population. But the numbers in the sample need to be accurate, reliable, and fairly collected.

Valid sampling

Being accurate

The apparatus should allow you to count a reasonably large number of the type of organism you are studying. For example, if you use a quadrat that is too small, then you will record fewer plants and animals. A small sample size is not very accurate and would not be reproducible.

Being reliable

Repeat readings make the data more reliable. If only one quadrat is recorded, then it might not represent the population accurately. The more quadrats you record, the more reliable your data will be.

Being fair

To be fair, all your readings should use the same equipment. They also need to be placed fairly. When recording distribution, quadrats can be placed at regular intervals along a transect. This avoids you choosing places that look promising, which would give biased readings. When estimating population size, quadrats should be placed randomly in an area, rather than choosing where to place them.

Factors affecting distribution

There are many factors that could affect where an organism lives. Collecting data about the distribution of particular species allows biologists to compare their distributions in relation to factors such as:

- temperature: for example, polar bears with adaptations to cold climates are found in the arctic
- nutrients: lions need gazelles to eat; both species are found together in the same area
- light: many plants live in sunny locations, or have adaptations to shady conditions
- availability of water: few species live in deserts
- availability of oxygen: plants and animals need oxygen to respire
- availability of carbon dioxide: plants grow well with plenty of carbon dioxide for photosynthesis.

Key words

distribution, population, community, relationship, sampling

A Describe what a quadrat is used for.

B How could a group of students record the distribution of limpets down a beach?

Exam tip

✔ Useful memory aids for sampling are 'Accuracy using Apparatus' and 'Reliability needs Repeats'.

Questions

1 What is a community? (E)

2 Describe what techniques you would use to estimate the population of daisies in a school field. (C)

3 Explain why you think that collecting data using sampling techniques gives only a rough estimate of population size. (A*)

Learning objectives

After studying this topic, you should be able to:

- know that environmental work can generate large amounts of data
- review some methods used to process data

Key words

mean, median, mode

▲ A researcher from Newcastle University monitoring water pollution. She will take several samples and analyse the data.

A Name three centralised values scientists generate from their data.

B Explain why centralised values are useful.

Numbers everywhere

Experiments generate lots of numbers. It can be hard to make sense of large collections of numerical data like this. Biologists process the data to try to look for meaningful patterns or relationships.

One common technique is to search for a centralised value that is typical of all the results and can be used to compare with other values. There are three ways of achieving this:

- The **mean** is the average value of the data. This is commonly used. For example, the researcher in the photograph might take several samples and quote the mean as the average level of pollution. A disadvantage is that the mean can be influenced by a rogue result which is very different from the other data.
- The **median** is the middle value of the data when arranged in rank order. So, the researcher could quote the middle value of all her levels of pollution. This is less affected by data points that are very high or very low compared with the others.
- The **mode** is the most common value of the data. Its advantage is that it is not affected by an extreme rogue result. The mode does not take account of the spread of the data.

Centralised values, or averages, like these are useful because they can give you a quick overview of what the data are showing. Biologists might use these centralised values and compare them with others, maybe from a different location, or to look for a relationship with other factors.

How scientists work

A case study: global warming

Scientists interpret data using centralised values. They analyse their data by looking for relationships. Any relationships that they find form their conclusions. However, other biologists might take the same sets of data and reach quite different conclusions from them.

One very important example is the evidence about environmental change.

Step 1: scientists make a hypothesis

People are concerned that global temperatures are rising, and that this is leading to habitat loss, such as the loss of the ice caps that are the habitat of the polar bear. Scientists think that rising levels of carbon dioxide in the atmosphere are causing this increase in temperature.

Step 2: scientists test their hypothesis

To test this idea, we need to collect data. Scientists take ice samples from the polar ice packs, formed over thousands of years. They record the carbon dioxide levels in air bubbles trapped in the ice. The structure of the ice also tells them about the temperature at the time it was formed. This gives a record of temperature and carbon dioxide levels over time.

▲ Scientists taking ice core samples

Step 3: analysing the data

From many samples, the scientists calculate the mean carbon dioxide level and temperature for each time period recorded. They then plot their data as a graph.

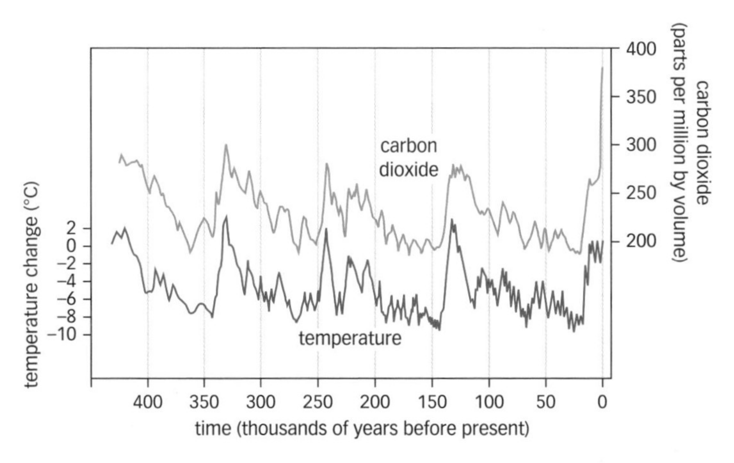

▲ Graph of carbon dioxide levels and temperature over the last 200 000 years

Step 4: interpreting the evidence

Some scientists are convinced that there is a direct relationship between the increase in global temperatures and the increase in the levels of carbon dioxide. They believe that human activity, including burning fossil fuels, is responsible for the increase in carbon dioxide. They also think that this climate change will have an impact on habitats and the distribution of many species.

Other scientists disagree with this interpretation. They feel that other factors, such as sun activity, have not been taken into account. At present the jury is still out!

Exam tip AQA

✔ When calculating a mean average, only quote one more decimal place than that of your data.

Questions

1. Why do scientists take more than one ice core sample for each time period?
2. Why do scientists calculate mean values from all the samples at each time period?

3. Why do scientists plot temperature and carbon dioxide levels on the same graph?

4. Explain why it is important that other scientists test the findings of an experiment.

Course catch-up

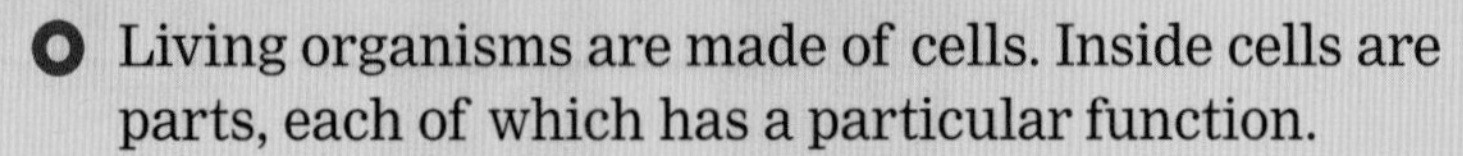

Revision checklist

- Living organisms are made of cells. Inside cells are parts, each of which has a particular function.
- There are similarities and differences between plant and animal cells.
- Fungal cells have a wall made from chitin.
- Bacterial cells have a wall, ribosomes, a membrane, and cytoplasm, but no nucleus.
- Cells have the same basic structure but they become specialised for different functions.
- Substances may move into and out of cells by diffusion. Diffusion is affected by distance, concentration gradient, surface area, and temperature.
- Groups of cells work together as tissues. Tissues work together as organs. Organs work together in systems in the organism.
- Different organs within each system have particular functions. The digestive system digests and absorbs food.
- In plants, roots, stems, leaves, and flowers are organs. Each has a specific function.
- Xylem and phloem are plant tissues. Xylem transports water and minerals; phloem transports dissolved food.
- Plants make their own food from carbon dioxide, water and light energy, by photosynthesis. The glucose they make can be converted to other sugars, starch, cellulose, fats and proteins (and nucleic acids). The waste product is oxygen.
- Leaves are organs adapted for photosynthesis. There are many of them, they are thin, have stomata for gaseous exchange, and have palisade cells with many chloroplasts.
- Rates of photosynthesis can be changed by levels of carbon dioxide, light intensity, and temperature.
- Humans can control the factors needed for photosynthesis and improve yields of commercially grown plants.
- Biologists use sampling techniques to investigate the biodiversity in different ecosystems, and the factors that affect animal and plant distribution. They analyse the data they collect.
- Scientists look for relationships from the analysed data. They then make a hypothesis and carry out experiments to test the hypothesis. Other scientists also repeat their investigations. If all their evidence supports the hypothesis, eventually it becomes accepted as a scientific theory.

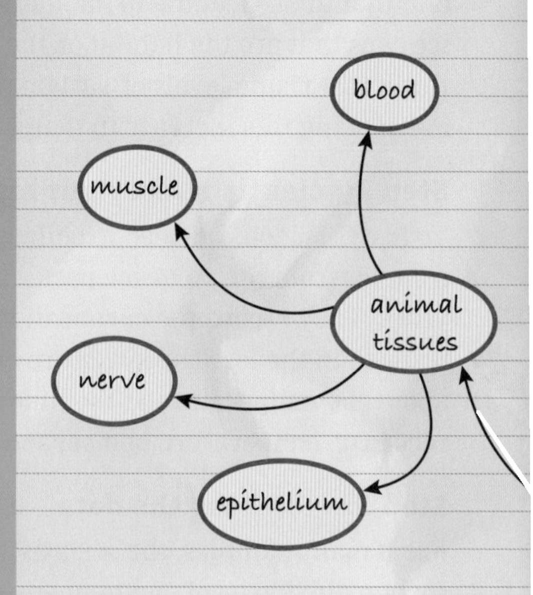

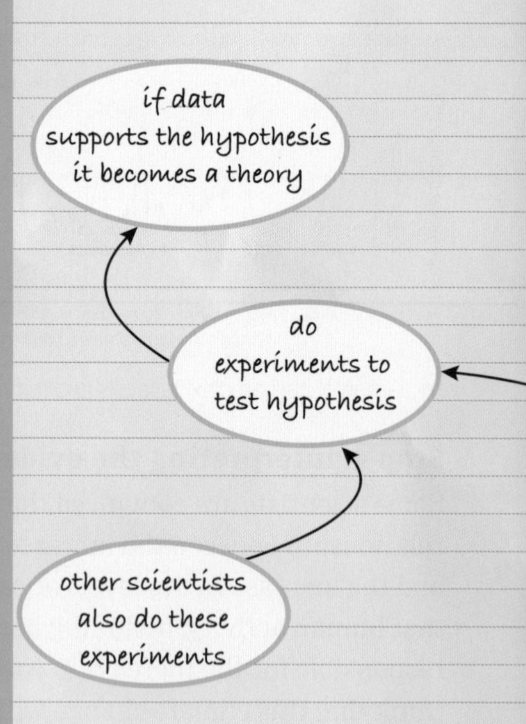

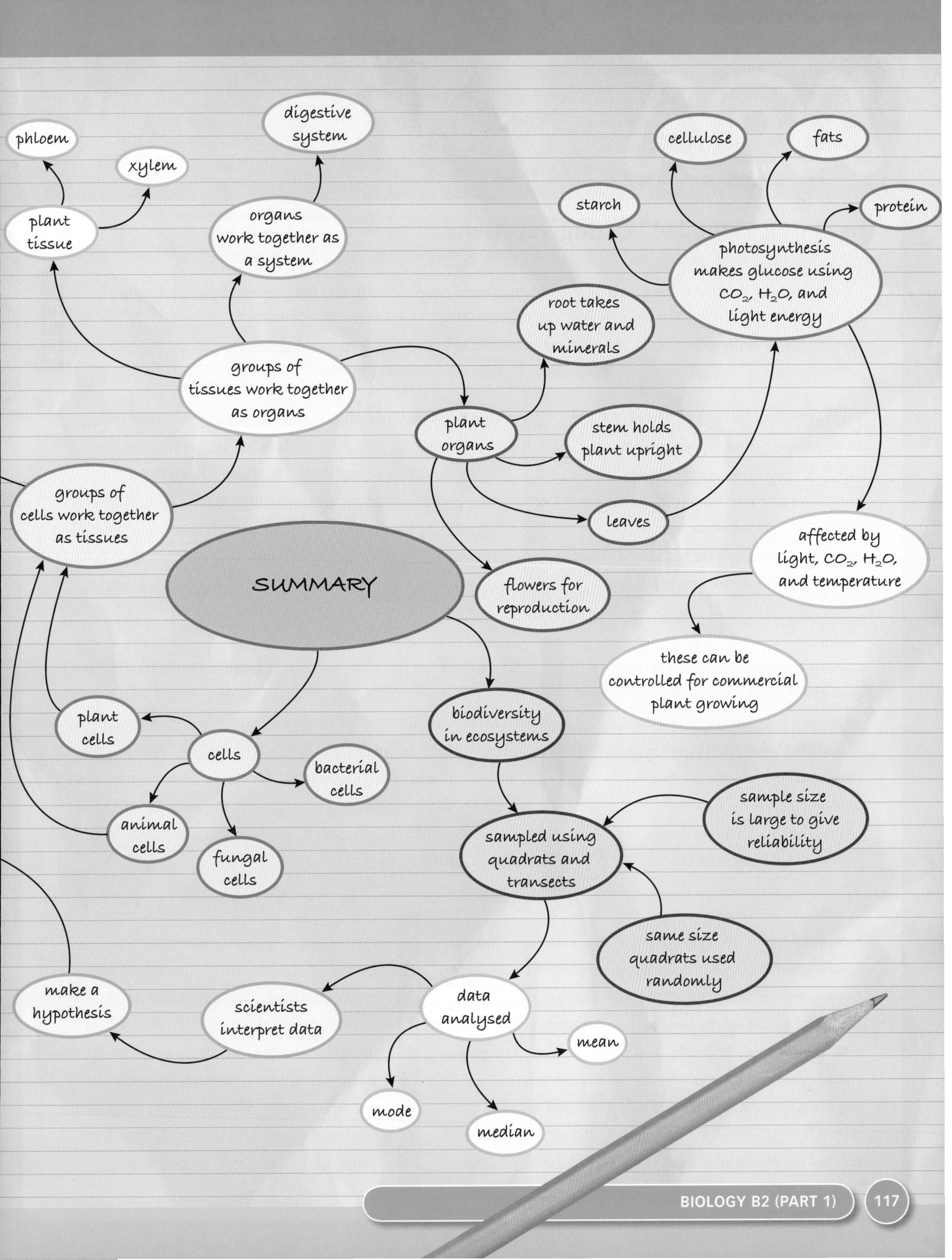

phloem
xylem
plant tissue
digestive system
organs work together as a system
cellulose
fats
starch
protein
photosynthesis makes glucose using CO_2, H_2O, and light energy
root takes up water and minerals
groups of tissues work together as organs
plant organs
stem holds plant upright
leaves
groups of cells work together as tissues
SUMMARY
flowers for reproduction
affected by light, CO_2, H_2O, and temperature
these can be controlled for commercial plant growing
biodiversity in ecosystems
plant cells
cells
bacterial cells
animal cells
fungal cells
sample size is large to give reliability
sampled using quadrats and transects
same size quadrats used randomly
make a hypothesis
scientists interpret data
data analysed
mean
mode
median

Answering Extended Writing questions

QUESTION

Since life began on Earth, many new species of organisms have evolved, and many species have become extinct. Explain how new species can evolve and how some species may become extinct.

The quality of written communication will be assessed in your answer to this question.

G–E

Living things change when the environment changes, so they can survive. they become new species. A long time ago on Earth lots of things became extinct when an asteroid crashed into us. Diseases and volcano's can also make animals go extinct.

Examiner: This answer wrongly suggests that living things make a decision to change, and shows no understanding of natural selection. There is no mention of genetic variation, or organisms with an advantage surviving and breeding to pass on favourable alleles. Three reasons for extinction are included, but only animals are mentioned. Two grammatical errors.

D–C

Living things have to compete for food and space. Some are fitter than others and they survive. Their children inherit the good genes and features. If their children become different enough they are a new species. Living things become extinct if humans kill too many.

Examiner: The term natural selection is not used. The candidate describes organisms as being 'fitter' rather than 'better adapted'. The answer does not specify that living things become new species when they can no longer interbreed successfully. One reason for extinction is given, but many living things became extinct before humans appeared on Earth. The spelling, punctuation, and grammar are good.

B–A*

New species can arise through natural selection. There's genetic variation within a species or population so some organisms are better adapted. These survive and breed and pass on the advantage to their offspring. Sometimes some animals in a population get separated from the others, by a mountain or a river. These animals change and then they can't breed with the original ones. Animals can become extinct if they are hunted too much or if there is a new disease or predator.

Examiner: Natural selection and isolation of a breeding population are well explained. Three reasons for extinction are described. However, the candidate only talks about animals. Plants, fungi, and bacteria also evolve or become extinct. This response is accurate, well organised, and fluent. The spelling, punctuation, and grammar are good.

Exam-style questions

1 The diagram shows a cell from the lung. Gases pass through this cell.

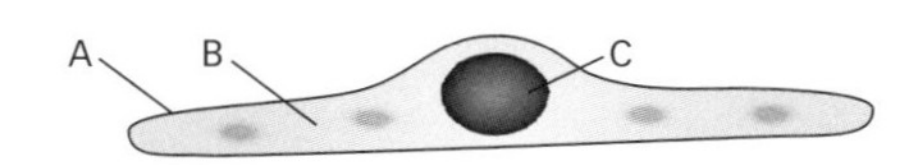

A01 **a** Name parts A, B and C

A01 **b** Which feature of this cell allows gases to pass through it?

- **i** it has a large nucleus
- **ii** it has many mitochondria
- **iii** it is thin

A01 **c** By what process does oxygen pass though this cell?

- **i** osmosis
- **ii** diffusion
- **iii** respiration

A01 **d** How is a bacterial cell different?

- **i** it has no membrane
- **ii** it has ribosomes
- **iii** it has no nucleus

2 A student investigated how a leaf makes starch. Diagram 1 shows how he treated the leaf. Diagram 2 shows where starch was present after 8 hours.

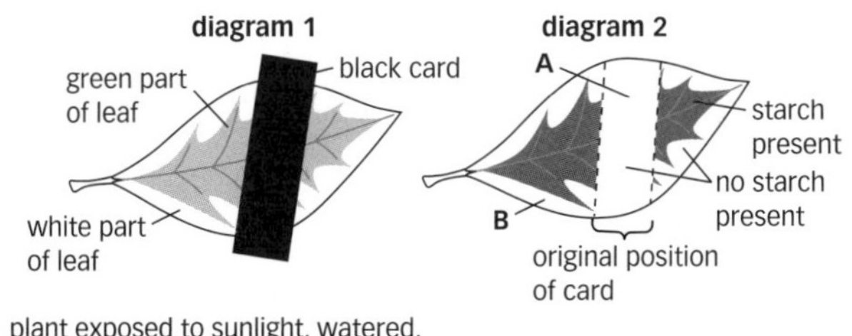

A01 **a** By what process did the leaf make starch?

A02 **b** Why was no starch found in

- **i** the part labelled A?
- **ii** the part labelled B?

A02 **c** Name the two independent variables in this investigation.

D–C

G–E

D–C

3 In an investigation potato chips were weighed before and after being placed in salt solutions for an hour.

salt concentration (M)	0.0	0.2	0.4	1.0	2.0
mass at start (g)	2.5	2.5	2.6	2.5	2.7
mass at end (g)	2.8	2.7	2.7	2.3	2.2
% change in mass	+12.0	+8.0		–8.0	–18.5

A03 **a** Fill in the missing value.

A03 **b** Why are the changes in mass expressed as a percentage change?

A01 **c** By what process do cells in the chips gain or lose water?

A02 **d** Name two factors that should be kept the same in this experiment to make it valid (fair).

A02 **e** How could you find out the strength of salt solution that causes no change to the mass of the chip?

B–A*

Extended Writing

4 A02 Describe how you would find out which plants are present on a school playing field.

G–E

5 A02 Explain how leaves are well adapted for photosynthesis.

D–C

6 A02 Explain how commercial plant growers can manipulate environmental conditions to increase the yield of crops grown in greenhouses.

B–A*

A01 Recall the science

A02 Apply your knowledge

A03 Evaluate and analyse the evidence

B2 Part 2

AQA specification match

Genes and proteins, inheritance, gene technology, and speciation

Why study this unit?

You learn things in school to help you understand the world around you. Some of the main areas of research in biology today are genes, ageing, and regenerative medicine. In this unit you will learn how the genetic code governs the making of proteins in your cells, and why proteins are important. Enzymes are proteins, and are useful outside the body as well as inside. You may use washing powder containing enzymes, and enzymes are also used to help make some foods. You will find out why you need energy, and how your cells respire to release energy from the food you eat.

You will learn how characteristics governed by genes are inherited, and how genetic disorders occur. You will also learn about genetic engineering and gene therapy, cloning, embryo screening, genetic fingerprinting, and stem cell research. Our knowledge of genetics can also be used to help understand how new species develop.

You should remember

1. There are two types of cell division: mitosis for growth and asexual reproduction; and meiosis for sexual reproduction.
2. Cells need energy for their chemical reactions (metabolism) and for division.
3. Respiration releases energy from the food you eat.
4. Genes, on chromosomes, determine your characteristics.
5. The cell that you developed from was made from your mother's egg and your father's sperm, and each contained your parents' genes.
6. There is variation within and between species.

In 1996, a team of scientists at the Roslin Institute in Scotland, led by Professor Ian Wilmut, cloned a sheep using genetic material taken from a cell of an adult sheep. Dolly became the most photographed and most famous sheep in history. She was not the first mammal to be cloned, but was the first to be cloned from an adult cell that had undergone differentiation into an udder (mammary gland) cell.

Professor Wilmut is now director of the Centre for Regenerative Medicine at Edinburgh University, and his team works on stem cells and their potential use in treating brain and spinal injuries, and diabetes. The use of stem cells is not in itself controversial, but sourcing these cells from human embryos is. Scientists have recently found some stem cells deep in the dermis layer of the skin, which may prove a less controversial source of stem cells for research.

14: Proteins

Learning objectives

After studying this topic, you should be able to:

- know that proteins are made of long chains of amino acids
- understand how the shape of a protein enables it to carry out its functions

Key words

proteins, antibodies, hormone, amino acids

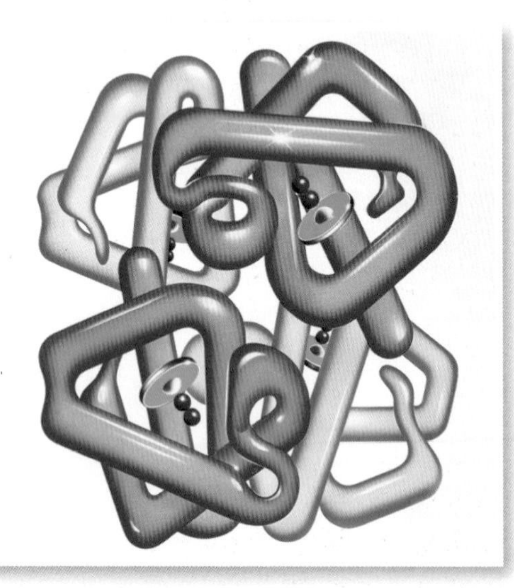

The structure of haemoglobin. The pink and blue areas are proteins. The green areas are where the iron is. Each iron atom can hold two oxygen atoms, shown as red spheres.

Did you know...?

About 75% of your dry mass is protein. Because your cells contain a lot of water, this means that about 25% of the mass of each of your cells is protein.

Why are proteins important?

All the basic structural material of your body is made of **proteins**. Your skin, muscles, bones, cartilage, ligaments, and cell membranes all have a lot of protein in them.

In addition, you also have many other proteins in your cells. These other proteins do important jobs. They include:

- **antibodies**
- **hormones** and their receptors on membranes of target cells
- channels in cell membranes
- catalysts (enzymes that increase the rate of chemical reactions)
- structural components of tissues such as muscle.

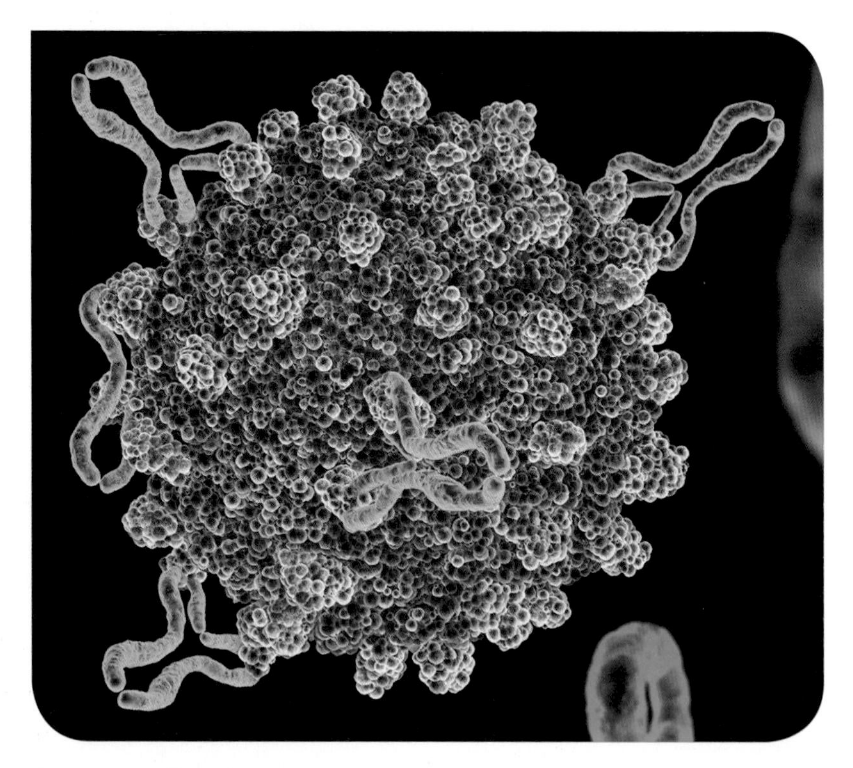

Antibodies surrounding a virus particle in the blood. Antibodies are proteins. Each antibody fits on to the particular antigen, also made of protein, on the particular virus coat.

A Name five parts of your body that are made of protein.

B Name three other types of protein that help your cells to work properly.

What are proteins made of?

All protein molecules contain the elements:

- carbon
- hydrogen
- oxygen
- nitrogen.

In addition, some also contain sulfur.

Proteins are big molecules. They are made of smaller molecules, called **amino acids**. The amino acids are joined together to make long chains. These long chains then fold up into particular (specific) shapes. Each type of protein has a specific shape. Another molecule with a particular shape can fit into it.

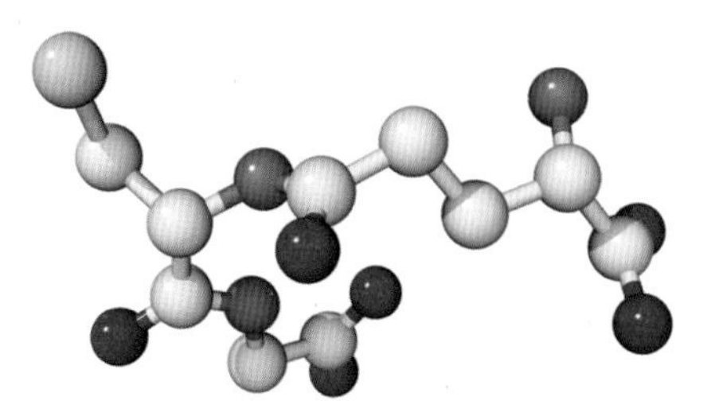

▲ Molecular model of an amino acid. The white balls represent carbon atoms; red represents oxygen; blue represents nitrogen; yellow represents sulfur. Hydrogen atoms are not shown.

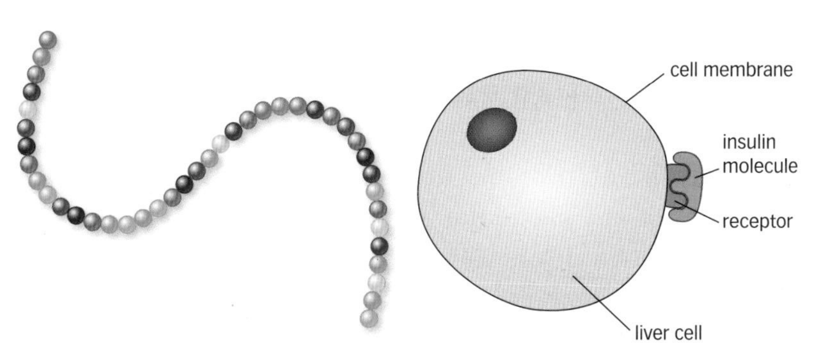

▲ A long chain of amino acids is called a protein

▲ The protein receptor on the surface membrane of a liver cell has a specific shape. The hormone insulin fits into it.

Exam tip

✔ Remember that if two things are shaped so that one can fit into the other, they are not the same shape. Scientists say they are complementary in shape, but you can just say that one fits into the other. Think of a boiled egg fitting into an egg cup. They are not the same shape. You couldn't get one boiled egg to fit inside another boiled egg.

Questions

1. Name a hormone that is made of protein.
2. What elements are present in protein molecules?
3. What smaller molecules are joined together to make proteins?

↓ E

4. Explain why the way each particular protein folds into a specific shape enables it to carry out its function (do its job).
5. A certain living organism contains 80% water. Some 75% of its dry mass is protein. How much of its total living mass is protein?

↓ C

6. Fertiliser for plants contains nitrates, a form of nitrogen. Why do you think plants need nitrogen to grow?
7. Why do you think you need to eat a certain amount of protein each day?
8. When you eat proteins from meat, eggs, or soya beans, your body uses them to make the proteins in your cells and tissues. How do you think this happens?

↓ A*

Learning objectives

After studying this topic, you should be able to recall that enzymes:

- ✔ are proteins that catalyse chemical reactions in living cells
- ✔ have high specificity for their substrates
- ✔ work best at particular temperatures and pH

Key words

enzyme, catalyst, substrate, specific, optimum, denatured

A What is a catalyst?

B Name three types of chemical reaction that enzymes speed up in living organisms.

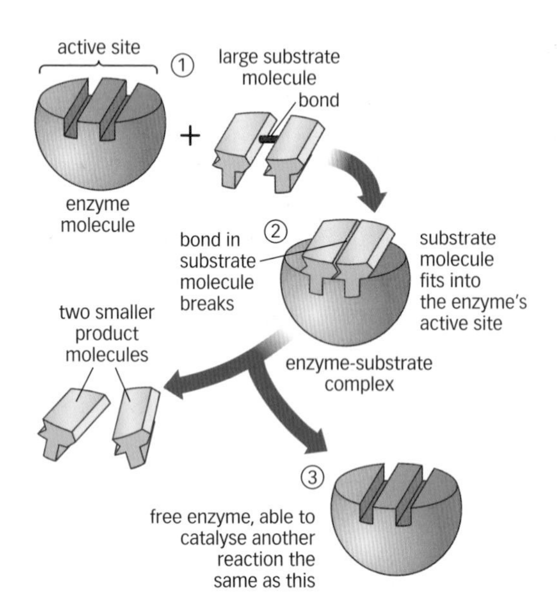

▲ The large substrate molecule fits into the enzyme's active site. A bond breaks and two product molecules are formed.

Enzymes are catalysts

Enzymes are **catalysts** because they speed up chemical reactions.

Most of these reactions, such as:

- photosynthesis
- respiration
- protein synthesis

take place inside living cells.

Enzymes can be used to catalyse the same type of reaction many times. This is like using one type of screwdriver to screw in many of the same type of screw, one at a time.

The shape of an enzyme is vital for its function

Enzymes, like all proteins, are folded into a particular shape. The shape of one particular area of the enzyme molecule, called the active site, is very important.

- The **substrate** molecules fit into the active site.
- This brings them together so they can form a bond.
- This makes a bigger molecule.

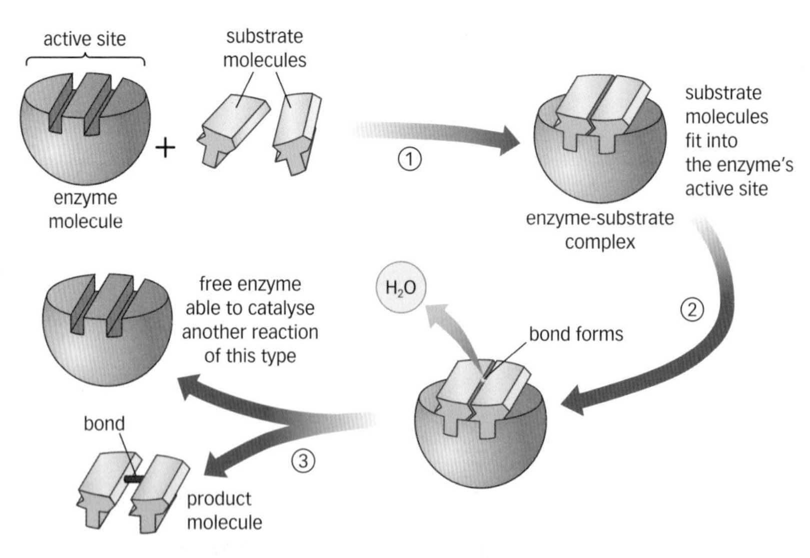

▲ The lock and key hypothesis is a hypothesis about how enzymes work. The two substrate molecules fit side by side into the enzyme's active site. A bond forms between them, and one large product molecule is formed.

In some cases (as shown on the left):

- A big substrate molecule fits into the active site.
- A bond breaks.
- Two smaller product molecules are made.

Enzymes have specificity for their substrate

- Only one particular type of substrate molecule can fit into an enzyme's active site. This is like the way only one type of key will fit into a particular lock.
- This means each enzyme is **specific** for its substrate molecules.

C What are enzymes made of?

D Why is the shape of the active site of an enzyme important?

E How is each enzyme specific for a particular substrate?

What makes enzymes work best?

Each enzyme works best at

- a particular temperature, known as its **optimum** (best) temperature
- its optimum pH.

Low temperatures

The enzyme and substrate molecules have less energy. They do not move very fast so they do not collide (bump into each other) very often. The rate of reaction is low.

High temperatures

As the temperature increases, the enzyme and substrate molecules move more quickly and collide more often. This gives a faster rate of reaction.

However, if the temperature becomes *too* high then:

- The shape of the enzyme's active site changes.
- The substrate molecule cannot now fit into the active site.
- The rate of reaction slows and eventually stops.

When the shape of the enzyme has changed in this way, it cannot go back to its original shape. The change is irreversible. The enzyme is **denatured**.

pH

Each type of enzyme works at an optimum pH.

If the pH changes very much, then:

- The shape of the active site changes.
- The substrate molecules cannot fit into it.
- The enzyme has been denatured.

Did you know...?

Many enzymes in the body could work more quickly at temperatures above 37°C. However, if we kept our bodies hotter than this, many of our other proteins would be damaged. At 37°C our chemical reactions go on fast enough to sustain life. But some bacteria can live in very hot places, and their enzymes work well at 100°C.

Questions

1 State two conditions that enzymes need to work best. (E)

2 Explain why increasing the temperature from 10°C to 25°C makes the rate of an enzyme-controlled reaction increase.

3 Explain why if the temperature increased to 60°C, the rate of the enzyme-controlled reaction would slow down and eventually stop. (C)

4 As well as having enzymes inside your cells, you also have them in your blood. Your blood pH needs to be kept very close to 7.2. Why do you think this is? (A*)

Exam tip AQA

✔ Always refer to the active site when you are explaining how an enzyme works.

Learning objectives

After studying this topic, you should be able to:

- ✔ understand that enzymes for digesting food work outside body cells
- ✔ know the roles of hydrochloric acid and bile in helping digestion

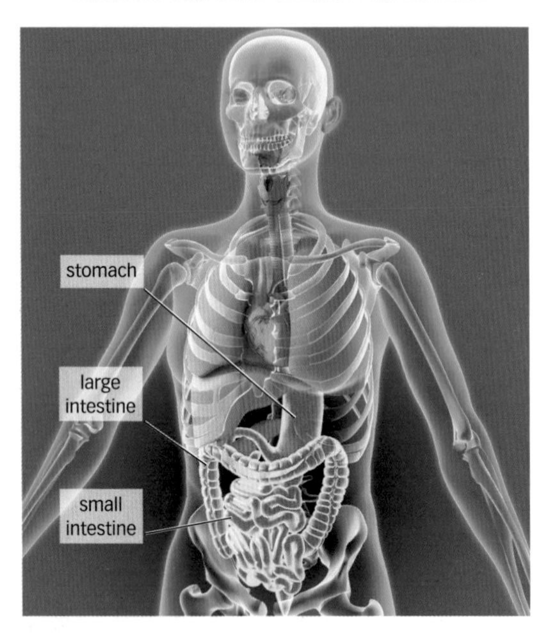

▲ Computer artwork showing the digestive tract (gut) inside the body

Did you know...?

You could digest your food without enzymes, but it would take several years to digest just one meal.

Exam tip AQA

- ✔ You should learn the information in the table on the next page. You need to know the names of these enzymes, where they are made and work, and what they do.

Why do you need to digest food?

It may seem a strange idea, but when food is in your gut it is still outside your body. Think of your body as being like a very long doughnut. The hole running through it is your gut. It is open to the outside at each end – the mouth and the anus.

Your gut or digestive tract runs from the mouth, via the gullet, stomach, small intestine, large intestine, and rectum to the anus.

The gut wall is part of your body of course, but the space in the middle of the gut is continuous with the outside of the body. So when the food you have chewed and swallowed is in your stomach and intestine, it is still outside.

You need to get it across the gut wall and into your bloodstream so it can go to your cells. The large food molecules have to be broken down into small molecules so they can diffuse across the gut wall and into the blood. This breaking down into smaller molecules is **digestion**.

A Explain why you need to digest your food.

How is your food digested?

You have special enzymes that catalyse the breakdown (digestion) of large food molecules to smaller molecules. They are called digestive enzymes.

Digestive enzymes

- are made in specialised cells in glands and in the lining of the gut
- pass out of the cells where they are made and into the gut
- come into contact with food molecules and catalyse the breakdown of large food molecules into smaller molecules.

B What is the substrate for the enzyme amylase?

C What is the product when amylase catalyses the breakdown of its substrate?

carbohy

Enzyme	Where it is made	What it does	Where it does it
Amylase	• In salivary glands • In the pancreas • In the lining of the small intestine	Catalyses the breakdown of starch molecules into sugar molecules.	In the mouth and in the small intestine.
Protease	• In the stomach • In the pancreas • In the lining of the small intestine	Catalyses the breakdown of protein molecules into amino acids.	In the stomach and small intestine.
Lipase	• In the pancreas • In the lining of the small intestine	Catalyses the breakdown of lipids (fat molecules) into fatty acid and glycerol molecules.	In the small intestine.

The best conditions

Enzymes need a particular pH to work best.

Some enzymes work well at acidic pH

Your stomach makes hydrochloric acid. This kills any bacteria that are in the food you eat. The proteases that work in your stomach can work well at low pH.

Some enzymes work best at alkaline pH

The enzymes that work in the small intestine work best in slightly alkaline conditions. When the food passes out of your stomach and into your small intestine, bile is released from the gall bladder. The bile enters the small intestine.

Bile

- is made in the liver
- is stored in the gall bladder
- neutralises the acid that was added to food in the stomach
- provides alkaline conditions in the small intestine for the enzymes there to work most effectively.

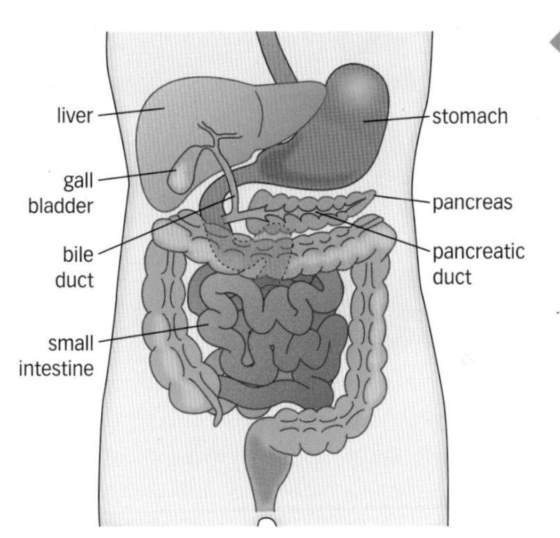

In this diagram of the digestive tract you can see how the bile from the gall bladder passes down a tube and into the small intestine. You can also see how enzymes made in the pancreas can pass along the pancreatic duct and into the small intestine.

Questions

1. Name three places in the body where amylase is made.
2. What types of molecules are made when lipase catalyses the breakdown of fats?
3. What is the function of bile?
4. Where is bile made?

↓ E

5. What is the function of hydrochloric acid in the stomach?
6. Does amylase work inside or outside body cells?

↓ C

7. Protease enzymes that work in the stomach work well at very low pH. Protease enzymes that work in the small intestine work best in slightly alkaline conditions. Do you think they are exactly the same type of enzyme? Explain your answer.

↓ A*

Key words

digestion, amylase, protease, lipase

Learning objectives

After studying this topic, you should be able to:

- ✔ know how enzymes obtained from microorganisms are used in biological detergents

▲ Fermenters (special large containers) used for growing large numbers of bacteria. The bacteria inside make enzymes and pass them out of their cells. The enzymes can be collected from the liquid inside the fermenter.

▲ False colour scanning electron micrograph of granules of biological washing powder. Inside the opened granules you can see capsules of enzymes (×40).

What are detergents?

Detergents are cleaning agents. Washing powders contain detergents. Some washing powders also contain enzymes.

Obtaining enzymes from microorganisms

Some microorganisms make enzymes that pass out of their cells. If bacteria or fungi are grown in large vessels (called **fermenters**), scientists can collect the enzymes that they make and pass out of their cells.

A What are detergents?

B How are bacteria grown on a large scale?

Biological washing powders

Biological washing powders were first made on a large scale in 1965. They contain soap and have enzymes added to them. The enzymes can break down stains on textiles such as clothes. The stains may contain:

- protein – such as egg yolk or blood
- fats – such as grease or sweat.

The enzymes added to washing powders are proteases and lipases.

▲ Assorted washing powders. The biological washing powder on the left contains enzymes to break down fats and proteins in stains. Non-biological powders do not contain enzymes. Some people say they have had allergic reactions to the enzymes in biological washing powders. However, skin specialists in hospitals have carried out investigations. They have found that biological washing powders do not irritate skin.

C What sort of stains on clothes do protease enzymes break down?

D What sorts of stains on clothes do lipase enzymes break down?

Washing powders are highly alkaline, so the enzymes added to them must be able to work well in solutions with high pH. They must also be able to work at a range of temperatures between 10 °C and 90 °C.

Some bacteria live in very hot places and have enzymes that work well at high temperatures. Some of the enzymes added to washing powders can be obtained from these bacteria.

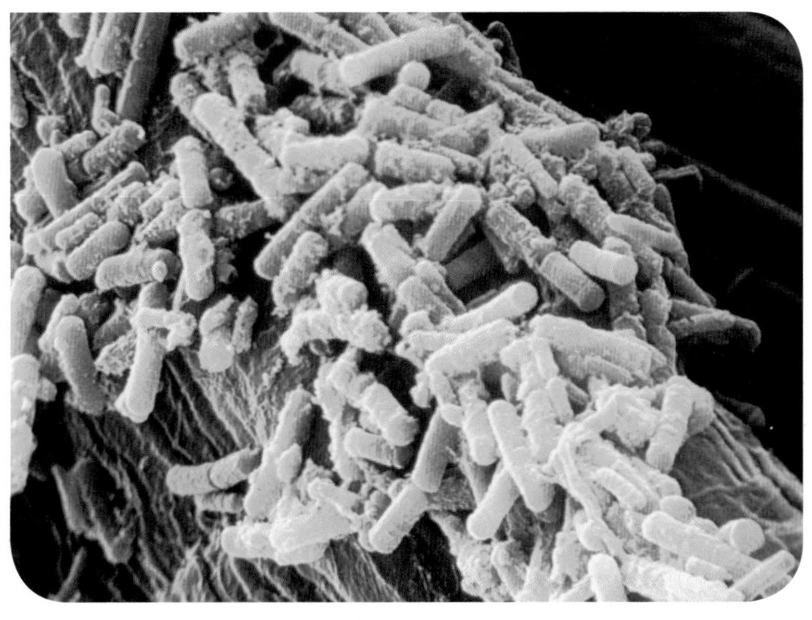

▲ These bacteria grow best between 55 °C and 75 °C and can survive at 90 °C. Their enzymes are not denatured at this high temperature (× 7200).

Low-temperature washing powders

Some biological washing powders are designed to be used at low temperatures. The enzymes added to them work well at low temperatures, so these washing powders remove stains better than other types of detergents at this temperature. Using cool water to wash clothes saves energy, and so saves money.

Some enzymes added to washing powders are denatured at high temperatures. These washing powders are designed to be used at temperatures up to 40 °C.

Key words

detergent, **fermenter**

Did you know...?

Some bacteria live near thermal vents on ocean beds at temperatures of 120 °C.

Questions

1. Why are enzymes for some biological washing powders obtained from bacteria that live in hot springs? (E)
2. What are the advantages of using biological washing powders at 20 °C?
3. What is the best pH for the enzymes added to washing powders? (C)
4. Explain what would happen to enzymes added to low-temperature washing powders if these powders were used in a hot wash at 80 °C. (A*)

Exam tip AQA

✔ Remember that not all enzymes become denatured at temperatures above 40 °C.

Learning objectives

After studying this topic, you should be able to:

- ✔ understand why enzymes are used in industry
- ✔ consider some examples of using enzymes in industry

▲ The protein in baby food has been pre-digested with enzymes, so the baby can get all the amino acids from it

A Why do babies need to absorb amino acids from their food?

B Where are the enzymes used in industry obtained from?

Key words

carbohydrase, starch, glucose, fructose, isomerase

Did you know...?

Sugar is added even to savoury processed foods such as soup and baked beans.

Why use enzymes?

Many reactions used in industrial processes can take place without enzymes. However, the reaction mixture would have to be heated to a high temperature and kept at high pressure. This would use a lot of expensive energy. It would also need expensive equipment that could stand the high temperatures.

When enzymes are used to catalyse the reactions, the reactions can take place at lower temperatures and pressures. This saves money because the equipment used is not so expensive. Also, not so much fuel is needed.

Scientists can get enzymes from microorganisms. Some microorganisms make enzymes that pass out of their cells. Scientists grow lots of these bacteria in big vessels in laboratories, and then extract the enzymes.

Properties of enzymes used in industry

Industrial enzymes must

- have a long shelf life
- be able to stand fairly high temperatures
- have a wider than usual pH tolerance
- be able to work in the presence of chemicals that usually stop enzymes working.

How enzymes are used in industry

Proteases

Protease enzymes break down proteins. Some are used to pre-digest protein in baby foods. The enzymes break down the protein molecules into smaller molecules. Babies' stomachs are not strong enough to digest bigger protein molecules. Pre-digesting protein in this way makes sure that the baby can absorb amino acids from the baby food.

Carbohydrases

Carbohydrases are enzymes that break down **starch** into sugar (**glucose**) syrup.

Many processed foods have sugar added to them for flavour or to make the food taste sweet. Cane sugar (the sort you may put in your tea or coffee) is expensive. Food manufacturers use carbohydrase enzymes, such as amylase, from microorganisms. They mix the enzymes with cheap starch obtained from things like corn stalks.

Corn stalks are a source of cheap starch

Glucose is added to some foods such as ice cream, and to some drinks. But it is not as sweet as cane sugar.

For some processed foods, a lot of glucose would need to be added. A type of sugar called **fructose** is much sweeter than glucose. Natural fructose is very expensive, but another type of enzyme from microorganisms, called an **isomerase**, can change glucose into cheap fructose.

Isomerase enzymes

An isomerase enzyme is mixed with the glucose syrup. This enzyme catalyses the changing of glucose molecules into fructose molecules. Both sugar molecules have the same number of atoms, but they are arranged slightly differently. Glucose and fructose molecules have slightly different shapes. Fructose tastes sweeter than glucose, so food manufacturers need to add less of it to their products. This is good for slimming foods.

Slimming products

Disadvantages of using enzymes in industrial processes

Although enzymes from microorganisms are tougher than other enzymes, at very high temperatures some are denatured. Also, the product may be contaminated with some enzyme molecules. However, if the enzymes are enclosed in special little capsules, this protects them from the high temperatures and stops them contaminating the product.

Exam tip AQA

✔ Some exam questions are meant to be hard. You need to think about the question and make a sensible suggestion based on what you know – in other words, apply your knowledge to a new situation.

Questions

1 Why do food manufacturers want to use fructose rather than glucose?

E

2 What are the advantages of using enzymes obtained from microorganisms in chemical reactions used in industry?

3 Describe how carbohydrase enzymes are used to obtain glucose sugar.

C

4 Describe how isomerase enzymes are used to turn glucose into fructose.

5 What are the disadvantages of using enzymes in industrial processes?

6 Fructose and glucose are both sugars. They have slightly differently shaped molecules. Why do you think they do not taste the same? (You have to think about this question and make a suggestion. Think about the shapes of protein molecules and taste receptors on your tongue.)

A*

19: Energy and life processes

Learning objectives

After studying this topic, you should be able to:

- ✔ know that the energy needed for all life processes is provided by respiration
- ✔ understand that respiration can take place aerobically or anaerobically

▲ Buffalo (*Bison bison*) grazing on grass

A State three reasons why living things need energy.

B Name three types of large molecules that are made in living cells using energy.

▲ This grey wolf (*Canis lupus*) needs energy to run

What is energy?

Energy is the ability to do work. All matter has energy. There are different forms, such as kinetic (movement), potential (stored), heat, sound, electrical, and light energy. Each form of energy can be transferred into another form.

- Plants trap sunlight energy and use it to make large molecules – proteins, fats, and carbohydrates. These molecules contain stored energy.
- Animals get these molecules, containing stored energy, by eating plants or eating other animals that have eaten plants.

Why do living organisms need energy?

All life processes in all living organisms (including plants as well as animals) need energy. The energy may be used

- to build large molecules from smaller ones
- for muscle contraction in animals
- to control body temperature in mammals and birds.

Building large molecules from smaller ones

- Plants use sugars, nitrates, and other nutrients to make amino acids.
- Amino acids are joined together in long chains during protein synthesis. All living things need to make proteins such as enzymes and parts of their structure.
- Plants join sugar molecules together to make starch.
- Animals join sugar molecules together to make glycogen, which is similar to starch.
- Living organisms join fatty acids and glycerol together to make lipids (fats).

Muscle contraction

Animals need to move, to find food or a mate, or to escape from predators. Muscle contraction needs energy and causes movement.

Controlling body temperature

Some organisms cannot control their temperature very well. As the surrounding temperature changes, their temperature may also change. They control it by moving into the shade or into a warmer place. Snakes and lizards are very slow and sluggish in winter, or at night, when it is cold.

Birds and mammals can be active at night and during the winter. This is because a lot of the energy from the food they eat is released as heat energy. This keeps their body temperature steady regardless of the external temperature. However, it means that they need to eat more food than animals such as fish, snakes, and lizards.

How is energy released from food molecules?

Respiration in living cells releases energy from glucose molecules. You get glucose when you digest carbohydrates that you eat.

Respiration is a process that involves many chemical reactions, all controlled by particular enzymes.

Aerobic respiration

Aerobic respiration uses oxygen. It happens continuously in the cells of plants and animals.

Anaerobic respiration

Anaerobic respiration is a different type of respiration, that takes place without oxygen. This does *not* happen continuously in plant and animal cells. It happens when cells are not getting enough oxygen.

Questions

1 Explain why animals need energy for movement.

2 What process in cells releases energy from food?

E

3 Explain why birds and mammals can be active at night when it is cold.

4 Explain the difference between aerobic and anaerobic respiration.

C

5 On a cold night in winter, a robin will lose a quarter of its body mass. Why do you think this is?

6 During the winter in the UK, many birds, such as swallows, cannot find enough food to eat to keep warm. How do you think they solve the problem?

7 During the winter in the UK, some mammals, such as hedgehogs, cannot find enough food to eat to keep warm. How do you think they solve the problem?

A*

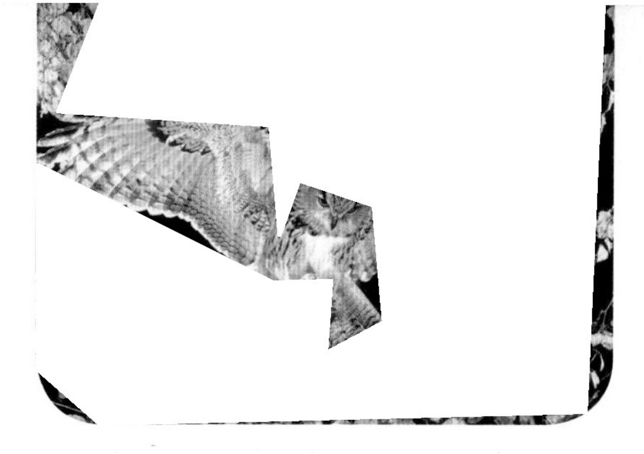

▲ Eagle owl (*Bubo bubo*) hunting at night

Did you know...?

The average temperatures of different species of birds and mammals vary.

Average body temperatures of some mammals and birds

Animal	Average body temperature (°C)
human	37.0
chimpanzee	37.0
dog	38.0
cat	39.0
rabbit	39.5
chicken	42.0
owl	38.5
eagle	48.0
penguin	38.0

Key words

respiration, aerobic, anaerobic

Exam tip AQA

✔ Remember that plants respire all the time, just as animals do.

Learning objectives

After studying this topic, you should be able to:

- ✔ know the summary equation for aerobic respiration
- ✔ know that most of the reactions in aerobic respiration take place in mitochondria
- ✔ understand the changes that take place in the body during exercise

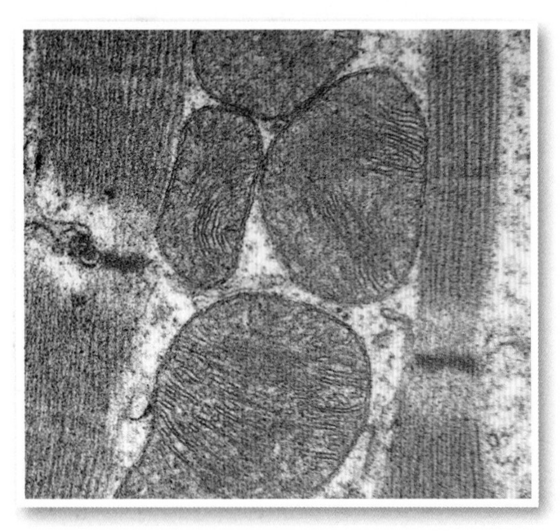

▲ Electron micrograph showing mitochondria, stained pink, in heart muscle (×78 800)

▲ This runner needs energy released from respiration. The energy enables her muscles to contract.

The equation for respiration

During aerobic respiration in living cells, there are chemical reactions which

- use glucose sugar and oxygen
- and release energy.

Although aerobic respiration involves a series of several chemical reactions, it can be summarised by the following equation:

glucose + oxygen → carbon dioxide + water (+ energy)

Mitochondria

Most of these chemical reactions take place inside **mitochondria**. These are sausage-shaped organelles (tiny organs) inside both plant and animal cells.

A Write down a word equation for aerobic respiration.

B Where in your cells do most of the chemical reactions involved in aerobic respiration take place?

Energy and exercise

The energy released by respiration is used by the organism. One of the things it may be used for is muscle contraction.

When you exercise, your muscles need to contract more. They need more energy. They have to carry out more respiration to release the extra energy.

More respiration means the muscle cells need

- more glucose
- more oxygen.

To meet those needs

- your rate and depth of **breathing** goes up, which gets more oxygen into the lungs and into the blood
- your **heart rate** goes up, so more blood with oxygen is quickly delivered to the muscle cells, and this also delivers more glucose to muscle cells
- your muscle cells break down more of their stored **glycogen** into glucose.

How do your muscles store glycogen?

When you eat carbohydrates, your enzymes digest them to sugar.

The sugar passes across the gut wall into the bloodstream.

Some of that sugar goes from the blood into muscle cells. Here it is changed into glycogen (a big molecule similar to starch) and stored.

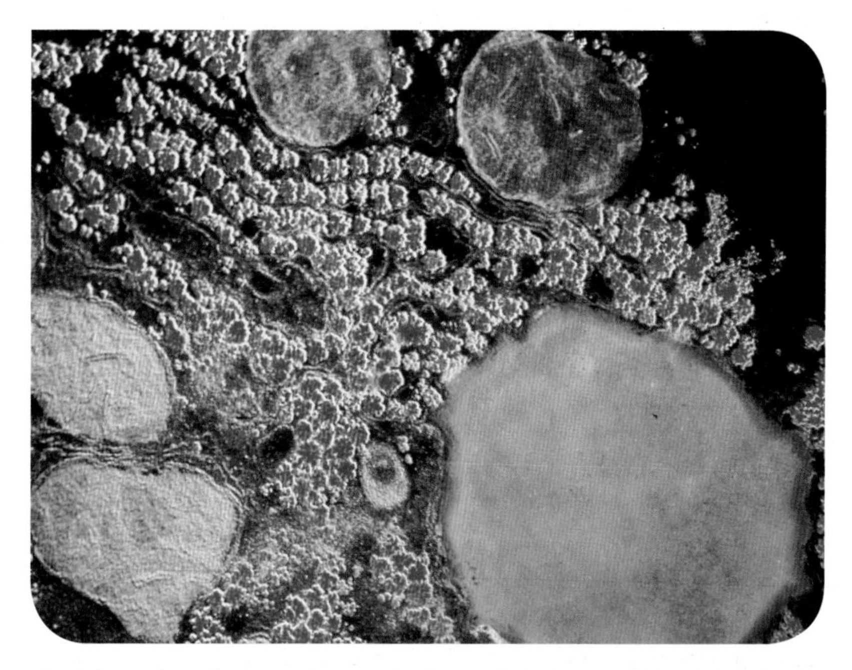

▲ False colour electron micrograph of stored glycogen (coloured pink) in a cell (× 9000)

When muscle cells respire more during exercise, as well as releasing more energy for their contraction they also release

- more heat
- more carbon dioxide.

Your body must not get too hot. The increased blood supply to and from your muscles carries away the extra heat to the skin. It also carries away the extra carbon dioxide to the lungs to be breathed out. Your heart rate and breathing rate stay high for a while after you finish exercising. This helps remove the extra carbon dioxide and heat.

Key words

mitochondria, breathing, heart rate, glycogen

Exam tip AQA

✔ Do not confuse breathing with respiration. Respiration happens in cells and releases energy from food. Breathing is your chest movements that get air into and out of the lungs.

Did you know...?

Your brain detects the extra carbon dioxide in your blood. It sends signals to the heart and lungs to make heart rate and breathing rate increase.

Questions

1 Why do your muscle cells respire more when you run?

2 Explain why your breathing rate increases when you play a game such as football.

E

3 Explain why your heart rate increases when you ride a bicycle.

4 Why do muscle cells store glycogen?

C

5 Why do you think your blood carries heat to your skin?

A*

Learning objectives

After studying this topic, you should be able to:

- know that muscles use anaerobic respiration when they do not receive enough oxygen
- understand why muscles become fatigued after a long period of vigorous activity

A Write a word equation for anaerobic respiration.

B Where, in cells, does anaerobic respiration take place?

Did you know...?

Some kinds of exercise use anaerobic respiration because they happen quickly and there is no time for your heart and breathing rates to increase. A golf swing, tennis serve, a shotput, and a 100 m sprint all use anaerobic respiration to provide the energy for muscle contraction.

When your muscle cells do not receive enough oxygen

You have seen that during exercise your muscle cells need to respire more. Your breathing and heart rates increase to try to meet that need. However, sometimes they do not deliver enough extra oxygen to respiring muscle cells during exercise.

When this happens, your muscle cells use another type of respiration as well. They use anaerobic respiration. Anaerobic respiration happens in the cytoplasm of cells.

During anaerobic respiration, glucose is incompletely broken down to **lactic acid** instead of carbon dioxide. A much smaller amount of energy is released per molecule of glucose than in aerobic respiration. However, this incomplete breakdown happens quickly. So, many molecules of glucose can be broken down to meet the muscle's extra needs for a while.

You cannot use anaerobic respiration for long because the build-up of lactic acid is toxic. It causes muscles to become **fatigued**. They stop contracting efficiently. As your blood flows through your muscles it carries away the lactic acid. The blood takes the lactic acid to your liver, to be processed.

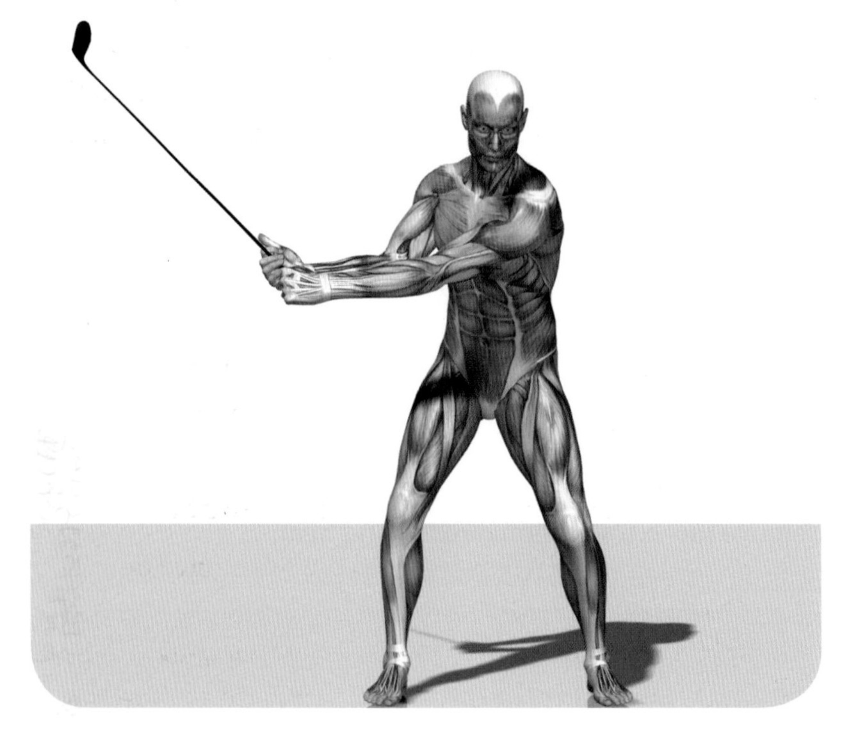

▲ Muscles involved during a golf swing use anaerobic respiration to release the energy needed

▲ This runner is suffering from a 'stitch', caused by lactic acid build-up

Key words

lactic acid, fatigue, **oxygen debt**

The oxygen debt

If you use anaerobic respiration to provide energy, then the muscle cells break down glucose to lactic acid. The lactic acid lowers the pH. This reduces the activity of the enzymes in muscles so they do not contract so efficiently. This is why you cannot exercise for long using anaerobic respiration.

When your heart rate and breathing rate have increased, you can use some of the extra oxygen to oxidise the lactic acid to carbon dioxide and water. This is called repaying the **oxygen debt**.

Questions

1. Which provides more energy for each molecule of glucose – anaerobic or aerobic respiration?
2. What causes muscle fatigue?
3. How does lactic acid affect the pH of your blood?
4. Why do you think you use anaerobic respiration if you run a 100 m sprint?
5. Why do you think an athlete's heart rate and breathing rate stay high for several minutes after running a 100 m sprint?

E C A*

Exam tip AQA

✔ Remember that aerobic means with oxygen. The prefix 'an' or 'a' means without. So anaerobic means 'without oxygen'.

Learning objectives

After studying this topic, you should be able to:

- know that mitosis is a type of cell division in body cells, producing two genetically identical cells
- know that organisms that reproduce asexually use mitosis

Key words

mitosis, chromosome, diploid, asexual reproduction, allele, growth

Exam tip

- Mitosis replaces damaged cells and repairs tissue. It does not repair cells.

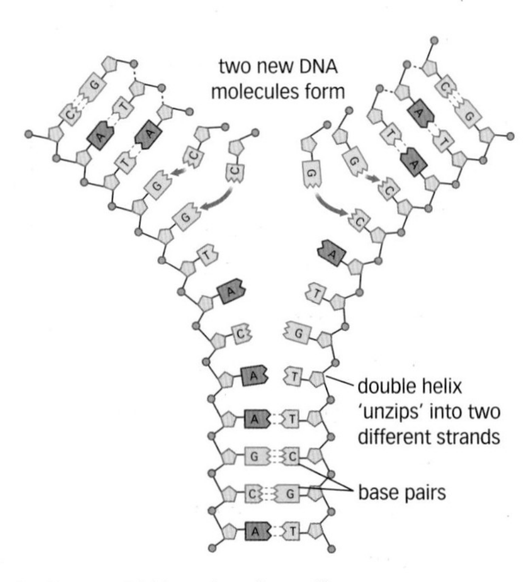

How a DNA molecule replicates

Why do body cells divide?

Body cells divide

- to replace worn-out cells
- to repair damaged tissues
- to grow by producing more cells.

Body cells divide by **mitosis**. Each cell produces two genetically identical daughter cells. This increases the total number of cells in a multicellular organism.

In the nucleus of most of your body cells you have two sets of **chromosomes**, arranged as 23 matching pairs of chromosomes. These cells are described as **diploid**.

This cell is about to divide. The chromosomes have coiled and become visible.

A Name a type of cell in your body that does not have any chromosomes in it.

Copying the cell's genetic material

Before a cell divides, its genetic material has to be copied. Each chromosome, made of one molecule of DNA (the genetic material), is copied. So, before a cell divides, each molecule of DNA copies itself. This is called DNA replication.

How DNA replicates

- DNA is a double-stranded molecule.
- The molecule 'unzips', forming two new strands.
- This exposes the DNA bases on each strand.
- Spare DNA bases in the nucleus line up against each separated strand of DNA.
- They only align next to their complementary DNA base, forming base pairs.
- One molecule of DNA has become two identical molecules.

How mitosis happens

- When each chromosome has made a copy of itself, these duplicated chromosomes line up across the centre of the cell.
- Then each 'double' chromosome splits into its two identical copies.
- Each copy moves to opposite poles (ends) of the cell.
- Two new nuclei form, each with a full set of chromosomes.
- The cell divides into two.
- Each cell is genetically identical to each other and to the parent cell.

Cells in the root tip of a hyacinth plant undergoing mitosis (× 180)

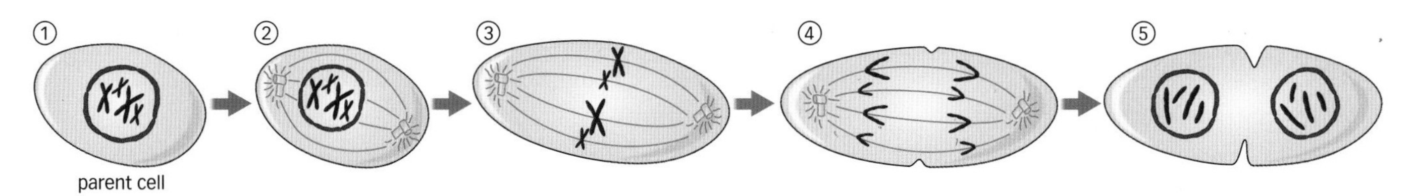

A cell dividing into two genetically identical cells by mitosis

Asexual reproduction

Some organisms can reproduce **asexually**. This type of reproduction also uses mitosis. The cells of the offspring produced by mitosis are genetically identical to the parent cells. They have the same **alleles** (versions of genes) as the parents.

Advantages of being multicellular

Early life forms on Earth were single-celled. There are still many simple single-celled organisms, such as the amoeba. Many organisms are now multicellular. Being multicellular means the organisms can be larger, can have different types of cells that do different types of jobs, and can be more complex.

Mitosis in mature organisms

In mature animals, cell division is mainly restricted to replacement of cells and repair of tissues. Mature animals do not continue to **grow**.

However, mature plants still have areas, such as root and shoot tips, where they can grow. The new cells made in these areas, by mitosis, can differentiate (become different and specialised) into many different types of plant cell.

Questions

1. Explain why a cell's genetic material has to be copied before it divides by mitosis.
2. What is DNA replication?

E

3. Explain why mitosis is used for asexual division.
4. What are the advantages to an organism of being multicellular?

C

5. How does the use of mitosis differ in mature plants and animals?
6. Explain how mitosis happens.

A*

Learning objectives

After studying this topic, you should be able to:

- ✔ know that gametes are made by meiosis for sexual reproduction
- ✔ know that gametes are haploid and combine to give a diploid zygote
- ✔ understand that meiosis produces genetic variation

Key words

gametes, meiosis, haploid, zygote, fertilisation

A What are gametes?

B Where in the body are female gametes made? Where in the body are male gametes made?

Gametes

Gametes are sex cells. They are involved in sexual reproduction. Each gamete has only one set of chromosomes.

- Egg cells are made in the ovaries and sperm cells are made in the testes.
- Gametes are made by a special kind of cell division called **meiosis**.

How meiosis happens

- Just before the cell divides by meiosis, copies of the genetic information are made, just as they are before mitosis.
- So each chromosome has an exact copy of itself.
- However, in meiosis, the cell divides twice, forming four gametes.
- In the first division the chromosomes pair up in their matching pairs.
- They line up along the centre of the cell.

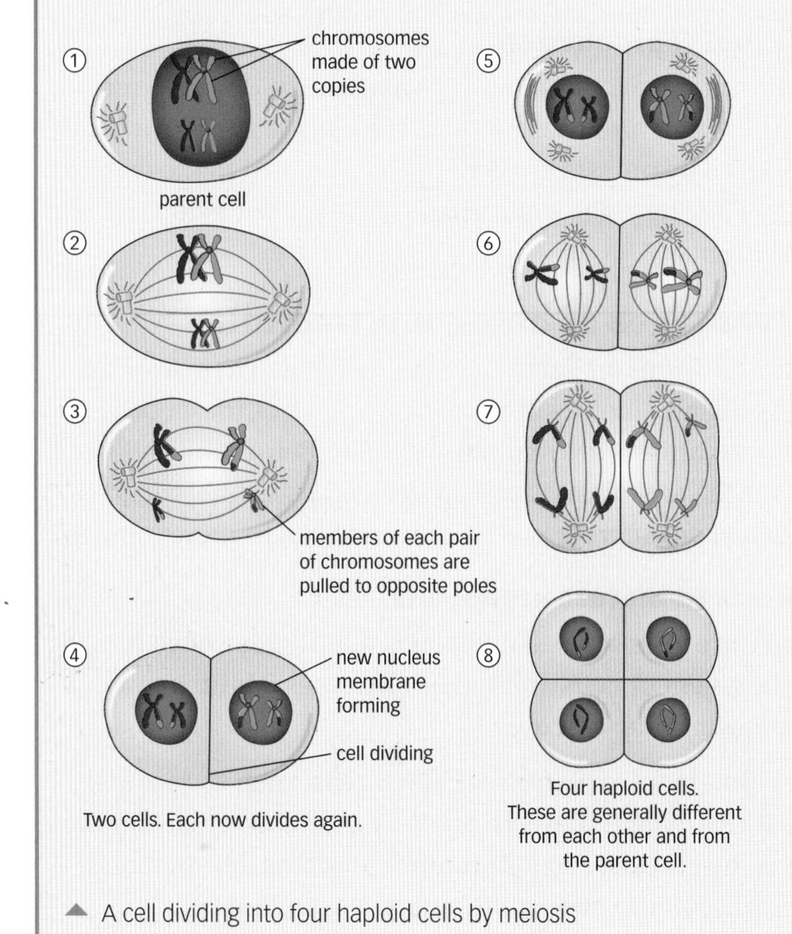

▲ A cell dividing into four haploid cells by meiosis

- The members of each pair split up and go to opposite poles (ends) of the cells.
- Now these two new cells each divide again.
- This time the double chromosomes split and go to opposite poles.
- Four cells, each having just one set of chromosomes, are made.

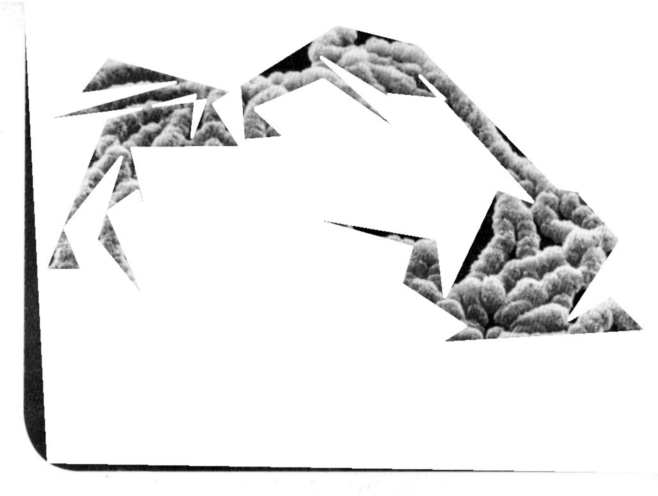

Coloured electron micrograph (×4100) showing chromosomes during meiosis

The cells made by meiosis are **haploid** gametes. They contain 23 chromosomes, *not* 23 pairs.

When two haploid gametes (an egg and a sperm) join, they produce a diploid cell called a **zygote**. This zygote will divide by mitosis into many cells and grow into a new individual:

- The joining of two gametes is called **fertilisation**.
- The combining of genetic material from two parents produces a unique individual.
- Half its chromosomes (and genes/alleles) have come from one parent and half from the other parent.
- It will have two sets of chromosomes.

A new individual will now develop from this cell, dividing many times by mitosis.

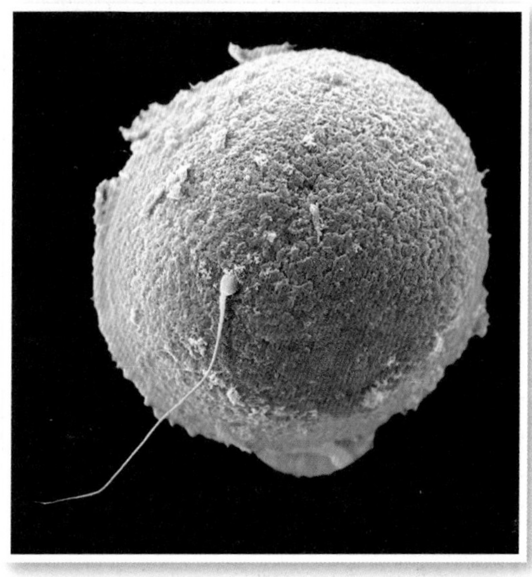

Coloured electron micrograph showing a human sperm (coloured blue) penetrating a human egg (×700)

Questions

1 How are gametes different from body cells?

2 Describe how male gametes are adapted to their function.

3 Describe how female gametes are adapted to their function.

4 Why do sperm cells need a lot of energy?

E

5 By what process will the mitochondria in sperm provide energy?

6 What do the following terms mean? (a) fertilisation, (b) haploid, (c) diploid, and (d) zygote.

C

7 Explain why sexual reproduction produces genetically unique new individuals.

8 What type of cell division do you think causes the zygote to develop into an embryo?

A*

Exam tip AQA

✔ Learn the spelling of meiosis. The word is similar to mitosis, so it has to be spelled correctly so that the examiner can be sure which type of cell division you are referring to.

24: Inheritance – what Mendel did

Learning objectives

After studying this topic, you should be able to:

- be familiar with the principles used by Mendel in his investigations of monohybrid inheritance

Gregor Mendel. Before he became a monk he was brought up on a farm where he gardened and kept bees.

Pea plants (*Pisum sativum*)

Who was Mendel?

Gregor Mendel was born in 1822 in what is now the Czech Republic. He became a monk and studied maths and science at the University of Vienna. He used his monastery garden to study variation and inheritance in pea plants. He published a paper about his findings in 1866, but not many people read it and no one else seemed to be able to understand it. His work did not achieve the recognition it deserved in his lifetime, but was rediscovered about 40 years after it was published.

A When was Gregor Mendel born?
B Where was he brought up?
C What did he study at Vienna University?

What did Mendel discover?

Before Mendel's plant-breeding experiments, people thought that characteristics were inherited by a blending mechanism. They thought, for example, that if a black dog and a white dog mated, then the puppies would be grey.

Mendel had access to the monastery garden and grew edible peas, *Pisum sativum*:

- He had several varieties of pea plants. He made sure, by sowing seeds from them and looking at the offspring, that each variety he used was true-breeding. For example, white-flowered plants always produced white-flowered offspring.
- He selected varieties with different characteristics.
- He cross-pollinated two different varieties. In one of his experiments he crossed tall-stemmed plants with short-stemmed plants.
- He took pollen (male gametes) from one variety of pea plant. He placed this on to the female part of the flowers of the other variety of pea plant.
- He then collected the seeds and grew them.
- He looked at the offspring (the **F_1 generation**) to see if they were tall- or short-stemmed.
- He saw they were all tall-stemmed. The short-stem characteristic seemed to have disappeared. He noted that they were not in between tall and short, so the blending idea did not seem to apply.

- He then allowed the plants of the F_1 generation to interbreed. He collected and grew their seeds.
- He looked at the offspring (F_2 generation) and saw that some were tall-stemmed and some short-stemmed.
- Then he did something very unusual for that time: he counted how many of each type there were.
- He found 787 tall plants and 277 short plants. He realised that there were three times as many tall plants as short plants.

He described the tall-stem characteristic as **dominant**. He described the short-stem characteristic as **recessive**.

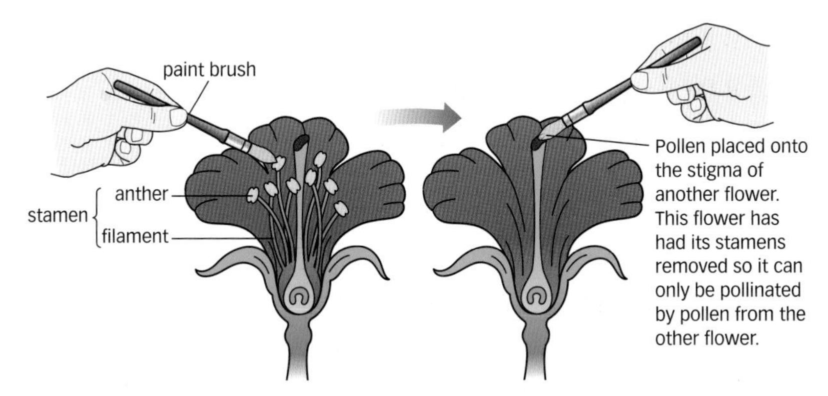

Cross-pollinating flowers – taking pollen from one flower and applying it to another flower to fertilise the egg cell and produce seeds. This technique allows researchers to carry out breeding experiments.

What did Mendel deduce?

Mendel said that inheritance was not by a blending mechanism but was due to **inheritance factors**. Factors for different characteristics are passed from parent to offspring separately.

He said:

- The cells in the pea plant each have two inheritance factors for every characteristic, such as stem length.
- Each pollen grain and each egg (female gamete) carries only one inheritance factor for a characteristic.
- Any pollen grain could fertilise any egg.
- If the offspring inherited one inheritance factor for tall stem and one inheritance factor for short stem, then it would show the dominant characteristic and be tall.
- If the offspring inherited two short-stem inheritance factors, then it would show the recessive characteristic; it would have a short stem.

Key words

F_1 generation, dominant, recessive, inheritance factor

Exam tip

✔ In his stem length experiment, Mendel was studying the inheritance of one characteristic. This is called monohybrid inheritance.

Questions

1 How did Mendel know if his peas were true-breeding?

2 Before Mendel did his inheritance experiments with peas, how did people at the time think that characteristics were inherited from parent to offspring?

3 When he crossed true-breeding tall pea plants with true-breeding short pea plants, what were the offspring like?

4 When he allowed these offspring to interbreed, what were the F_2 offspring like?

5 What conclusions did Mendel draw from the results of these experiments?

Learning objectives

After studying this topic, you should be able to:

- know how other scientists rediscovered Mendel's work
- understand that there are different forms of each gene, called alleles
- be able to interpret genetic diagrams of monohybrid inheritance

Key words

monohybrid inheritance, DNA (deoxyribonucleic acid), gene

A What is monohybrid inheritance?

B If a pea plant has inheritance factors **Tt**, is it tall- or short-stemmed?

Exam tip AQA

- Always follow the conventions for drawing genetic diagrams.

male gametes / female gametes	T	t
T	TT tall	Tt tall
t	Tt tall	tt short

▲ Punnett square showing the monohybrid inheritance of crossing the F_1 generation

Genetic diagrams to explain Mendel's results

Genetic diagrams show how characteristics are inherited according to Mendel's laws. In each diagram we are looking at the inheritance of just one characteristic, such as the tall- or short-stemmed pea plants. The inheritance of one characteristic is described as **monohybrid inheritance**.

All of the first generation, the F_1 generation, were tall-stemmed. You can see that each plant had one dominant inherited factor and one recessive inherited factor. They all showed the dominant characteristic.

Mendel then interbred these **Tt** plants to form the F_2 generation.

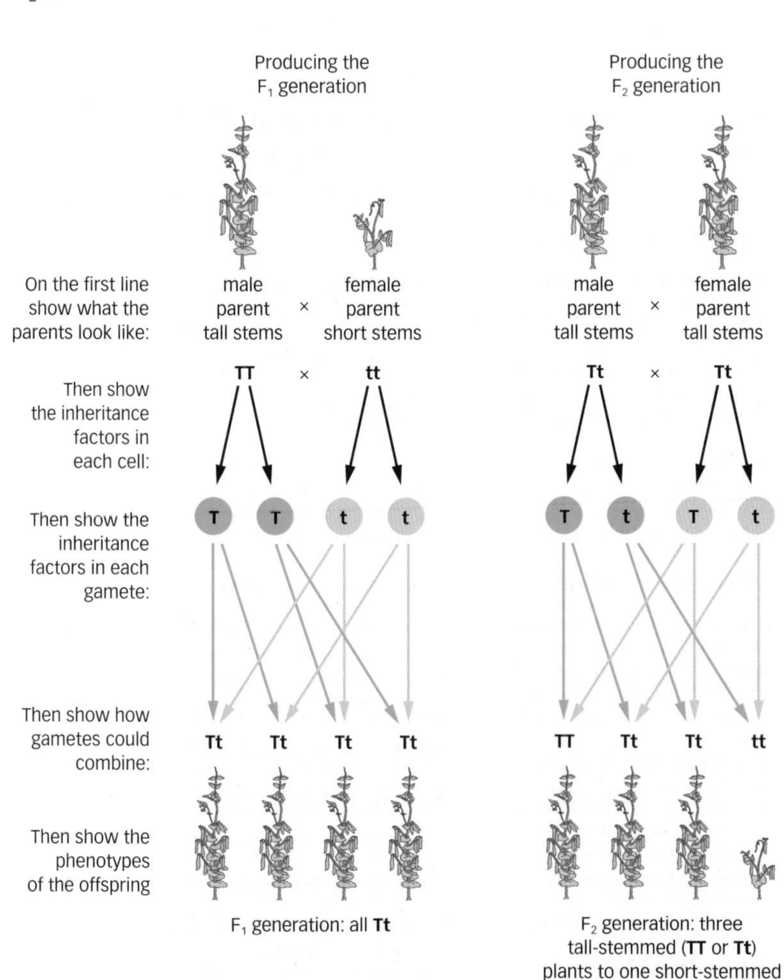

Instead of a genetic diagram, it may be clearer to use a Punnett square (see left) to show the inheritance.

You can see from this diagram that three out of the four possible combinations of factors produce tall-stemmed plants. One out of four possible combinations of factors produces short-stemmed plants. This gives a ratio of three tall-stemmed plants to one short-stemmed, which is what Mendel saw in his results.

Mendel did not get any prizes

By 1900 scientists had realised that the idea of blending inheritance was probably wrong. They were seeking a successful theory about separately inherited factors. Two biologists, independently of each other, duplicated the breeding experiments that Mendel had done about 40 years previously. When they read Mendel's account and conclusions they were able to understand their own findings.

Chromosomes, genes, and alleles

Scientists now know that the chemical that governs inheritance of characteristics is **DNA (deoxyribonucleic acid)**. DNA carries the genetic information:

- It is coiled into structures called chromosomes, which are in the nucleus of the cell.
- Each small section of DNA, called a **gene**, on a chromosome codes for a particular characteristic.
- Each pair of chromosomes contains the same genes, although each gene may have different forms called alleles.
- Each allele of a gene codes for the same characteristic, but for a slightly different version of it.

In the case of Mendel's tall- and short-stemmed pea plants, the characteristic was height. This was controlled by a single gene. The gene had two alleles:

- One, the dominant allele, had instructions for tall stems. This controls the development of the visible characteristic even if it is only present on one of the chromosomes.
- One, the recessive allele, had instructions for short stems. This controls the development of the visible characteristic only if the dominant allele is not present.

Questions

1 The diagram shows the results of crossing a pea plant having alleles **Tt** with a pea plant having alleles **tt**.

parents	tall stem	short stem
parents' genes	Tt	tt
gametes	T t	t t

gametes \ gametes	T	t
t	Tt	tt
t	Tt	tt

(a) What do the offspring with **Tt** alleles look like?

(b) What do the offspring with **tt** alleles look like?

E

(c) If the original parent plants produced 300 seeds, how many would you expect to have the **Tt** alleles, and how many would you expect to have **tt** alleles?

C

2 Use a genetic diagram to predict the outcome of crossing a pea plant that has **TT** alleles with a pea plant that has **tt** alleles.

A*

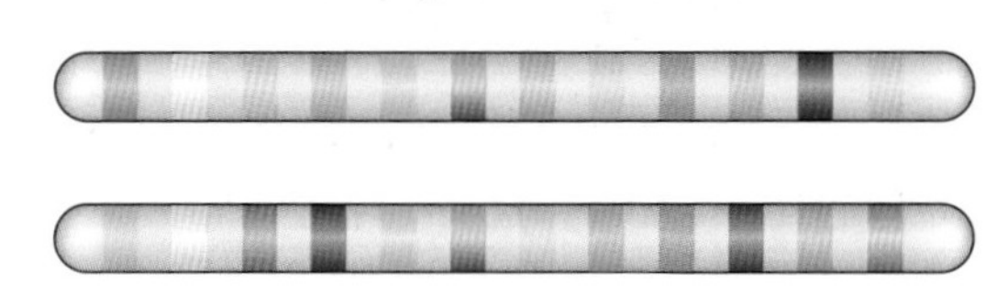

A pair of chromosomes and their genes. Many of the genes have different alleles on the two chromosomes.

26: The inheritance of sex

Learning objectives

After studying this topic, you should be able to:

- ✔ know that one of the 23 pairs of human chromosomes carries the genes for sex (gender)

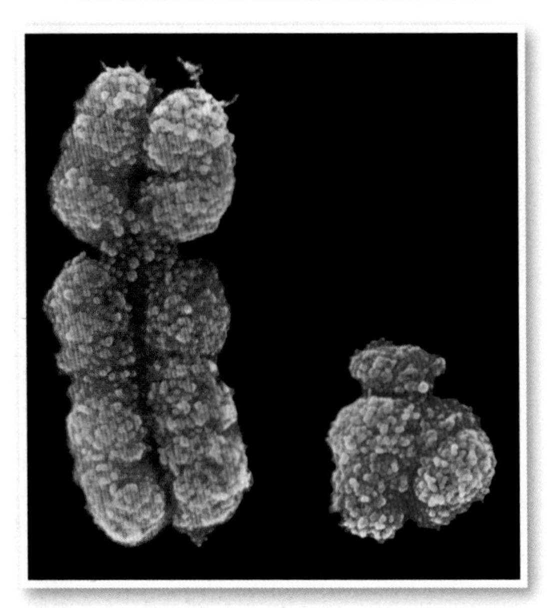

▲ XY chromosomes. The Y chromosome (blue) is much smaller than the X chromosome (pink). One small part of the X chromosome matches a small part of the Y chromosome, so that these two chromosomes can pair up for meiosis.

A Which of his sex chromosomes does a father pass to his son?

B Which of his sex chromosomes does a father pass to his daughter?

Did you know...?

Not all living organisms have the same mechanism as humans for determining sex. Most male mammals are XY, but male birds and butterflies are XX.

What makes us male or female?

You know that you have 23 pairs of chromosomes in the nucleus of all your body cells. One of these pairs of chromosomes determines sex. If, in this pair of **sex chromosomes**, you have two large X chromosomes, you are female. If you have one large X chromosome and a smaller Y chromosome, you are male.

Inheritance of sex

We can use a genetic diagram to show how sex (gender) can be inherited.

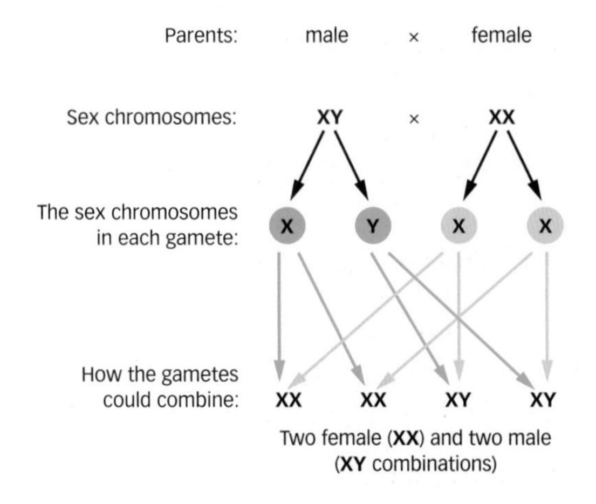

▲ There is an equal chance that each baby will be a girl or a boy

You can see that half the male gametes (sperm) have an X chromosome and half have a Y chromosome. However, all the female gametes (eggs) have an X chromosome.

Fertilisation is random, and either type of sperm could fertilise an egg. At each pregnancy there is a 50:50 chance of conceiving a girl or a boy. In a large population, there will be equal numbers of male and female offspring.

◀ Non-identical twins: boy and girl. Twins like these are rare. The mother produced two eggs at the same time. One was fertilised by a Y-bearing sperm and the other by an X-bearing sperm.

XX or XY?

The key gene, called SRY, which triggers a human embryo to develop into a male, is on the Y chromosome. This gene triggers the development of the testes. The absence of an SRY gene makes an embryo develop into a female.

Genetic diagrams

You may be asked to construct a genetic diagram for a monohybrid cross. You must follow the conventions:

- Show the characteristic of the parents.
- Show the alleles present in the parents' cells.
- Use upper case letters to represent a dominant allele.
- Use the lower case version of the same letter to show a recessive allele.
- Put gametes in circles.
- Show all the different possible combinations of alleles at fertilisation.
- Put an 'x' to show a cross (mating).

Here is a genetic diagram to explain a monohybrid cross between a father who has two alleles for free earlobes and a mother who has two alleles for attached earlobes.

Parents' characteristics: free earlobes × attached earlobes

Parents' alleles: (EE) × (ee)

Gametes: all (E) × all (e)

Offspring's alleles: all (Ee)

Offspring's characteristics: all have free earlobes but have one dominant and one recessive allele

Always make your upper case letters really big and the lower case letters small. And if you are told to use certain letters in your answer, then use them.

Key words

sex chromosomes

Questions

1. How many pairs of chromosomes do you have, in each cell nucleus, that do not play a part in determining your sex?

E

2. Explain how sex is determined in humans.
3. A couple have three children – all girls. What are the chances of them having a boy at the next pregnancy? Show how you worked out your answer.

C

4. A mother with blue eyes (recessive) and a father with brown eyes (dominant) have three children. Two of them have brown eyes and one has blue eyes. Construct a genetic diagram to explain how these parents can produce these offspring.
5. Some humans have XXY chromosomes in their cells. Do you think they will be male or female? Explain your answer.
6. Some genetic diseases, such as red–green colour blindness, are due to a faulty recessive allele on the X chromosome. Why do you think this disorder is more common in males than in females?

A*

Learning objectives

After studying this topic, you should be able to:

- ✔ know that chromosomes are made of DNA
- ✔ know that a gene is a small section of DNA
- ✔ understand that each gene codes for a particular protein

Key words

DNA bases

DNA

DNA stands for deoxyribonucleic acid. This chemical carries coded genetic information. Each molecule of DNA has a double helix structure. In the nucleus of all your cells you have 46 large molecules of DNA. Each of your chromosomes is one molecule of DNA.

You will remember from studying mitosis and meiosis that before a cell divides, all the molecules of DNA make copies of themselves. Then they condense into tightly coiled chromosomes so that they can move from the centre of the cell to the ends of the cell as the cell divides. When the cell has finished dividing into new cells, these chromosomes unravel. In that state they can govern the making of particular proteins.

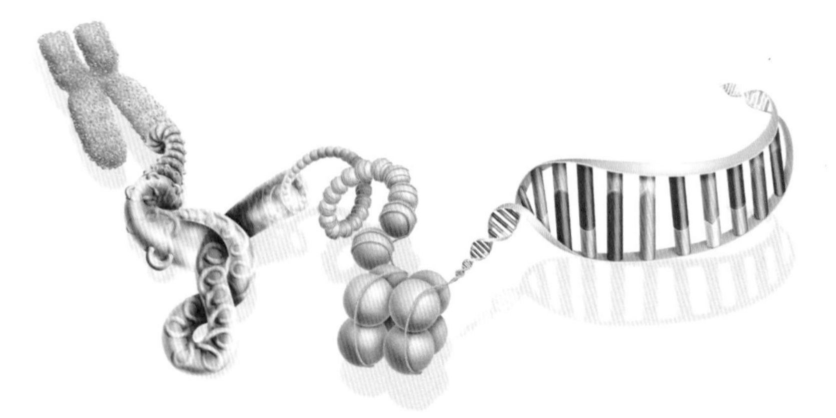

▲ How DNA condenses and coils into a chromosome. This DNA molecule has made a copy of itself so that when it forms a chromosome, the chromosome consists of the original and an identical copy. They are joined at a region near the middle. This gives the classic shape of visible chromosomes. It is only when chromosomes are coiled up like this that they take up stains and you can see them under a light microscope.

Did you know...?

Scientists are now finding that some genes code for more than one protein.

A What does DNA stand for?

B How many molecules of DNA are there in a chromosome?

Genes

Each chromosome is one large molecule of DNA. Within each DNA molecule, shorter sections of DNA form specific genes.

Each gene has coded genetic information. It codes for a particular combination of amino acids that makes a specific protein.

How genes code for proteins

Each section of DNA has a sequence of **DNA bases** in it. There are four bases, A, T, G, and C. You do not need to know their names.

1. These bases form a code. They are 'arranged' in groups of threes, or triplets.
2. Each triplet specifies a particular amino acid. So ATC will specify a different amino acid from ACT.
3. The sequence of base triplets on a section of DNA specifies the sequence of amino acids in a protein.
4. In turn, the sequence of amino acids in the protein governs how the protein will fold up into a particular shape.
5. Each different type of protein has a specific shape and can fit another molecule. This is how enzymes each fit just their own specific substrate molecule.

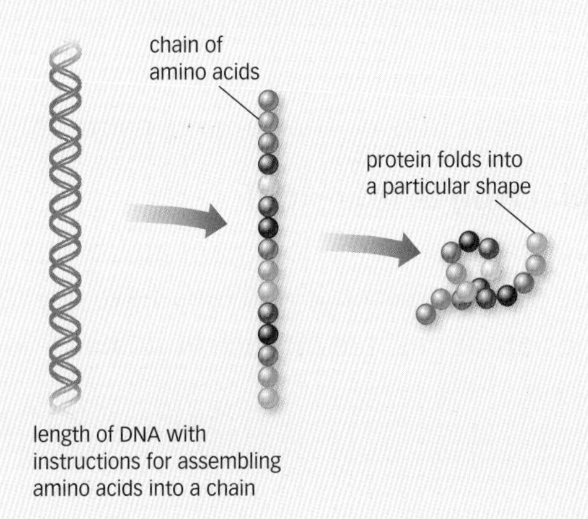

▲ Simplified diagram to show how the coded information in a gene determines the shape and the function of a protein

All proteins have a specific shape, and this allows them to carry out their function. The characteristics that you inherit involve proteins. These characteristics may rely on the help of enzymes and hormones to develop, or they may directly involve a protein.

You have channels in your cell membranes that allow certain chemicals, such as chloride ions, into and out of the cells. If you have cells with faulty chloride ion channels in the membranes, you will suffer from cystic fibrosis (see spread B2.28).

Exam tip AQA

✔ Remember that genes are on chromosomes. Each gene is a small section of DNA. Each chromosome is one huge molecule of DNA.

Questions

1. Why does DNA condense and coil into chromosomes before a cell divides? (E)
2. Why does your DNA have to be unravelled when the cell is not dividing? (C)
3. Humans have about 20 000 genes. The two members of each pair of chromosomes have the same genes on them. So, on average, about how many genes do you think there are on each of your chromosomes? (A*)
4. Explain how different genes code for specific proteins.

Learning objectives

After studying this topic, you should be able to:

- know that some genetic disorders are inherited
- consider two genetic disorders – polydactyly and cystic fibrosis

Key words

digit, mucus, tract

▲ An extra finger

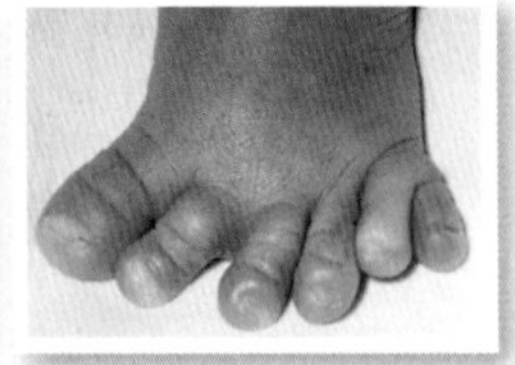

▲ An extra toe

Exam tip

- Study family trees carefully. There are lots of clues to help you answer questions, but you have to think.

Gene and chromosome defects

There are about 5000 genetic disorders caused by a defect in a single gene. There are also genetic disorders caused by whole chromosomes being abnormal. You need to consider just two examples of a genetic disorder.

Polydactyly

This simply means having more **digits** (fingers or toes) than usual (*poly* means many).

About 1 baby in every 500 births has one extra digit. This type of polydactyly is also called hexadactyly (*hexa* means six). The extra digit is usually on the little-finger side of the hand, or little-toe side of the foot. However, it may be on the thumb side as in the picture on the left. The extra digit causes no harm, but it is usually removed surgically at birth. Sometimes seven or nine digits are formed on a hand or foot.

Polydactyly is more common amongst black African and African American children than among white children.

How polydactyly is inherited

The condition is caused by a dominant allele. This means you have the condition even if you only have one faulty allele. If one parent has it, then each child has a 50% chance of inheriting it.

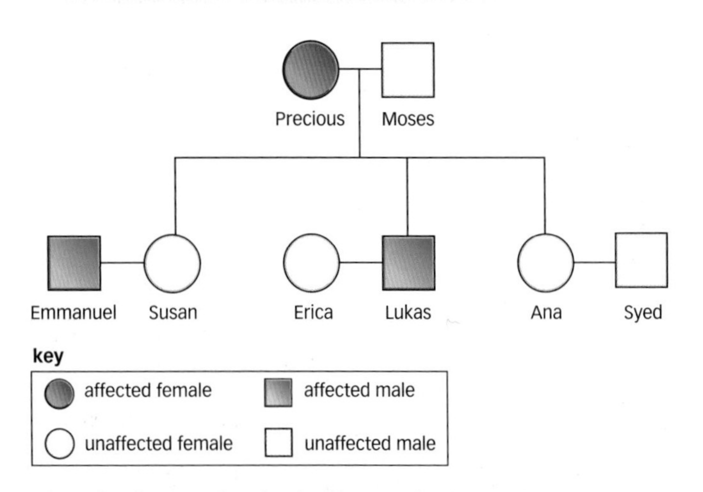

▲ A family tree showing incidence of polydactyly in a family

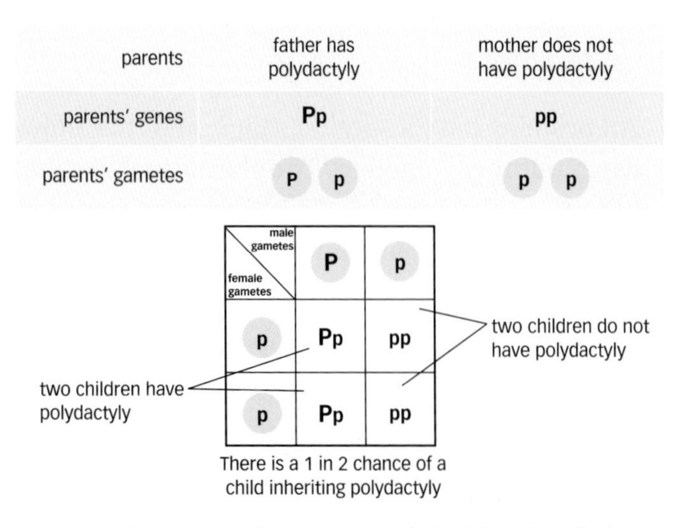

▲ A genetic diagram showing how polydactyly is inherited

Cystic fibrosis

Cystic fibrosis (CF) is a disorder of cell membranes. The membranes of cells lining your airways and pancreas have faulty chloride ion channels. As a result you will have thick sticky **mucus** in these tubes or **tracts**. You get scarring (fibrosis) and cysts (cystic) in the pancreas. Other symptoms are

- frequent lung infections and difficulty breathing
- failure to thrive, as children cannot properly digest food because enzymes from the pancreas do not pass into the gut
- it shortens life, although antibiotics, physiotherapy, and lung transplants have extended life expectation. Many people with CF are now in their thirties and expect to live longer
- men and women may be infertile. Men can make sperm but do not have a sperm duct. Women have thick mucus that blocks their cervix.

How cystic fibrosis is inherited

CF is caused by a faulty, recessive allele of the gene for the cell-membrane-channel protein. A person with the disorder has to inherit two copies of the recessive allele – one from each parent. The parents may not have the disease. They may be symptomless carriers. They will each have one recessive allele and one normal, dominant allele. So they have enough functioning cell-membrane channels.

At each pregnancy for two carrier parents, there is a 25% chance that the child will have CF.

Questions

1 Look at the family tree for polydactyly. What are the chances that:

(a) Syed and Ana will have a child with an extra digit?

(b) Erica and Lukas will have a child with an extra digit? In each case explain how you arrived at your answer.

E

2 Look at the CF family tree.

(a) What are the chances of Julia and Michael having a child with CF?

C

(b) Joan and David are both carriers of CF. How can you tell this from the family tree?

(c) If Joan and David have another child, what are the chances that it will have CF? In each case, show how you work out your answer.

A*

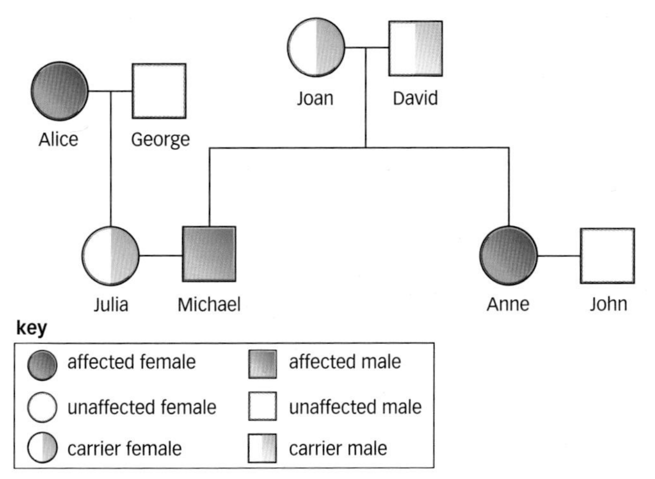

▲ A family tree showing incidence of CF within a family

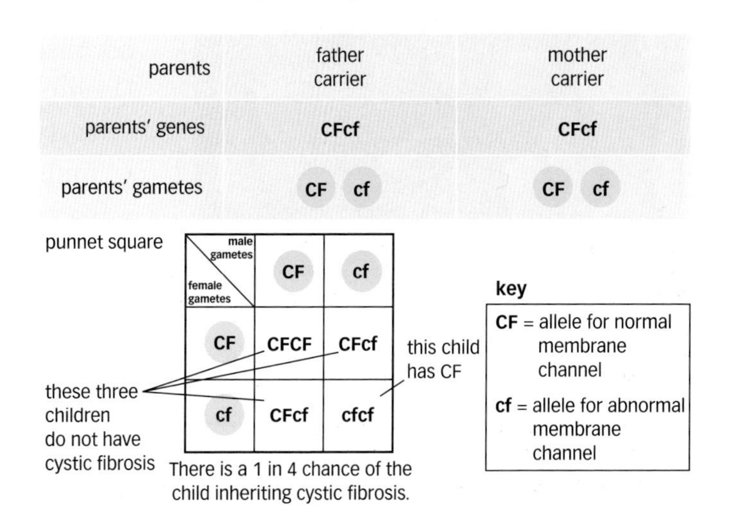

▲ A genetic diagram showing how CF is inherited

Learning objectives

After studying this topic, you should be able to:

- ✔ know that embryos can be screened for alleles that cause genetic disorders
- ✔ understand that stem cells from embryos can be used for medical research
- ✔ explain how DNA fingerprinting can be used to identify individuals

Key words

stem cell, DNA fingerprinting

A Who might use embryo screening during pregnancy?

B List two advantages of using embryo screening.

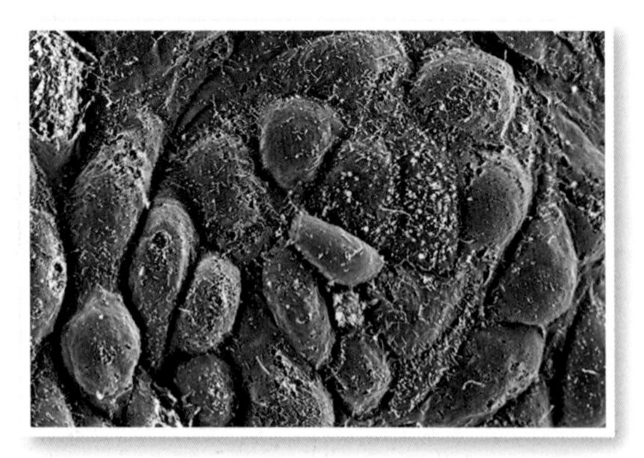

▲ Coloured scanning electron micrograph of human embryonic stem cells (× 1500). These were grown at the Centre for Life, Newcastle upon Tyne. Scientists can make them develop into any kind of human cell (and we have 220 different cell types) by giving them an appropriate chemical signal.

Embryo screening

If both parents know they are carriers of a serious genetic disorder, such as cystic fibrosis (CF), they may decide to have children in the following way:

- undergo in vitro fertilisation (IVF) to produce embryos
- have the embryos tested to see if they have CF (if they have two copies of the abnormal allele)
- only implant embryos that are free of CF.

There are several advantages of this:

- Their children will not have CF. Although CF can be treated, people with it have many health problems and may live shorter lives.
- Their children will not have the CF allele so will not be able to pass on the disorder to their own children.
- Although IVF is expensive, in the long term, money will be saved by the NHS as the children born will not have to be treated for CF. Also, they will not need a heart–lung transplant later in life.

However, as a result:

- Some embryos are formed that do not get the chance to develop into people.
- Some people who have a genetic disorder think this procedure is discriminating against them.

Stem cells

Stem cells are useful in medical research and treatments because they can develop into many different kinds of cell. Doctors can get stem cells from

- early embryos
- umbilical cord blood cells
- some types of adult tissues, such as bone marrow
- and more recently from amniotic fluid.

Using stem cells

Early embryo cells are the most useful because they still are able to differentiate into any type of cell. In mature animals, the ability of cells to differentiate is restricted to repair and replacement.

Medical research is developing ways of using stem cells to

- treat people with Parkinson's disease
- repair spinal-cord injuries
- grow tissues or organs for transplanting
- treat people with type 1 diabetes.

Scientists are allowed to keep spare embryos created by IVF alive for up to 14 days. Single cells taken from them are used as stem cells, to develop into different types of cells, such as nerve cells, blood cells, and muscle cells.

This raises ethical issues, as the spare embryos are not allowed to develop into people. However, even without stem-cell research these embryos would still be discarded.

DNA fingerprinting

In 1984 Professor Sir Alec Jeffreys, at the University of Leicester, discovered that certain parts of your DNA (not the genes, but other bits called VNTRs) vary a lot between people. However, they are similar within families. He used this knowledge to make **DNA fingerprints** of people.

Uses of DNA fingerprinting

The technique was used in a famous court case. A mother wanted to bring her child to the UK, but needed to prove that the child was hers. Alec Jeffreys did DNA fingerprints and proved that the woman was the mother of the girl.

Leicestershire Police heard about the technique and used it in 1986. It showed that a person who had confessed to a murder had not done it. Using DNA fingerprinting, the force caught the real murderer, Colin Pitchfork. He was the first criminal convicted in the UK using DNA evidence.

DNA fingerprinting is used to establish

- paternity and maternity
- criminal convictions for rape and murder
- whether a person is innocent of a crime after they were convicted
- family relationships.

The remains of some bones found in Russia were shown to be those of the Russian royal family, executed in 1917.

Did you know...?

Scientists can now take one cell from an eight-cell embryo and use that cell to grow more stem cells. The remaining seven-cell embryo can develop normally into a baby.

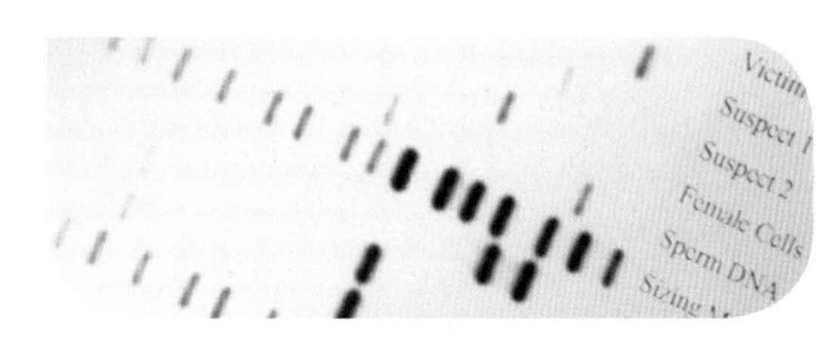

▲ DNA fingerprint results from a rape investigation. Suspect 1's DNA matches that of the sperm found in the victim.

Exam tip 

✓ When you answer questions like 2–5 below, you must be objective. Do not rant about your own personal opinions, but put forward reasons for and reasons against.

Questions

1 What are stem cells? (E)

2 Think of reasons for and reasons against the following, writing down your ideas:
 (a) Having a national database with a sample of everyone's DNA on it.
 (b) Using embryo stem cells for medical research.
 (c) Embryo screening. (C)

3 What sort of division do you think is used when a stem cell divides? (A*)

Learning objectives

After studying this topic, you should be able to:

- know that organisms have changed over time
- appreciate that there is evidence for evolution, such as fossils
- know how fossils were formed
- evaluate fossil evidence

Key words

evolution, fossil

Evidence for evolution

The theory of **evolution** says that all the organisms around today have developed from previous life forms. According to the theory, organisms change gradually over time. Any proposed theory requires evidence to support it before it can be accepted.

Any scientific observation that supports the idea of evolution is part of the body of evidence for evolution. There are several types of evidence, but perhaps the most significant is the fossil record.

Fossils

Fossils are the preserved remains of living things from many years ago. There are many ways that fossils have been preserved.

Fossil types	
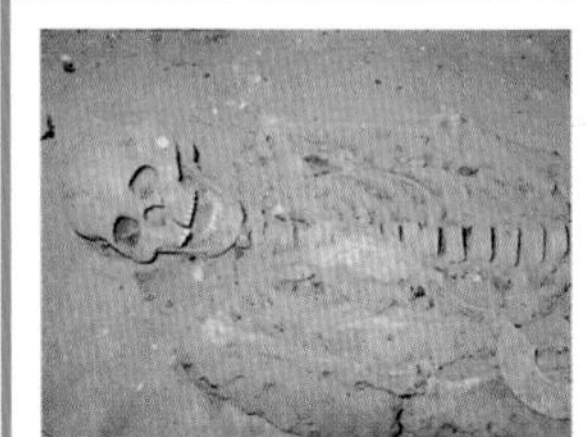	Remains of hard parts of animals such as bones and shells that do not decay easily.
	Parts (most often hard parts) of the organism are replaced by other materials such as minerals as they decay.
	Parts of organisms that have not decayed because one or more of the conditions for decay are missing. For example, if an animal is sealed in amber, then it is preserved because there is no oxygen for the microbes that cause decay.
	Evidence of an organism, such as footprints, burrows, or traces of plant parts are preserved.

Exam tip

- Remember that fossils give good evidence for evolution, but they do not give a complete picture. You may be asked to evaluate how good the evidence is.

A What is a fossil?

B What are the four main types of fossil?

How are fossils formed?

Perhaps the best known of the fossil types are those found in rock, formed as the hard parts of the organism are replaced by minerals. These fossils are formed like this:

- The plant or animal dies, and falls into soft mud or silt, often at the bottom of a lake or the sea.
- The body becomes covered in silt or mud.
- This gradually turns to rock, encasing the dead body.
- Over millions of years the hard parts of the body of the plant (leaves) or animal (bones and shells) are replaced by minerals. Soft parts of bodies do not always fossilise well, as they decay quickly.
- If earth movements make the land rise then the remains may become exposed at the surface.

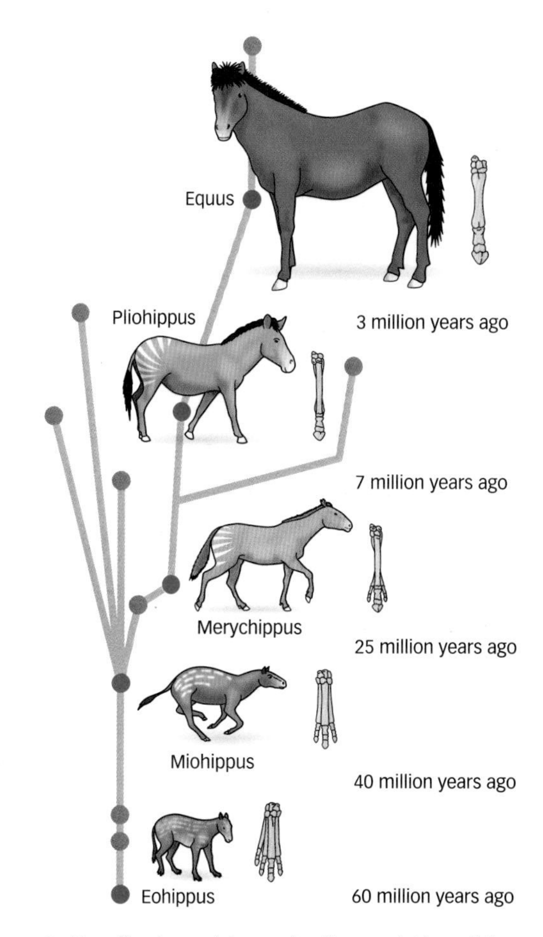

▲ Fossils give evidence for the evolution of the horse

Studying fossils

Biologists and palaeontologists study fossils, date them, and put them into date order. This can reveal the gradual change of one type of plant or animal into another over time. This can give evidence for the steps in evolution.

One of the best studied fossil records is that of the horse. From the record it can be seen that:

- The horse has grown, originally being the size of a dog.
- It now stands on long legs with only one digit.
- Its teeth are now adapted for eating grass.

C Describe how the fossils of dinosaur bones formed in rocks.

D How do scientists work out how organisms have changed over time?

How useful are fossil records?

It is rare for fossils to form, because the conditions must be right. Many fossils that have formed have not been found. Some fossils have been damaged, so they are less reliable. The evidence of past organisms in fossils is not complete.

Soft-bodied organisms and single-celled organisms, such as the earliest organisms, do not fossilise well. Any traces of such small organisms are often destroyed by geological processes such as erosion. This means that scientists are not absolutely certain about how early life forms evolved.

Questions

1 Explain how fossils help us to describe how an animal has evolved. E

2 (a) How do we know that dinosaurs existed? C

(b) What evidence would scientists use to show that the dinosaurs were closely related to modern day reptiles?

3 Why do you think scientists can explain how the horse evolved, but have difficulty in reliably explaining how birds evolved? A*

Learning objectives

After studying this topic, you should be able to:

- understand extinction
- offer reasons for the causes of extinction
- understand what is meant by an endangered species

Key words

endangered, **extinction**

A What does the word 'extinction' mean?

B What is the difference between an endangered species and an extinct species?

Where are they now?

Evolution leads to new species being formed. These species evolve because they are better adapted to survive in their environment. Some existing species may not be able to compete or survive as well as the newer ones. The result is that they reduce in number, eventually becoming **endangered** as their numbers become low. If conditions do not improve for them, then all the members of the species will die out. This is called **extinction**. In this way, new species replace old ones in the cycle of life. One well known example of extinction is the dinosaurs.

The causes of extinction

There are many possible causes of extinction. Several factors may combine to bring about an extinction.

Changes to the environment over geological time

If the environment changes, for example the temperature increases, some organisms will not be as well adapted to the new conditions. They may be outcompeted by better-adapted rivals. The woolly mammoth became extinct about 10 000 years ago, partly because the climate became warmer.

▲ The woolly mammoth was widespread in northern regions of America and Europe, but became extinct about 10 000 years ago

▲ Drawing of the extinct dinosaur *Tyrannosaurus rex*

Major catastrophic events

Major global events can affect the environment so much that some species cannot survive. Examples include massive volcanic eruptions and collisions with asteroids – one theory is that an asteroid impact resulted in reduced sunlight levels, which meant that plants could not photosynthesise so effectively. Many scientists believe that this led to the extinction of the dinosaurs.

New predators

The arrival of a new efficient predator might cause the extinction of some prey species. Humans are a good example – they can hunt as a team, and are very effective. Humans have been responsible for the extinction of many animals, such as the dodo.

The dodo lived on the island of Mauritius. The birds became extinct in the seventeenth century, not long after humans first landed there. Humans hunted them, and rats released by humans ate their eggs.

Dutch elm disease has wiped out more than 25 000 trees in the UK since 1967

New diseases

Some diseases are so virulent that they can wipe out an entire species before it has time to develop immunity. The elm tree population is almost extinct due to a fungal disease called Dutch elm disease.

New competitors

The arrival of a new and more effective competitor can cause problems for a resident species. An example is the introduction of the grey squirrel into the UK. This species outcompetes its relative the red squirrel, which is now in serious decline and is restricted to small areas of Scotland, Cumbria, Anglesey, and the Isle of Wight.

The red squirrel has been largely replaced in the UK by the grey squirrel, which is more efficient at gaining resources such as food

Questions

1. What caused the dodo to become extinct? (E)
2. Global warming is causing the icecaps to melt. What will be the effect on polar bear populations, and why?
3. Suggest why an asteroid impact might have led to the extinction of many plants at the time of the dinosaurs. (C)
4. Explain how the introduction of the grey squirrel might lead to the eventual extinction of the red squirrel. (A*)

Learning objectives

After studying this topic, you should be able to:

- know that during evolution new species form
- understand the mechanism by which a new species develops

Key words

speciation, **isolation**, **mutation**

Alfred Russel Wallace

Did you know...?

Wallace was in a position to publish his ideas at about the same time as Darwin. If he had, a £10 note may now have a picture of Wallace instead of Darwin!

More and more new species

During evolution new species are formed. But how does this happen? Looking at the fossil record, it is clear to see that often one original ancestor species forms two new species. So the number of species increases.

Darwin knew that this must happen, but when he wrote *The Origin of Species* he could not explain how it happened. Modern biologists have explained the process.

Alfred Russel Wallace

Alfred Wallace was a Welsh biologist who worked at the same time as Darwin. He was also interested in explaining the evolution and formation of species. Although he did a lot of work collecting evidence, he was not as famous as Darwin. He found more evidence which contributed to Darwin's theory of evolution.

Alfred Wallace worked in South America and Malaysia. He was particularly interested in the ideas of geographical barriers leading to the formation of new species. He was perhaps the first biologist to suggest how isolating two populations might lead to **speciation**.

Forming new species

Sometimes two different populations of the same species change in different ways. Each group gradually changes over time, maybe to adapt to different environmental conditions. This then forms two new species. The process by which this happens is called speciation. It can be explained in four key steps.

A What is speciation?

Step 1: isolation

The population becomes separated or **isolated** by some kind of barrier. The barrier could be a mountain range, a river, or the sea between different islands. Individuals in the two isolated populations of the species can no longer meet and interbreed, as they cannot cross the barrier.

Step 2: genetic variation

In each of the populations there will be variation between individuals. This variation is caused by a wide range of alleles. These are alternative versions of genes, leading to different characteristics. In addition, different **mutations** will occur in the separated populations. This increases the variation.

Step 3: natural selection

Over time, each separated group of the population evolves differently on each side of the barrier. Some of the variants will be better adapted to survive the conditions. Because conditions will be slightly different in the two areas, different variants will survive in each area. Each successful variant will pass on its own alleles to the next generation. The longer they are separated, the more different the two groups become.

Step 4: speciation

The two sub-populations have changed so much that they can no longer interbreed. They have formed two new closely related species.

▲ Modern humans and chimpanzees both evolved from a common ancestor over 5 million years ago in Africa. The ancestral population split into two groups, one standing upright and evolving into the modern human, the other becoming the modern chimp.

B What are the two main causes of variation in a species that, in the right conditions, could lead to speciation?

Questions

1 List three ways in which populations can become isolated. (E)

2 Explain why two new species might not develop without a physical barrier.

3 Wallace discovered two species of monkey on either side of the river Amazon in South America. How could Darwin and Wallace have explained how they evolved from a common ancestor? (C)

4 Over 65 million years ago, ancestral primates reached the island of Madagascar from mainland Africa.
(a) Why did primates evolve into new primate species (even humans) in Africa, but into lemurs on Madagascar?
(b) What might have led to the evolution of more than 90 species of lemur on Madagascar? (A*)

Exam tip

✔ Learn the steps in the formation of a new species in sequence.

Course catch-up

Revision checklist

- Proteins make up much of the structure of living organisms and perform vital functions in cells.
- Enzymes are proteins that catalyse chemical reactions in cells. Each enzyme catalyses a specific reaction and works best at a particular temperature and pH.
- Enzymes are crucial for digestion.
- Biological washing powders have enzymes to break down stains.
- Enzymes are used in industry. They make baby food easier for babies to digest; can make sugar from cheap starch, for use in processed foods; some can turn glucose into a sweeter sugar called fructose.
- Respiration in cells provides the energy needed for an organism's life processes.
- Aerobic respiration requires oxygen.
- Anaerobic respiration does not require oxygen.
- Cells divide by mitosis for growth and asexual reproduction. Mitosis produces cells that are genetically identical to each other and to the parent cell. Living organisms made of many cells can be large, complex, and can have different types of cell for different functions.
- Certain cells divide by meiosis to make gametes for reproduction. Meiosis produces genetic variation.
- Gregor Mendel discovered how characteristics are inherited.
- Mendel's ideas explain many inheritance patterns. We use genetic diagrams to explain the results of genetic crosses.
- You are female if you inherit two X chromosomes, and male if you inherit one X and one Y chromosome.
- Genes are made of DNA and are found on chromosomes. Genes control characteristics because they contain coded information for making particular proteins.
- There are different versions of each gene. Some versions make abnormal proteins that lead to genetic disorders.
- Differences in people's DNA can be used to identify them. Embryos may be screened for genetic defects. Stem cells may be used to grow any type of cell to treat some illnesses.
- Fossils provide evidence for evolution and allow us to see how organisms have changed over time.
- Fossils show us what extinct organisms looked like.
- Mutations, isolation of populations, and competition all contribute to the formation of new species.

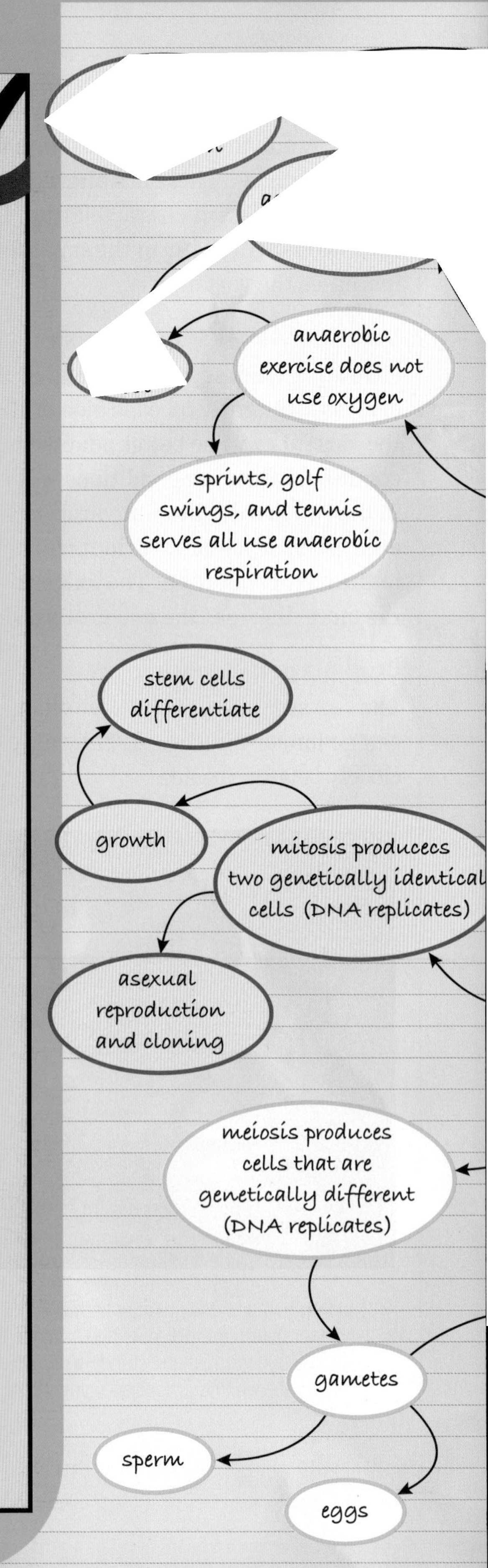

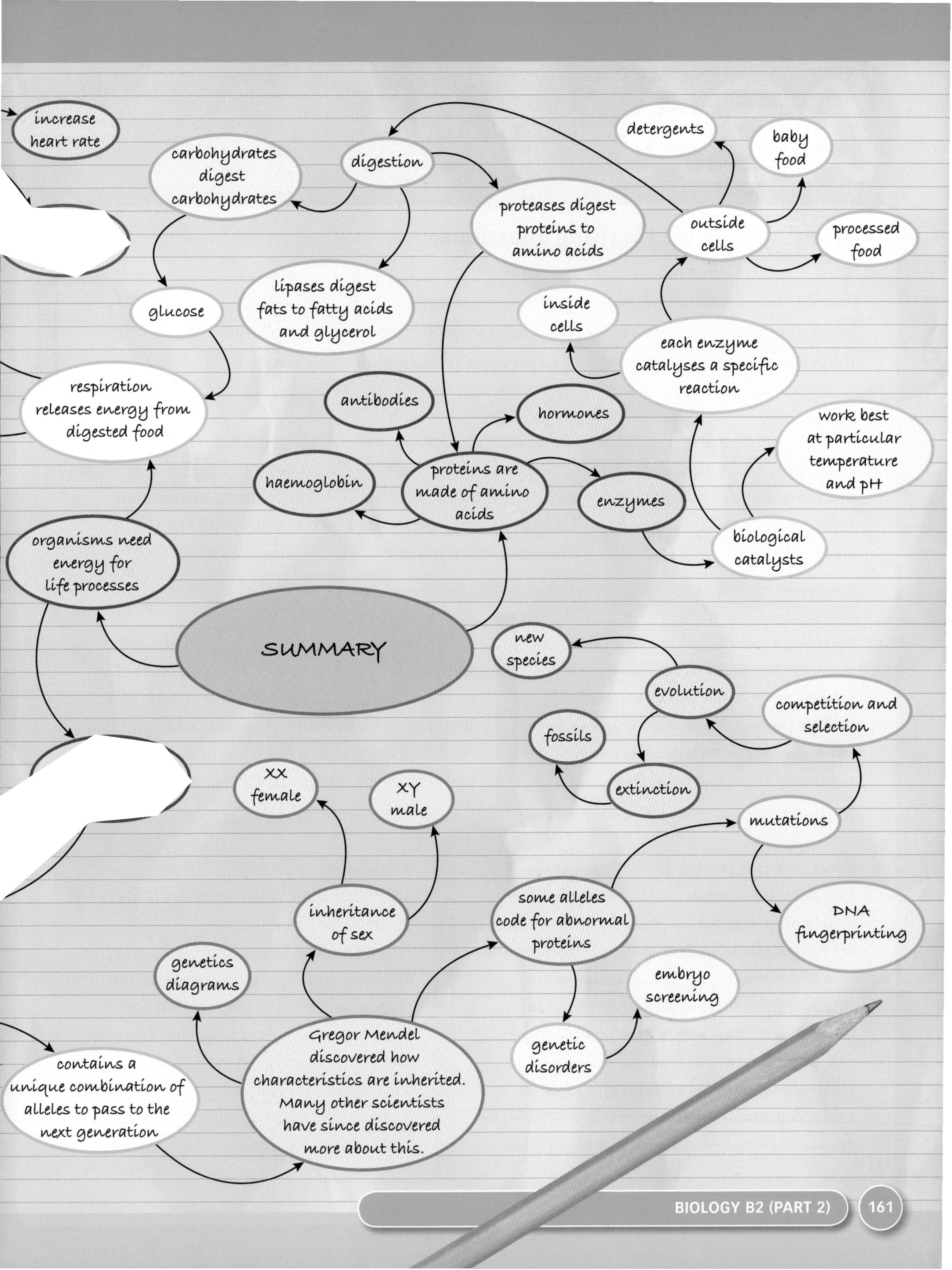
increase heart rate
carbohydrates digest carbohydrates
digestion
proteases digest proteins to amino acids
detergents
baby food
outside cells
processed food
glucose
lipases digest fats to fatty acids and glycerol
inside cells
each enzyme catalyses a specific reaction
respiration releases energy from digested food
antibodies
hormones
work best at particular temperature and pH
haemoglobin
proteins are made of amino acids
enzymes
biological catalysts
organisms need energy for life processes
SUMMARY
new species
evolution
competition and selection
fossils
extinction
XX female
XY male
mutations
inheritance of sex
some alleles code for abnormal proteins
DNA fingerprinting
genetics diagrams
embryo screening
Gregor Mendel discovered how characteristics are inherited. Many other scientists have since discovered more about this.
genetic disorders
contains a unique combination of alleles to pass to the next generation

Answering Extended Writing questions

QUESTION

Describe how enzymes can be used in the home and in industry.

Evaluate the advantages and disadvantages of using enzymes in the home and in industry.

The quality of written communication will be assessed in your answer to this question.

G–E

Enzymes help get cloths clean. they are in detergents. Enzymes are used in the food industry. Enzymes are made of protein.

Examiner: This answer is vague. There is a reference to detergents, but the term 'food industry' is too vague as an example of enzyme use. The candidate has included some irrelevant information. Although the last sentence is a correct statement, it will not score any marks as it is not relevant to the question. One spelling mistake.

D–C

Washing powders have enzymes to remove stains better. On some you cant use a hot wash as the enzyme would be denatured. Enzymes can change cheap starch into sugar for sweets. they are also used in slimming foods and baby food.

Examiner: Four uses of enzymes are briefly mentioned, but no details and no examples of enzymes are given. The candidate implies that better removal of stains is an advantage and that not being able to use a hot wash in some instances is a disadvantage. There are a few grammatical errors.

B–A*

Some washing powders for clothes contain enzymes like protease and lipase to get rid of blood and grease stains. These enzymes mean the clothes can be washed at a lower temperature and this saves energy. Some enzymes are obtained from bacteria and work well at high temperatures. They aren't denatured like most enzymes. But they are expensive to get. Protease enzymes are added to baby food so babies can easily digest the proteins. Isomerases are used to make fructose for slimming foods.

Examiner: Specific types of enzymes are described, along with examples of their use in the home or in industry.
One advantage, using lower temperatures and saving energy, is clearly described. The answer implies that the fact that enzymes from bacteria work at high temperatures and are not denatured is an advantage.
This answer is clear and uses technical terms correctly. The spelling, punctuation, and grammar are good.

Exam-style questions

1 Scientists compared the performance of a new detergent with an existing detergent.

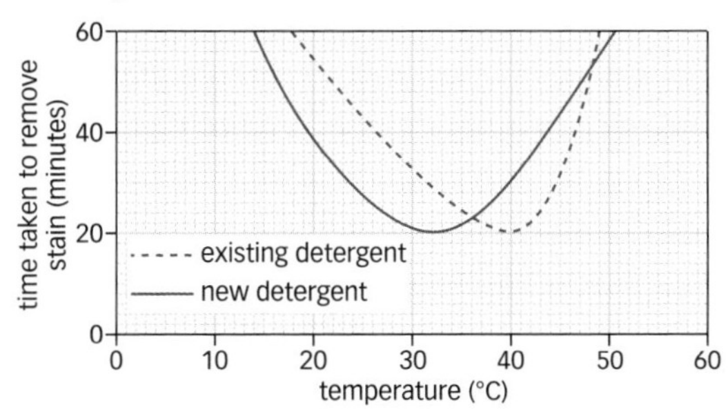

G–E

A03 **a** Describe the effect of increased temperature on the time taken by the existing detergent to remove the stain.

A03 **b** Which product works best at 30 °C?

A02 **c** Explain why neither detergent works well at 60 °C.

A03 **d** Is the new detergent likely to be more environmentally friendly than the existing detergent? Give a reason for your answer.

D–C

2 **a** Why do living organisms need energy?
A01

A01 **b** Write a word equation for aerobic respiration.

A01 **c** Where in cells does most aerobic respiration take place?

A01 **d** Write a word equation for anaerobic respiration.

A01 **e** Which type of respiration releases more energy from each molecule of glucose?

A01 **f** Which type of respiration causes muscle fatigue?

A02 **g** Explain why breathing rate and heart rate increase when you run.

B–A*

3 A student repeated one of Mendel's plant breeding experiments. She crossed pea plants grown from green seed with pea plants grown from yellow seed. All the first generation offspring produced yellow seeds. She crossed some of these with plants grown from green seed. The offspring from this second generation contained 19 plants with yellow seeds and 21 plants with green seeds.

A03 **a** Which seed colour is the dominant characteristic?

A03 **b** What is the ratio of yellow: green seeds in the second generation?

A02 **c** Seed colour in peas is controlled by a single gene that has two alleles. Draw a genetic diagram to show why this ratio of yellow seeds to green seeds was produced in the second generation. Use A to represent the dominant allele and a to represent the recessive allele.

A02 **d** How could the student have made the results more reliable?

Extended Writing

G–E

4 Describe how enzymes in your digestive
A01 system digest food.

D–C

5 Explain how fossils provide evidence
A02 for evolution.

B–A*

6 Describe how mitosis and meiosis are
A02 (a) similar (b) different.

A01 Recall the science
A02 Apply your knowledge
A03 Evaluate and analyse the evidence

Exchange and transport

AQA specification match		
B3.1	3.1	Osmosis
B3.1	3.2	Sports drinks
B3.1	3.3	Active transport
B3.1	3.4	Exchange surfaces
B3.1	3.5	Gaseous exchange in the lungs
B3.1	3.6	Exchange in plants
B3.2	3.7	Rates of transpiration
B3.2	3.8	The circulatory system
B3.2	3.9	The heart
B3.2	3.10	Arteries and veins
B3.2	3.11	Capillaries
B3.2	3.12	Blood
B3.2	3.13	Blood cells
B3.2	3.14	Transport in plants
Course catch-up		
AQA Upgrade		

Why study this unit?

The bodies of the human and the flowering plant are exquisitely balanced machines. It is vital that the cells of these bodies are supplied with a constant supply of life-supporting molecules like water, oxygen, carbon dioxide, and nutrients. In this unit you will discover how these molecules reach those cells.

You will study the ways in which molecules get into and out of cells. Keeping our bodies well hydrated and well fuelled is important. Larger plants and animals have needed to develop special exchange surfaces, like gills, lungs, and leaves, in order to pick up useful molecules from the outside. Once molecules have entered a plant or animal, they need to be moved around. The movement of water in the plant will be studied in this unit. In animals, the development of the circulatory system has allowed transport of molecules around the body. The circulatory system includes blood, blood vessels, and the living pump, the heart.

You should remember

1. The process of diffusion, allowing some molecules to enter cells.
2. The plant organs, and their structure and function.
3. How plant and animal organ systems compare.
4. The structure of cells.

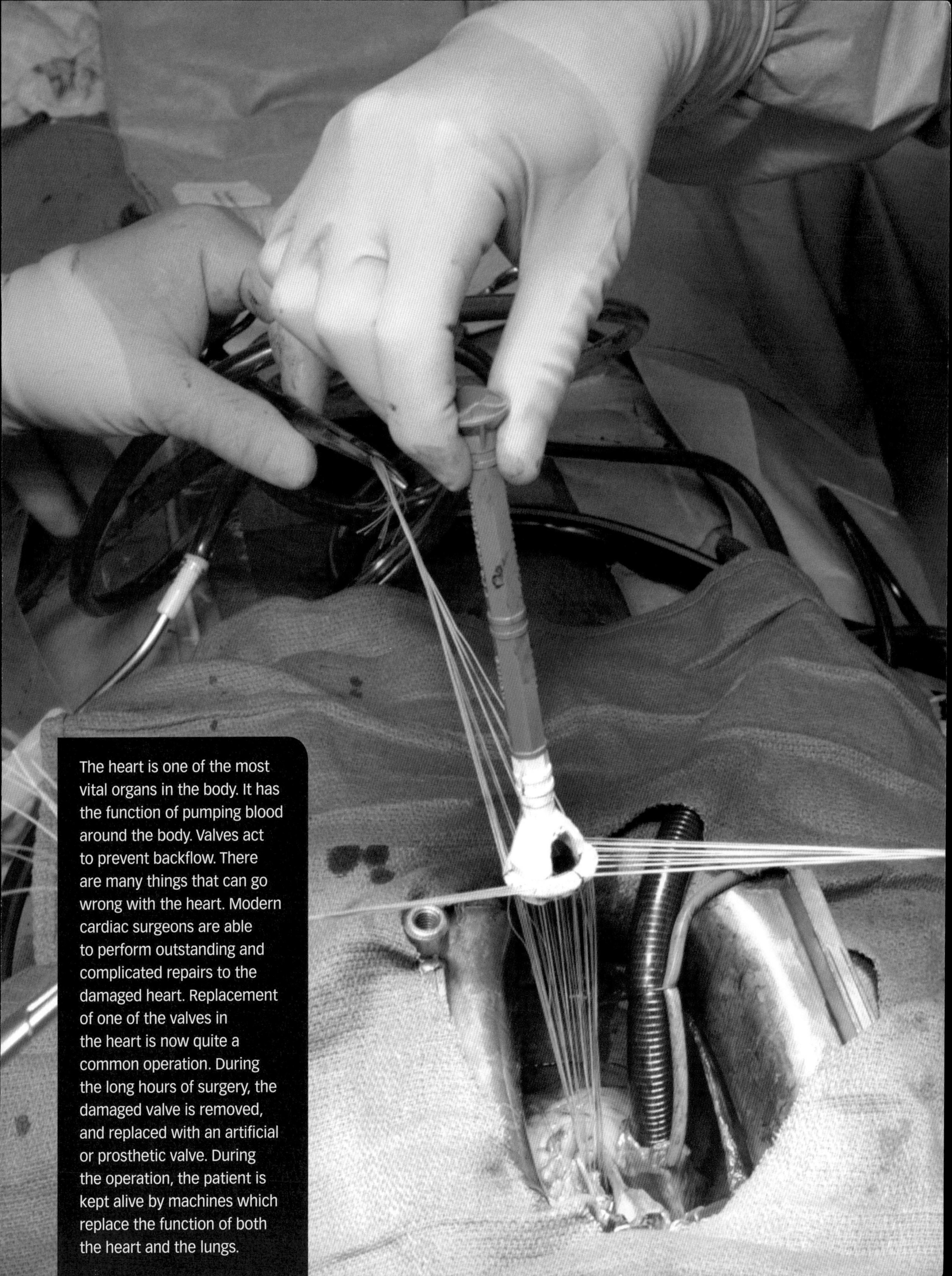

The heart is one of the most vital organs in the body. It has the function of pumping blood around the body. Valves act to prevent backflow. There are many things that can go wrong with the heart. Modern cardiac surgeons are able to perform outstanding and complicated repairs to the damaged heart. Replacement of one of the valves in the heart is now quite a common operation. During the long hours of surgery, the damaged valve is removed, and replaced with an artificial or prosthetic valve. During the operation, the patient is kept alive by machines which replace the function of both the heart and the lungs.

Learning objectives

After studying this topic, you should be able to:

- ✔ understand the process of osmosis
- ✔ know the effects of osmosis on plant and animal cells

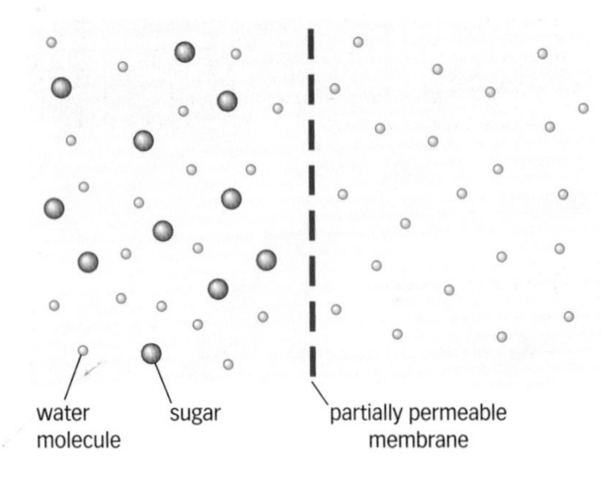

▲ A partially permeable membrane has pores that allow water molecules through, but not larger sugar molecules

A In which direction do water molecules move in osmosis?

B What is a partially permeable membrane?

▲ A wilted coleus plant. The cells have lost water by osmosis.

Moving molecules

There are three processes by which substances can move into or out of cells. Dissolved substances (molecules or ions) move by

- diffusion
- active transport.

Moving water

Osmosis is a special kind of diffusion. Water moves into and out of cells by osmosis.

The cell membrane has tiny holes called **pores**. Larger molecules such as sugars and proteins are too big to pass through the pores, but very small molecules, including water, can pass through. This type of membrane is called a **partially permeable membrane**, because only some molecules can pass through it.

The diagram on the left shows a dilute sugar solution separated from pure water by a partially permeable membrane. The sugar molecules are too big to pass through the pores. Water molecules pass from the pure water to the sugar solution by diffusion. This dilutes the sugar solution.

In osmosis, water will move from an area of high water concentration (pure water or a dilute solution) to an area of low water concentration (a more concentrated solution) across a partially permeable membrane.

Osmosis and cells

Water can move into or out of cells by osmosis. This movement of water is important for both plants and animals, because it keeps their cells in balance.

When plant cells take up water by osmosis, the cells become firm. The cell contents push against the inelastic cell wall. When plant cells are firm like this, they help support the plant. If they lose water, the cells become soft and the plant wilts.

Osmosis is also important in animals. There is no cell wall in animal cells, so they are very sensitive to water concentrations. If they take in or lose too much water, the cells are damaged and can die.

Osmosis and plant cells

water water

Surroundings are a less concentrated solution than cell contents (higher water concentration).

Surroundings have the same concentration as cell contents.

Surroundings are a more concentrated solution than cell contents (lower water concentration).

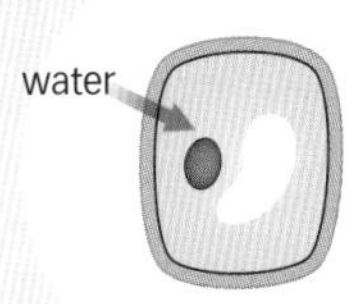

Cell placed into a dilute solution. It takes up water by osmosis. The pressure in the cell increases; this is called turgor pressure. The cell becomes firm or turgid.

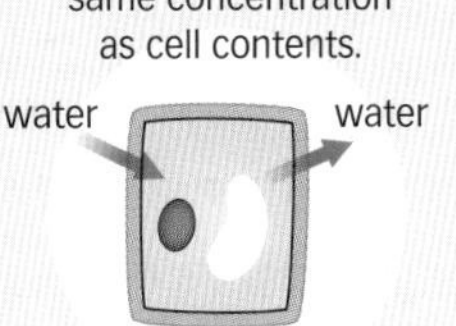

Cell placed into a solution with the same concentration as its contents. There is no net movement of water. The cell remains the same.

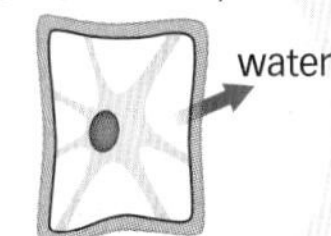

Cell placed into more concentrated solution. It loses water by osmosis. The turgor pressure falls and the cell becomes flaccid (soft). Eventually the cell contents collapse away from the cell wall. This is called a plasmolysed cell.

▲ Water movement by osmosis in plant cells

Osmosis and animal cells

isotonic solution

water water

Surroundings are a less concentrated solution than cell contents (higher water concentration).

Surroundings have the same concentration as cell contents.

Surroundings are a more concentrated solution than cell contents (lower water concentration).

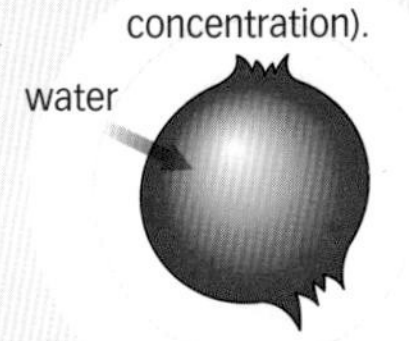

Cell placed into a solution that is more dilute than its contents. It takes up water, swells, and may burst. This is called lysis.

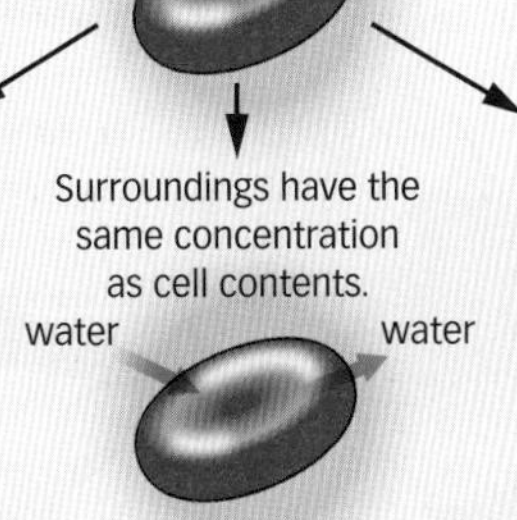

Cell placed into a solution with the same concentration as its contents. There is no net movement of water. The cell remains the same.

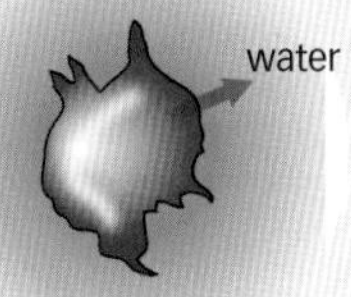

Cell placed into a more concentrated solution. It loses water by osmosis. The cell becomes crenated (it crinkles).

▲ Water movement by osmosis in animal cells

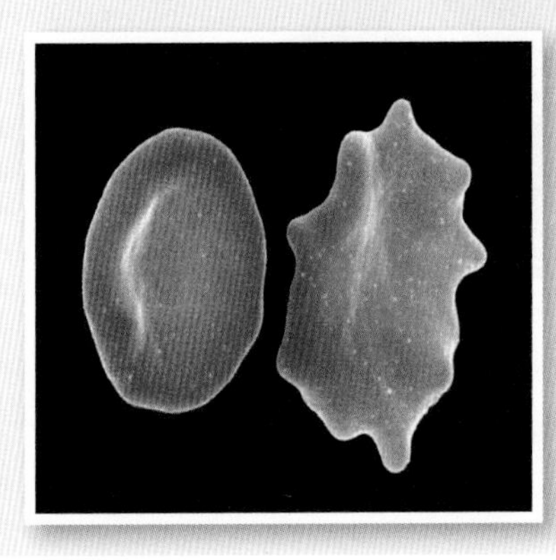

◀ A normal red blood cell and a crenated red blood cell

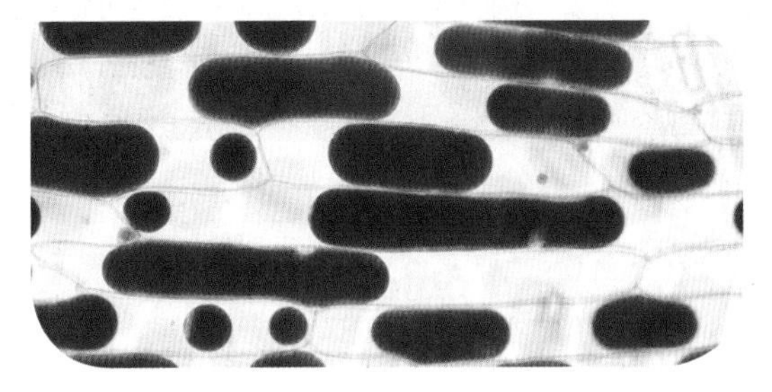

▲ Soft plant cells

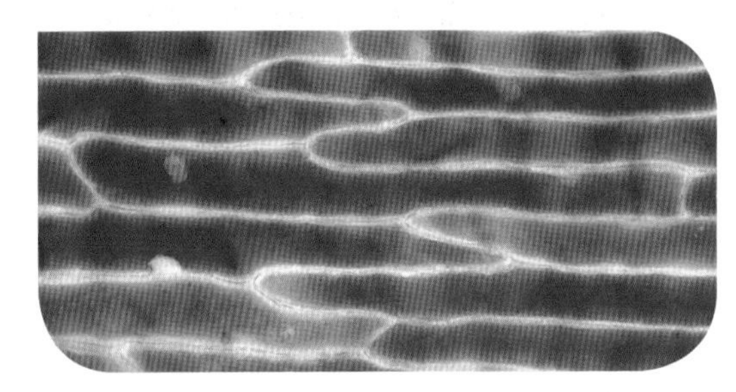

▲ Firm plant cells

Key words

osmosis, pore, partially permeable membrane

Questions

1 Name three ways in which substances can move into or out of cells. (E)

2 A piece of potato is weighed and then placed into a concentrated sugar solution. After 24 hours it is removed, dried, and weighed again:
   (a) Describe what would happen to the mass of the potato piece after 24 hours.
   (b) Explain why this has happened. (C)

3 A casualty from a road accident has lost blood. Why are they given a transfusion of blood, not water? (A*)

2: Sports drinks

Learning objectives

After studying this topic, you should be able to:

- know that water keeps the body hydrated
- understand that exercise increases water and ion loss from the body
- know the composition of soft drinks and sports drinks

Key words

dehydrated, **isotonic**

▲ You sweat more during strenuous exercise

A What is the recommended water intake for an adult?

B Name two substances that are lost from the body during sweating.

C Explain why water is important in keeping muscles healthy.

A body of water

The human body contains about 60–70% water. The water gets into your cells by osmosis. If the level of water in your body falls, your cells become **dehydrated** and they do not function properly. It is important to replace lost water by drinking fluids:

- Inactive people need two to three litres of fluid per day to remain hydrated.
- You need more fluid during exercise.
- You can sweat up to two litres of water per hour on a hot day.
- When you sweat, you lose not only water but also ions from your body. The body's balance of water and ions is now disturbed.

The importance of hydration

The effect of exercise

When you exercise, you use energy. The energy is released from sugars during respiration. Respiration also generates heat. To help cool your body down, sweating occurs. Most exercise sessions result in some sweating. You must replace the fluids lost to stop you dehydrating. Good tips are:

- Always start an exercise session fully hydrated.
- Drink 400–600 ml of water during the two hours before exercise.

Sports drinks

Sports drinks contain

- water to hydrate the body
- carbohydrates such as glucose for respiration, to provide energy
- mineral ions to keep the muscles healthy
- caffeine to make you feel awake.

There are three main types of sports drink:

Type of drink	What it contains	How it helps the body
Hypotonic drinks	In these drinks, the concentrations of dissolved substances are lower than in the body. There is usually less than 4 g of carbohydrate per 100 cm^3 of the drink.	Being more dilute than the blood, a lot of water will move from the gut quickly into the blood by osmosis. This type of sports drink is mainly for hydration.
Isotonic drinks	These drinks have concentrations of dissolved substances at about the same level as in the body. They usually have a carbohydrate concentration of between 4 and 8 g per 100 cm^3 of the drink.	Being the same concentration as the blood, some water will move from the gut into the blood by osmosis. Some sugar is also absorbed. This type of drink provides both hydration and fuel replacement.
Hypertonic drinks	In these drinks, the concentrations of dissolved substances are higher than in the body. They typically have sugar levels above 8 g per 100 cm^3 of the drink.	These drinks provide high sugar levels for absorption. They are often called 'power' drinks. They are mainly for supplying fuel to the muscles. They can cause a sudden sugar rush, followed by a crash, which can cause problems.

▲ Sports drinks keep your body hydrated, and also replace glucose and mineral ions

Questions

1 List four ingredients in a typical sports drink. (E)

2 Explain why long-distance runners would need isotonic sports drinks.

3 Why are hypertonic drinks often referred to as 'power' drinks? (C)

4 Explain how the water in a hypotonic sports drink is able to get into the blood from the gut. (A*)

Learning objectives

After studying this topic, you should be able to:

- know that active transport is a method of moving dissolved molecules into or out of a cell
- give some examples of active transport

Key words

concentration gradient, active transport, protein carrier

Against the flow

Diffusion and osmosis are not the only ways that particles can get into or out of cells. Some dissolved molecules or ions need to move from a low concentration to a high concentration, against a **concentration gradient**. This can happen by a process called **active transport**.

> **A** Active transport is one process by which molecules can move into or out of a cell. Name two others.

An example of active transport is the movement of sodium ions out of nerve cells in the human body.

1. The ion attaches to the protein carrier.

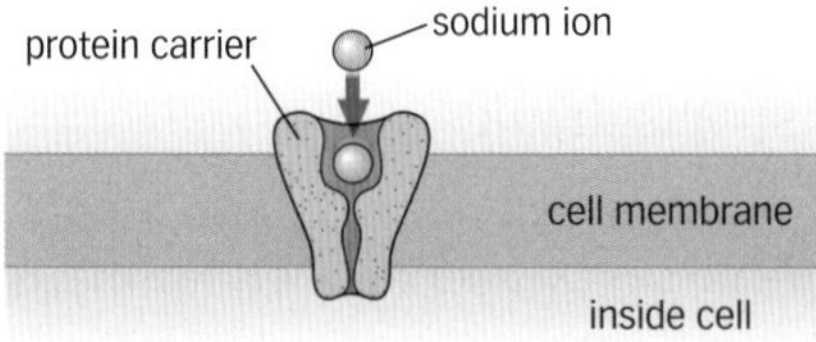

There is a high concentration of sodium ions on the outside of the nerve, and a low concentration on the inside.

There are **protein carriers** in the nerve cell membrane. Sodium ions fit into these carriers.

2. The protein uses energy to change shape.

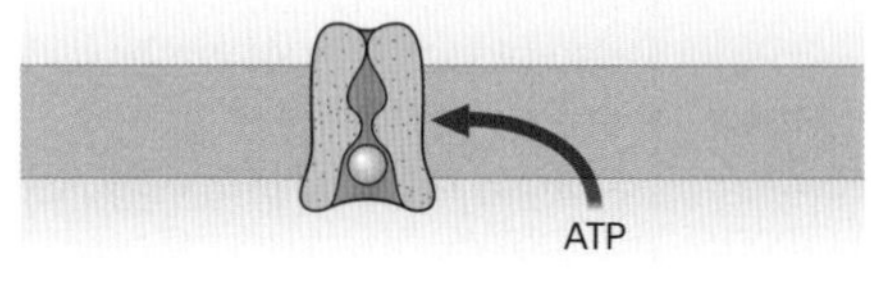

The proteins can change shape. This uses energy from a molecule called ATP, which is made in respiration.

3. The ion moves to the inside of the cell.

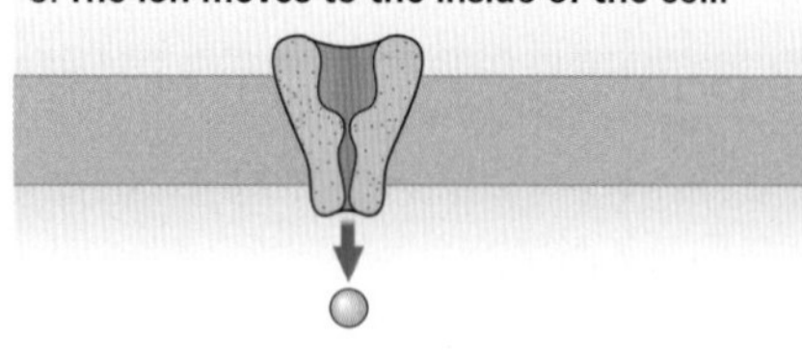

As the protein changes shape, it moves the sodium ion from the inside of the membrane to the outside.

The sodium ion falls off the protein carrier.

The protein immediately returns to its normal shape.

▲ Protein carriers in the membrane bring about active transport. Unlike diffusion and osmosis, the process requires energy.

Here are some key features of active transport:

- It pumps substances against a concentration gradient (from low to high concentration).
- It requires energy from respiration, which it obtains from the molecule ATP.
- It needs a protein carrier in the membrane.

> **B** What are the three distinctive features of active transport, which make it different from the other methods?

Examples of active transport

Active transport is involved in:

- the absorption of mineral ions by cells, for example in roots
- the absorption of sugars by cells.

Plants need mineral ions to maintain healthy growth. Mineral ions are at low concentrations in the soil – lower than their concentrations in the plant's cells. The minerals therefore cannot move into the plant by diffusion. Root hair cells on the outsides of roots are highly adapted for the uptake of minerals by active transport. They have long, fine extensions that fit between soil particles, and many protein carriers in their cell membranes which take up mineral ions from the soil. Each protein carrier is specific to one ion.

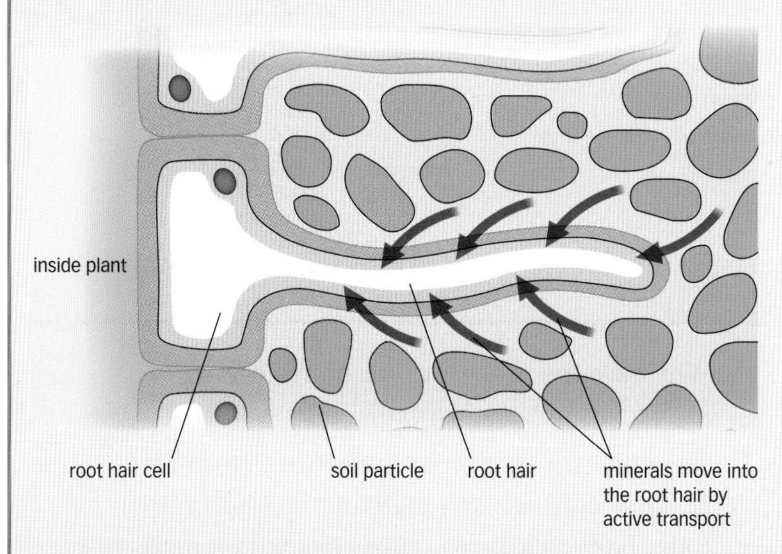

▲ Dissolved mineral ions are taken into plants through the root hairs. They move against a concentration gradient, by active transport.

Glucose can also be taken up by active transport. Glucose is absorbed by the cells lining the gut by active transport. There is often more glucose in the gut cells than in the fluid in the gut. For example, a hypotonic sports drink does not contain a lot of glucose. The cells lining the gut have membrane proteins that take up the glucose by active transport. These proteins absorb both glucose and sodium at the same time. Sports drinks often contain sodium ions, to help ensure glucose uptake.

Questions

1. What is energy used for in active transport? (E)
2. Explain why active transport is needed in roots to absorb minerals from the soil.
3. Suggest why a poison such as cyanide, which stops respiration, will stop active transport. (C)
4. Why is active transport affected by high temperatures?
5. Explain why root hairs need more than one type of carrier protein. (A*)

Learning objectives

After studying this topic, you should be able to:

- ✔ understand the need for exchange surfaces in larger organisms
- ✔ know the common features of exchange surfaces
- ✔ understand the adaptations of the villus as an exchange surface

Key words

exchange system, villus

▲ The axolotl is an aquatic amphibian. It has external gills for gaseous exchange in water.

The problem with being big

As organisms get bigger, they reach a size at which the surface area of their body is no longer large enough to allow sufficient materials to be exchanged with their surroundings. Important molecules such as oxygen cannot get into the body quickly enough to keep the organism alive.

As cells or organisms get bigger, their surface area to volume ratio gets smaller. As the ratio falls, exchange over the body surface becomes inefficient. To overcome this problem, organisms have developed specialised exchange surfaces.

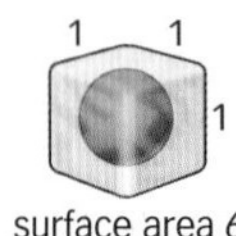

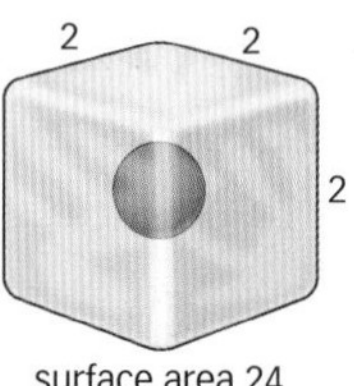

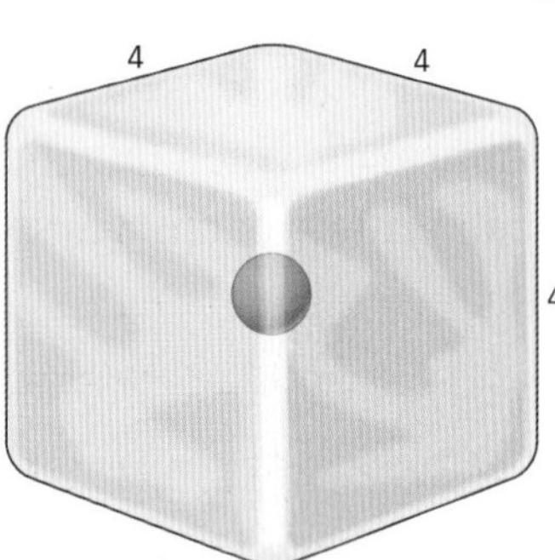

surface area : volume ratios

6 : 1
24 : 8 = 3 : 1
96 : 64 = 3 : 2 or 1.5 : 1

▲ The surface area to volume ratio falls as a cell, or organism, gets bigger

A Why does the process of exchange become inefficient in larger organisms?

B The diagram shows the surface area to volume ratio for cube cells with a length of 1, 2, and 4 units. Calculate the ratio for cube cells with lengths of 3 and 5 units.

Exchange systems

Larger organisms have developed many organ systems that are involved in exchange. These **exchange systems** include

- the surface of the lungs in land animals, for gaseous exchange
- the surface of the gills in aquatic animals, for gaseous exchange
- the digestive system of mammals and other animals, for uptake of nutrients
- the leaf, for gaseous exchange in plants
- the roots, for uptake of water in plants.

Features of exchange systems

At first, the various exchange systems of different organisms and for different functions look very different. But they all have common features:

- Large surface area: this provides a larger area over which diffusion can occur. This will increase the rate of diffusion.
- Thin surface: this provides a very short distance over which diffusion has to occur. This will also increase the rate of diffusion.
- Blood supply: a good blood supply over the exchange surface allows any particles that are taken up to be moved away quickly. This will maintain a greater concentration gradient.
- Turnover: the molecules or ions being exchanged need to be constantly replaced. For example, the air inside the lungs is constantly changed by ventilation – breathing in and out – and food is regularly supplied to the gut. This turnover maintains a high concentration gradient.

The villus – efficient by design

The small intestine is the major site of absorption of digested foods by both diffusion and active transport. It has many adaptations for absorption.

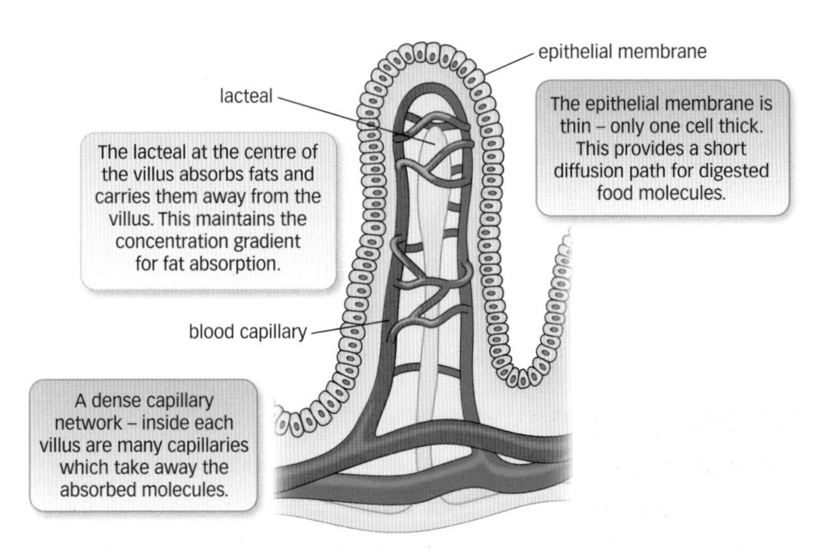

▲ Exchange across the villus. Each villus increases the surface area for both active transport and diffusion.

It has a large surface area for exchange – the small intestine is long (about 6.1 m), and its internal surface area is increased further by the presence of millions of finger-like projections called **villi**.

Exam tip 

✓ Whenever you see a question about exchange surfaces, remember to look for the features listed on the left.

C Identify the two features of exchange surfaces in the axolotl that are clearly evident.

D Explain how breathing helps to maintain the concentration gradient in the lungs.

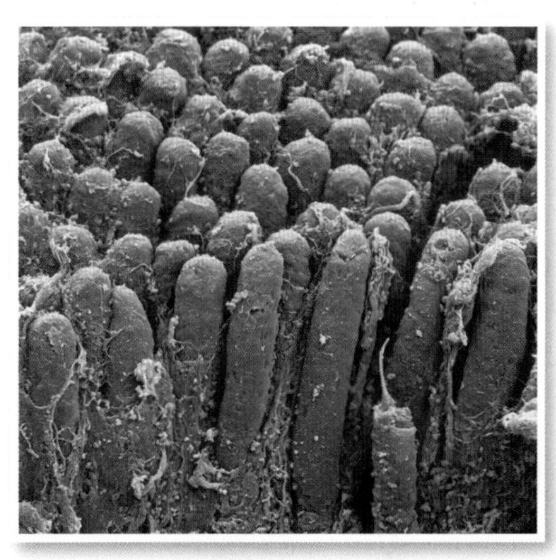

▲ The villus is well adapted for the absorption of digested food – it shows all the features of an exchange system

Questions

1. List three adaptations of a villus for exchange. (E)
2. For each adaptation named above, describe how it helps maintain exchange. (C)
3. Explain why a specialised surface is needed in humans for absorbing food. (A*)

5: Gaseous exchange in the lungs

Learning objectives

After studying this topic, you should be able to:

- ✔ know that the lungs are the organs of gas exchange in the human
- ✔ explain the process of gas exchange in the lungs
- ✔ explain the sequence of steps in the ventilation of the lungs

Key words

lung, alveolus, ventilation

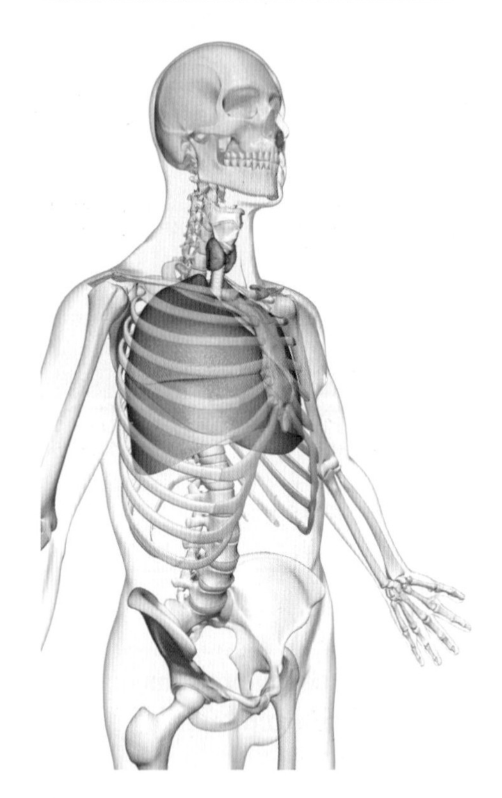

▲ The lungs are located in the thorax, surrounded by the ribs

A Where precisely does gaseous exchange occur in the lungs?

A breath of fresh air

The **lungs** are the specialised organs of many land animals for gaseous exchange. They are located in the chest or thorax, surrounded by the ribcage. The ribs protect the lungs and are also used during the process of breathing.

The thorax is separated from the abdomen by the muscular sheet called the diaphragm. This encloses the lungs in the thorax, and also plays a role in breathing.

Exchanging gases in the alveoli

You breathe air in through your nose and mouth. Here it is warmed, and many microbes are filtered out. This warmed air passes down your windpipe or trachea, which branches right and left into each of two lungs.

The airways continue to branch many times, getting smaller and smaller each time. Eventually, the tubes end in millions of tiny air sacs called **alveoli**. The alveoli have an excellent blood supply. It is in the alveoli that gaseous exchange occurs.

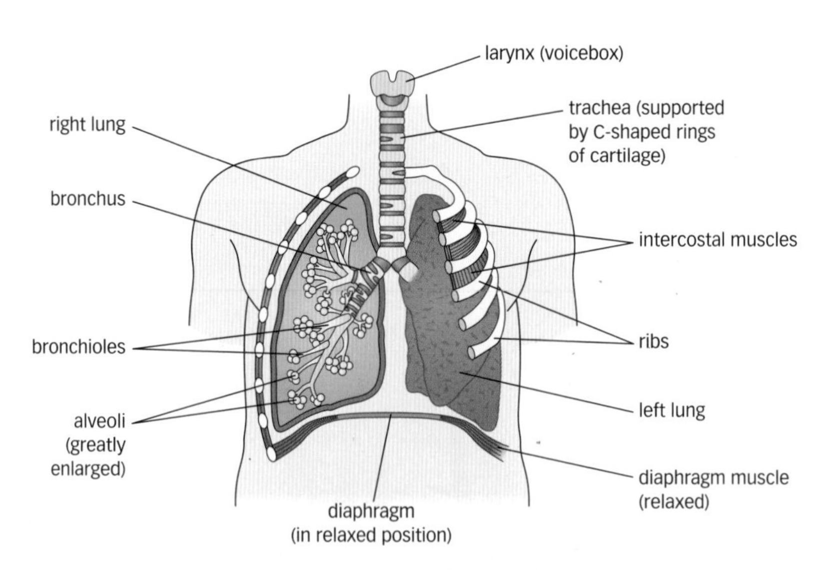

▲ This diagram shows a surface view of the left lung, and a section through the right lung showing the airways and air sacs inside

Breathing in brings air that is high in oxygen into the alveolus. Oxygen diffuses from this area of high oxygen concentration into the blood, where the oxygen concentration is low. At the same time, the carbon dioxide concentration is high in the blood arriving at the alveolus, and low in the air inside the alveolus. So carbon dioxide diffuses out of the blood into the alveolus. The stale air inside the alveolus becomes lower in oxygen and higher in carbon dioxide, and is breathed out.

The alveolus greatly increases the surface area for gas exchange. This maximises the rate of diffusion.

The wall of the alveolus is very thin – just one cell thick, and that cell is flattened, making it even thinner. This makes the diffusion pathway very short.

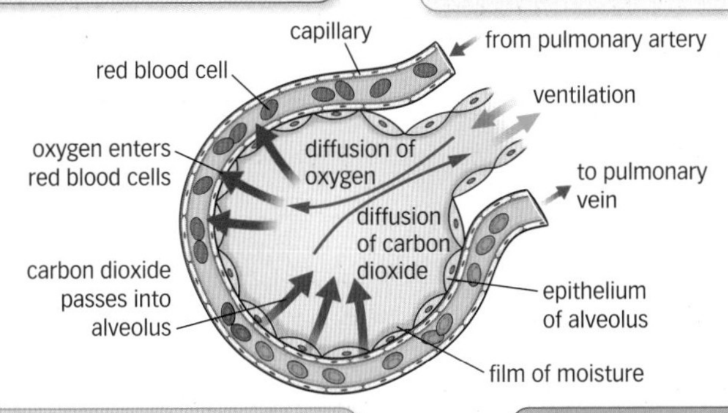

There is a dense blood supply to take away the absorbed gases. This maintains the concentration gradient.

The lining of the alveolus is moist. This allows dissolved gases to diffuse.

▲ How the alveoli and capillaries in the lungs aid gaseous exchange

The composition of inhaled and exhaled air

Gas	Inhaled air	Exhaled air
nitrogen (%)	78	78
oxygen (%)	21	17
carbon dioxide (%)	0.04	4
other gases (%)	1	1
water vapour	little	saturated

Ventilation

Ventilation is the process of breathing in (inhaling) and breathing out (exhaling) as shown in the diagrams.

Questions

1 Explain what causes the ribcage to move up and down during ventilation.

2 As you breathe in, describe the role of the diaphragm.

E

3 Describe the adaptations of the alveolus that make it an effective exchange surface.

C

4 Sketch a graph to show the pressure changes inside your lungs for two breaths in and out. Label the parts when you are inhaling and exhaling.

A*

B Which gases are exchanged in the lungs?

Breathing in – inhaling

1. The intercostal muscles between the ribs contract, lifting the ribcage up and out. This expands the thorax.
2. The diaphragm muscle contracts, flattening the diaphragm. This also expands the thorax.

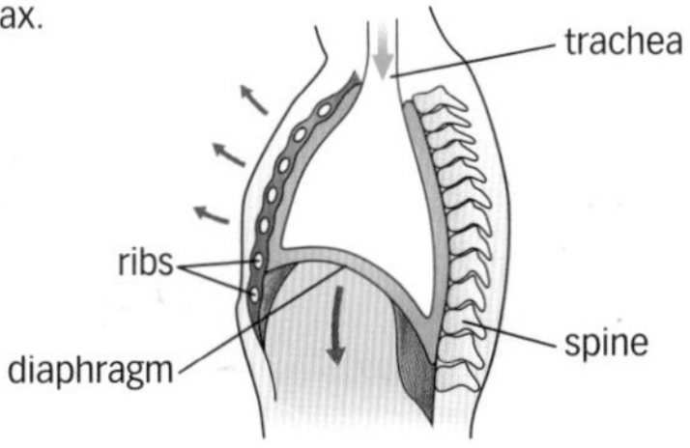

3. The volume inside the lungs has increased, and the pressure inside has decreased.
4. Air rushes into the lungs due to the lower pressure.

Breathing out – exhaling

1. The intercostal muscles relax, and the ribs fall, reducing the volume of the thorax.
2. The diaphragm muscle relaxes, and arches up. This also reduces the volume.

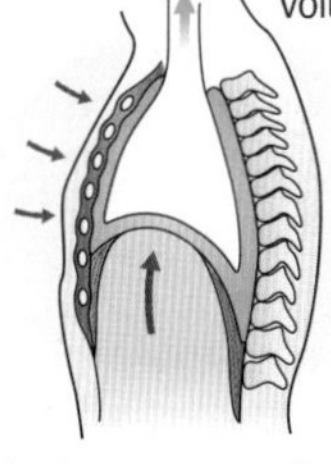

3. The volume inside the lungs has decreased, increasing the pressure in the lungs.
4. The higher pressure forces the air out.

Exam tip

✔ Ventilation is a sequence of events. Learn the four steps in order.

Learning objectives

After studying this topic, you should be able to:

- ✔ know that plants exchange substances across their roots and leaves
- ✔ understand the process of transpiration
- ✔ describe gaseous exchange in the leaf and the role of the stoma

Key words

transpiration, transpiration stream, stomata

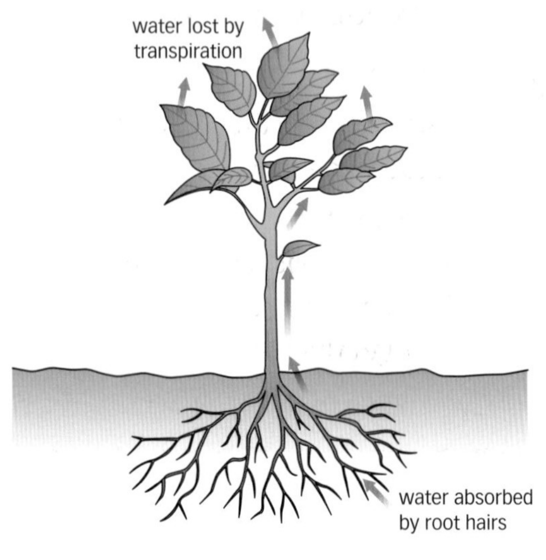

▲ The transpiration of water through a plant

In and out of plants

Plants also need exchange surfaces. There are two major exchange surfaces in plants:

- roots
- leaves.

Plants need water

Water is important for plants for a number of reasons:

1. Water is needed for the process of photosynthesis.
2. When water evaporates from the leaf it has a cooling effect.
3. Water enters the cells of the plant by osmosis, and makes the cells firm. This helps to support the plant.
4. As water moves through the plant, it transports dissolved minerals.

Transpiration

Plants take up water and minerals from the soil through their root hairs, which extend into the soil. The root hairs greatly increase the surface area of the root for the absorption of water and dissolved mineral ions.

The water flows up the stem and into the leaves. Water exits the plant by evaporation through the leaves in a process called **transpiration**. The flow of water from the roots to the leaves is called the **transpiration stream**. Leaves are also well adapted for exchange – their flattened shape gives a large surface area, and they have many internal air spaces.

A By what process does water move into the plant?

B Where does water:
(a) enter a plant (b) leave a plant?

C State two uses of water by the plant.

Controlling water loss: stomata

Leaves are highly adapted to be efficient at photosynthesis. A consequence of these adaptations is that the leaves can lose a lot of water by transpiration. To help reduce this, the leaf can control water loss through pores in the leaf, called **stomata**.

- Most stomata are on the lower leaf surface. There are very few on the upper leaf surface. More water would evaporate through the stomata on the upper surface because this area is warmed by the Sun.
- Each stoma can be opened or closed. When the plant is photosynthesising the stomata are open. The stomata are closed at night. When the stomata are closed, this reduces water loss.
- If there is little water, a plant is in danger of losing water faster than it can be replaced. The stomata do not open when the plant is short of water, and this reduces water loss. This prevents the plant dehydrating and wilting.

There are two special cells, called guard cells, on either side of the stoma. When there is plenty of light and water, the guard cells take up water by osmosis, swell, and become firm. This causes them to bend and open the stoma. Water then leaves via the stoma. If there is little water, then the guard cells cannot become firm. Then they do not open the stoma.

Gaseous exchange in the leaf

The leaf is the site of gaseous exchange in the plant. When the stomata are open, gases can both enter and leave.

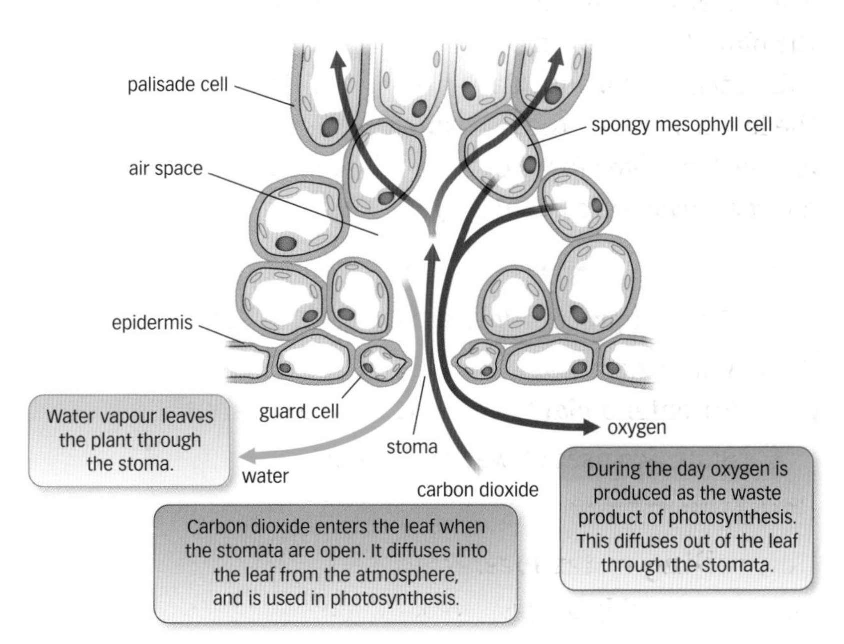

▲ Gaseous exchange in the leaf

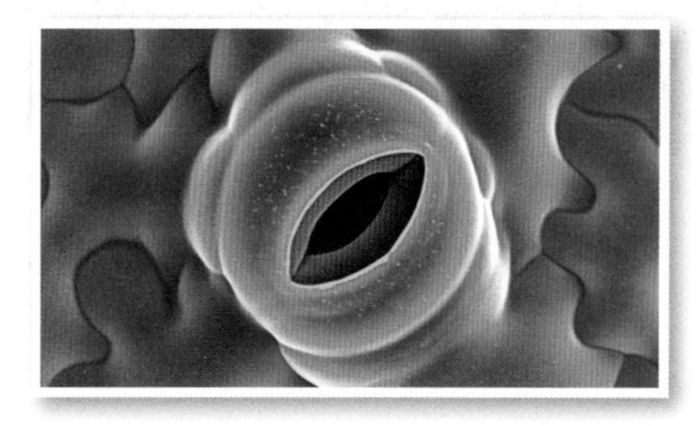

▲ When conditions are good for photosynthesis, the guard cells are turgid, opening the stoma. Carbon dioxide can enter the leaf and water can leave.

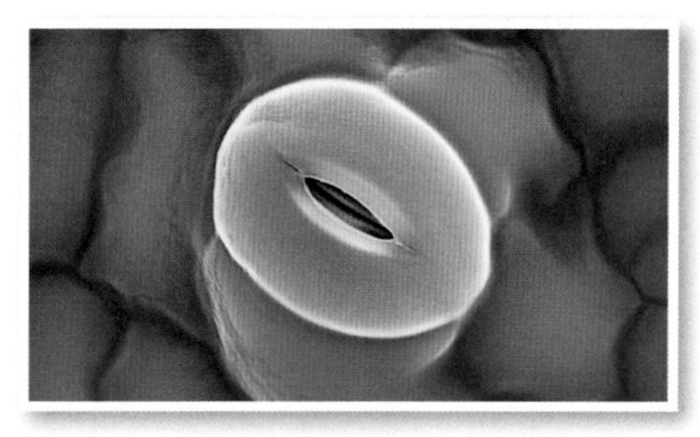

▲ When conditions are not good for photosynthesis, the guard cells close the stoma. This reduces water loss from the plant.

D List two ways that a plant can reduce water loss.

E Describe three occasions when osmosis plays a part in the movement of water through the plant.

Questions

1. Which gases are exchanged through the stomata during the day? ↓ E
2. Describe the process by which stomata can open. ↓ C
3. Wilting is a response of the plant to dehydration. Suggest how wilting helps prevent further dehydration. ↓ A*

7: Rates of transpiration

Learning objectives

After studying this topic, you should be able to:

- know that the rate of transpiration can change
- describe and explain how environmental factors can change the rate of transpiration

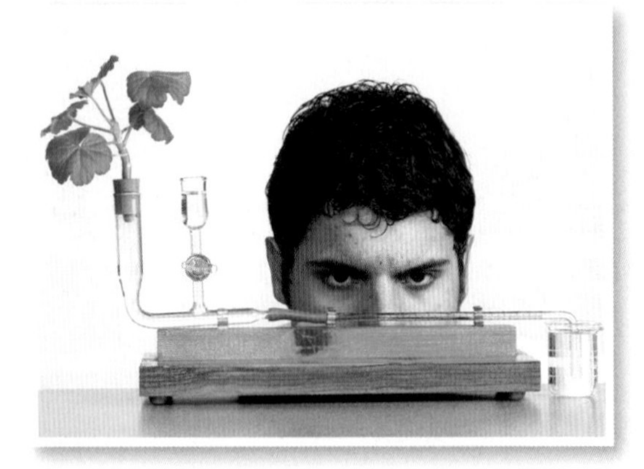

A bubble potometer

A A rate is a speed, which is distance divided by time. What two measurements would you need to take in an experiment using a potometer, in order to calculate the rate of transpiration?

B Increasing the light intensity will increase the rate of transpiration. How would you notice this using the bubble potometer?

C When comparing the rate of transpiration in two plants, why is it important to conduct the experiments at the same time of day?

Factors affecting the rate of transpiration

There are four main factors in the environment that can affect the rate of evaporation of water. Anything that affects evaporation will affect how quickly water moves through the plant – the **rate of transpiration**. The following factors make the rate of transpiration faster:

- an increase in light intensity
- an increase in temperature
- an increase in air movement
- a decrease in humidity.

Biologists use a piece of apparatus called a bubble **potometer** to measure the rate of transpiration. You can change a factor such as the light level, or temperature, and note the change in the rate of transpiration using a bubble potometer, by measuring how fast a bubble moves along a glass tube. The bubble shows how quickly water is moving through the plant.

Increasing the rate of transpiration

Higher light intensity

Stomata close in the dark and open in the light. When the light intensity is greater, more stomata will open. This allows more water to evaporate, so the rate of transpiration will be faster.

A higher light intensity increases the rate of transpiration. The stomata open to allow oxygen into the leaves for photosynthesis.

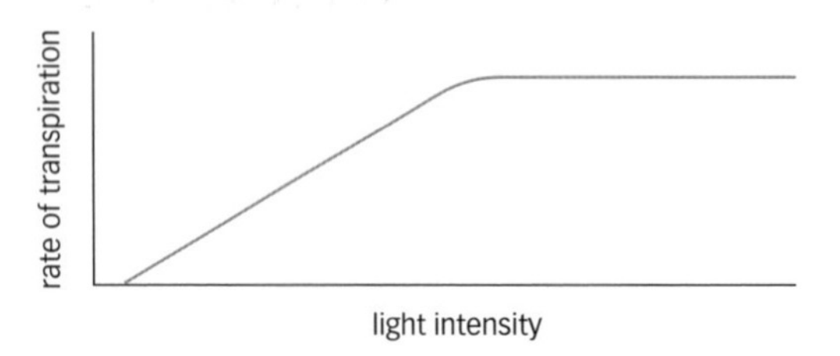

Graph of transpiration rate against light intensity. The rate increases until all the stomata are open, and transpiration is at a maximum.

Increase in temperature

The higher the temperature, the faster the particles in the air will move. This means that water molecules move faster and evaporate from the leaf quicker. So the rate of transpiration will increase.

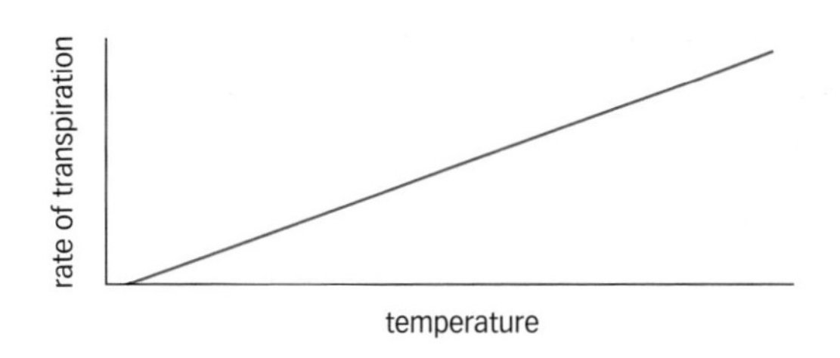

▲ A warmer temperature increases the rate of transpiration

Increased air movement

When air moves over the leaf, it moves evaporated water molecules away from it. The faster the air movement, the quicker the water will be moved away. This increases the diffusion of water out of the leaf, because water molecules do not build up in the air outside the leaf. The concentration of water outside the leaf is kept lower, keeping a high concentration gradient between the inside of the leaf and the air outside. So the rate of transpiration increases.

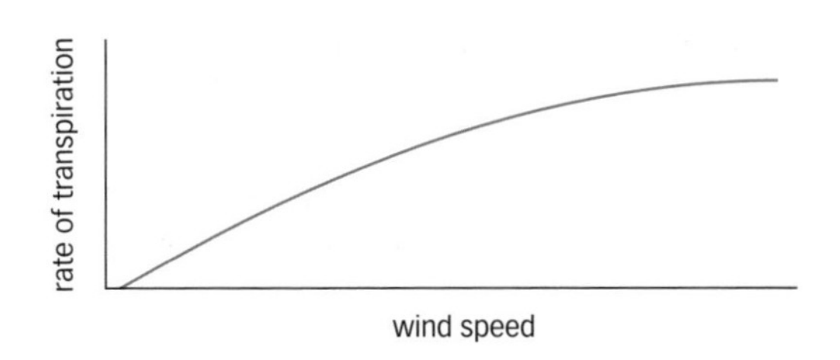

▲ The rate of transpiration is higher on a windy day

Decreased humidity

The less humid the air, the less water there is in it (it is drier). This again makes for a greater concentration gradient between the inside and the outside of the leaf. Water molecules will diffuse out more quickly, increasing the rate of transpiration.

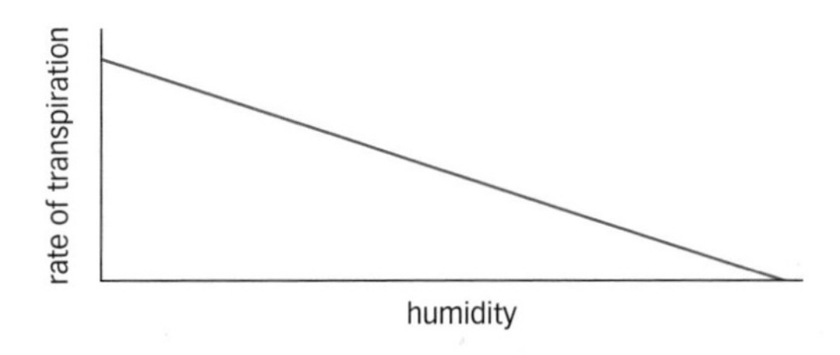

▲ The rate of transpiration is higher when the air is less humid

Key words

rate of transpiration, potometer

Exam tip AQA

✔ Try to remember the factors affecting the rate of transpiration by thinking of the best conditions for drying clothes.

Questions

1. Why do gardeners need to water their plants more in the summer? (E)
2. Explain why plants on a sand dune will lose water faster than plants in a woodland. (C)
3. Why do florists spray ferns with water to help keep them healthy? (A*)

Learning objectives

After studying this topic, you should be able to:

- ✔ understand that substances are transported around the body by the circulatory system
- ✔ know that circulatory systems can be medically assisted

Key words

blood, plasma, trauma

A Explain why single-celled organisms do not need a transport system.

Did you know...?

On 7 July 2005 a bomb exploded on a London bus. The bus had by chance been diverted, and was outside the British Medical Association (BMA) headquarters when the bomb exploded. Many senior doctors were on hand but they could not remember how to make up saline, as they had not done this for a long time. However, a policeman delivered bags of saline and the doctors infused these into the casualties.

Why do you need a transport system?

Diffusion in single-celled organisms

Small organisms such as amoebae do not need a circulatory system. They have a large surface area compared to their volume. They are surrounded by the water they live in. Dissolved oxygen diffuses from this water into the cell through the cell membrane. Waste material diffuses out of the cell. There is no need for a special transport system.

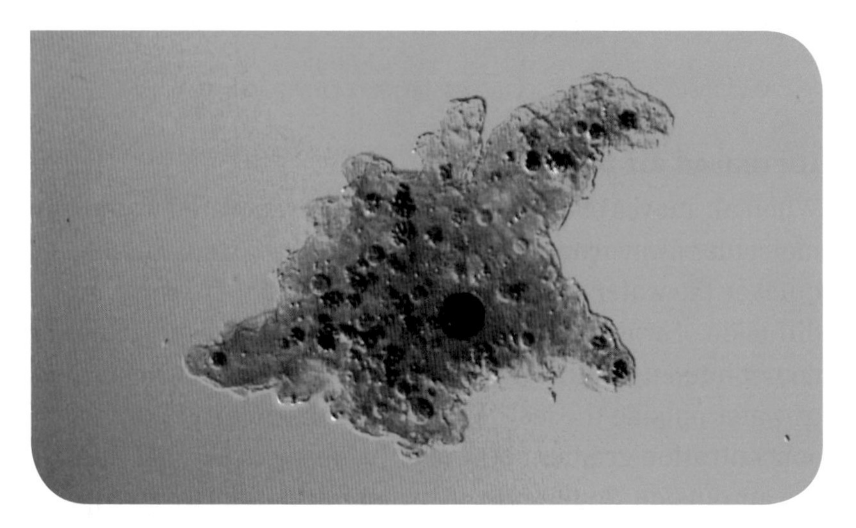

▲ *Amoeba proteus,* a single-celled organism. It lives at the bottoms of ponds and puddles. It is between 0.2 and 0.5 mm long.

Larger animals need a circulatory system, because diffusion alone cannot efficiently transport substances to and from their cells. They need blood to do this.

The human transport system

The **blood** is your main transport system. It takes oxygen and nutrients, such as glucose and amino acids, to cells. It collects and removes waste, such as urea and carbon dioxide, from cells.

Blood is a tissue. Red blood cells, white blood cells, and platelets are suspended in the fluid **plasma**.

B Name three substances that blood carries to your body cells.

C Name two substances that your blood carries away from your body cells.

Artificial blood

If people have lost a lot of blood, or have a problem with their blood, doctors may infuse their blood with extra 'artificial blood'. Various artificial blood products are available, but none contain red or white cells or platelets, so they are really substitutes for blood rather than artificial blood. Some are for increasing the blood volume, and some are for carrying oxygen.

Increasing the blood volume

These volume expanders are used by paramedics in emergency situations. If a patient has lost a lot of blood, but there is no real blood of the correct type to infuse, saline (salt) solution can be used instead. Some volume expanders also contain sugars and proteins. The volume expanders can maintain normal blood pressure, and the remaining haemoglobin in the patient's blood will carry enough oxygen to the body tissues to sustain a motionless patient.

Oxygen carriers

If the blood loss was such that more than two-thirds of your red cells were lost, you would need artificial blood that can carry oxygen. Some types contain chemicals that can carry and release oxygen. Other types contain encapsulated haemoglobin. Haemoglobin cannot be used free in the blood without being in capsules, because it would be filtered out in the kidneys and would damage them. These types of oxygen carriers are undergoing trials on emergency patients.

Uses of artificial blood products

Unlike real blood, artificial blood products do not have to be matched to patients. They could be useful

- for treating war casualties
- in countries where blood transfusions may not be safe, as blood may not be screened for disease
- to rapidly treat **trauma** patients who have serious injuries, often as a result of violence or an accident
- because they can be stored for one to three years at room temperature
- because they immediately restore full oxygen-carrying capacity to recipients, whereas this takes 24 hours with real blood.

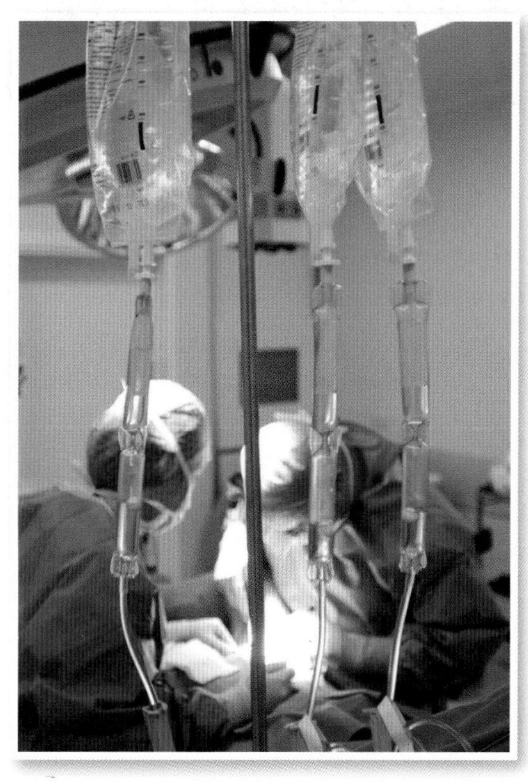

▲ Saline drips replace fluid lost by a patient during an operation

Questions

1. What are blood expanders, and when might they be used?
2. What are oxygen carriers?

↓ E

3. Why might the remaining haemoglobin in a trauma patient's blood be enough to sustain a motionless patient, but not someone who is mobile?

↓ C

4. Evaluate the usefulness of artificial blood products.

↓ A*

Exam tip

✔ If you are asked to evaluate something, try to think of some benefits and some risks.

Learning objectives

After studying this topic, you should be able to:

- ✔ know the names and locations of the four main chambers of the heart
- ✔ understand that there are two separate circulation systems, one to the lungs and one to the rest of the body
- ✔ understand the functions of ventricles and heart valves

Key words

ventricles, atria, valves

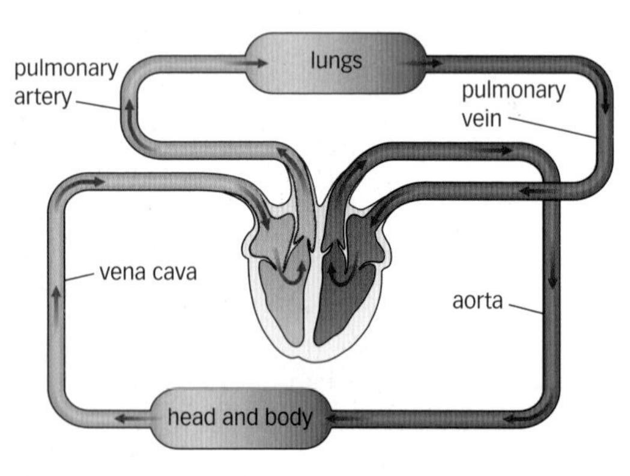

▲ The human double circulatory system

A The double circulatory system allows mammals to be very active. Why do you think this is?

B Draw a simplified diagram showing the double circulatory system. Label the atria, ventricles, aorta, vena cava, pulmonary artery, and pulmonary vein.

Two separate circulatory systems

Humans and other mammals have a double circulatory system. There are two circuits from the heart.

Blood passes

- from the heart to the body organs and tissues
- back to the heart
- to the lungs to remove carbon dioxide and collect oxygen
- back to the heart before going to the body again.

Because the blood makes two circuits from the heart, the heart needs four chambers. It is a double pump. Blood in a double circulatory system is under high pressure and so it transports material more quickly around the body.

The heart

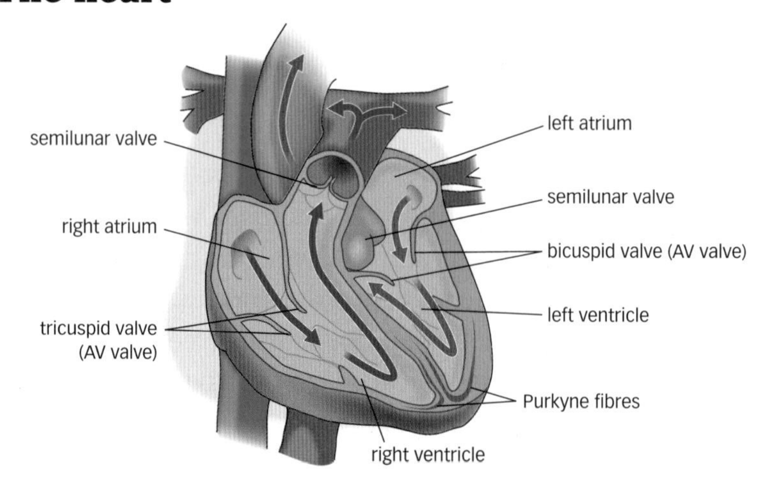

▲ Section through the human heart. Red arrows show the path of oxygenated blood. Blue arrows show the path of deoxygenated blood.

Your heart is an organ. Its function is to pump blood around the body. The wall of the heart is made of muscle. It is a specialised type of muscle called cardiac muscle.

The walls of the **ventricles** (lower chambers) are thicker than the walls of the **atria** (upper chambers). The left ventricle wall is thicker than the right ventricle wall. Blood leaving the left ventricle goes all around the body. Blood leaving the right ventricle goes to the lungs:

- Blood enters the atria of the heart, from veins.
- The atria contract and force blood into the ventricles.
- The ventricles contract and force blood out of the heart, into arteries.
- Blood flows from the heart to the organs, in arteries.
- Blood returns from the body organs to the heart, in veins.

Artificial parts for hearts

Heart valves

The **valves** in the heart prevent the blood from flowing backwards. Surgeons can replace diseased or damaged heart valves. The valves used may be synthetic, or taken from other animals such as cows or pigs. There is no risk of rejection of these valves, as heart valves have no capillary blood supply. This means that white blood cells do not patrol the heart valves.

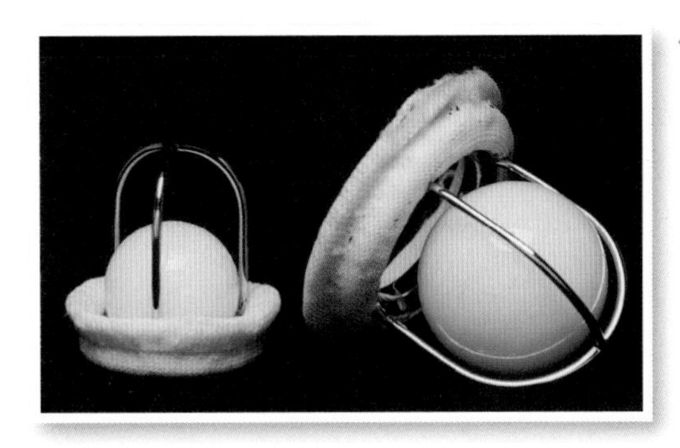

Artificial heart valves. The one on the left is closed and the one on the right is open. They are made of inert plastic and metal that will not react with any chemicals in blood.

Artificial hearts

The Jarvik-7 artificial heart is made from polyurethane and titanium. It was developed by Dr Robert Jarvik and a biomedical engineer, Dr Lyman. The inside is smooth and seamless so it does not cause blood clots that would cause strokes. Many people have had this type of artificial heart implanted while waiting for a heart transplant. However, the wires protrude through the skin.

In France in 2008, Professor Alain Carpentier, a cardiac surgeon, and engineers from the group that makes the Airbus aircraft, developed a new artificial heart that has been tested on calves and will soon be tested on humans.

Questions

1. Explain why artificial hearts and valves are made of inert materials. (E)
2. Why do you think it is important for biomedical engineers to continue to develop better artificial hearts? (C)
3. Explain why the risk of rejection would be a serious problem if heart valves had a capillary blood supply. (A*)

Did you know...?

In the future, it may be possible to use artificial hearts instead of heart transplants. This would be useful, because there is a shortage of donors for heart transplants. Many recipients of heart transplants are young and otherwise healthy, so using artificial hearts may one day save many lives.

The Jarvik-7 artificial heart

Exam tip

- ✔ You need to learn the names and functions of the aorta, pulmonary artery, vena cava, and pulmonary vein.

Learning objectives

After studying this topic, you should be able to:

- describe the structural features of arteries and veins
- understand how the structures of arteries and veins enable them to carry out their functions
- describe how stents may be used

Key words

elastic, muscle, lumen, stent

Did you know...?

A five-year-old child could crawl through the aorta of a blue whale.

Arteries

Every time your heart beats, the ventricles contract and force blood out of the heart, into arteries. The two arteries leaving the heart are the aorta and pulmonary artery.

The aorta:

- leaves the left ventricle
- carries oxygenated blood to your body tissues
- smaller arteries branch off from it to take blood to the head and brain, and to your other body organs.

The pulmonary artery:

- leaves the right ventricle
- takes deoxygenated blood to the lungs.

Because blood is forced out of the ventricles of the heart at each beat, it enters the arteries in high-pressure spurts. Artery walls contain

- a lot of **elastic** fibres to allow them to stretch, when each spurt of blood enters, and then to recoil. This pulsation helps smooth the flow of blood
- a lot of **muscle** fibres to withstand and maintain the high pressure
- a folded inner lining that can expand as the walls stretch with each high-pressure spurt
- a narrow **lumen** (the space inside the artery).

Arteries have a pulse. You can feel your pulse where an artery crosses over a bone and/or is near to your skin. Measuring your pulse tells you your heart rate, as each pulse corresponds to each beat of your heart.

A Explain why the lining of arteries is folded.

B Explain why measuring your pulse tells you how fast your heart is beating.

▲ Light micrograph of transverse section of an artery (left) and a vein (right) (×30). Notice the thick wall and small lumen of the artery, and the thin wall and wide lumen of the vein.

Veins

- Veins carry blood back from body tissues to the heart.
- Blood in your veins is under very low pressure.
- Their lumen is wide to allow low resistance to blood flow.
- The walls are thin as they do not have to withstand high pressure.
- Their walls contain less muscle and fewer elastic fibres.
- The inner lining is smooth.
- Veins are surrounded by skeletal muscles. When your leg muscles contract, this helps push blood through the veins, up towards your heart.
- They have valves to prevent backflow of blood.

The vena cava brings deoxygenated blood back to the right atrium. The pulmonary veins bring oxygenated blood from the lungs to the left atrium.

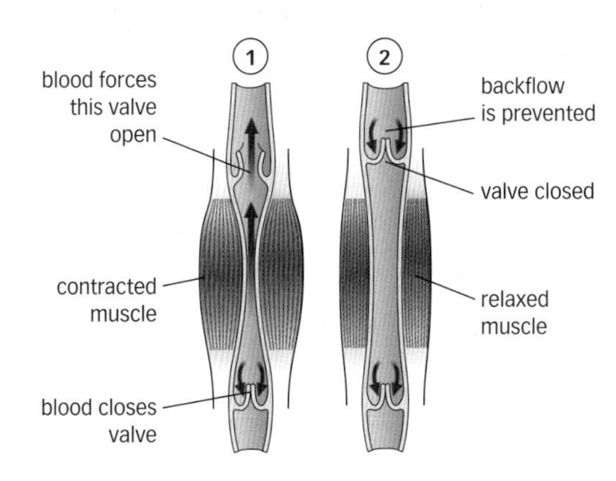

▲ How the valves in veins prevent the backflow of blood

What happens if your arteries become narrow?

As you age, and especially if you eat a lot of saturated fat, your arteries can become narrower. Fatty deposits form under their linings, and this obstructs the lumen. This could reduce blood flow and even cause a clot. If this happened in a coronary artery, supplying the heart muscle, you would have a heart attack.

A **stent** can be inserted into a narrowed or blocked artery. It makes the lumen wider again and eases the flow of blood. A stent is a narrow mesh tube that is inserted into the blocked artery. As it expands, it widens the artery lumen.

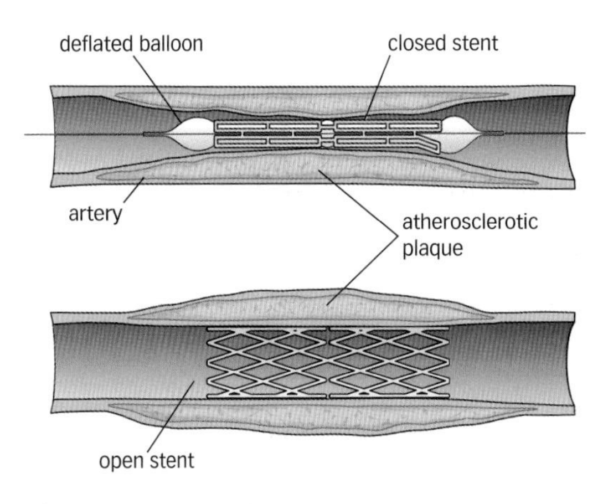

▲ A coronary stent

Questions

1. Make a table to compare arteries and veins.
2. Name the artery that carries deoxygenated blood.
3. Name the vein that carries oxygenated blood.

↓ E

4. Explain how blood gets back to the heart in veins, even though there is no pulse and the blood in veins is under very low pressure.

↓ C

5. Some hospitals treat patients who have had a heart attack within 24 hours, and insert a stent into the blocked coronary artery. This minimises damage to the heart muscle. Explain how a stent widens a blocked artery.

↓ A*

Exam tip

✔ Remember, arteries take blood from ventricles, and veins take blood to atria.
Always V and A.

Learning objectives

After studying this topic, you should be able to:

- ✔ understand the role of capillaries at tissues
- ✔ describe the structure of capillaries

Key words

capillaries

▲ Light micrograph of a capillary network branching off from an arteriole (×200)

Exam tip AQA

- ✔ Remember that blood plasma is forced out through capillary walls because blood is at higher pressure in arterioles than in capillaries.

Capillaries: where exchange takes place

As arteries reach the body tissues, they branch into narrower vessels called arterioles. Arterioles divide into smaller vessels called **capillaries**. Capillaries allow substances to be exchanged between blood and tissues. Capillaries then join to form venules, and these venules join to make veins. Veins take blood back to the heart.

The structure of capillaries

Capillaries are the smallest of your blood vessels:

- They have a wall that is made of just one layer of flattened cells. There are tiny gaps between the cells that make up the capillary walls. These gaps allow blood plasma to leave the capillaries.
- Blood in capillaries is at low pressure and so will not damage their thin walls.
- Their lumen diameter is just wide enough to let through red blood cells, usually one at a time in single file.
- Your blood flows slowly in capillaries.
- Capillaries form vast networks at the tissues of each of your organs.
- This gives a large total surface area for the exchange of materials.

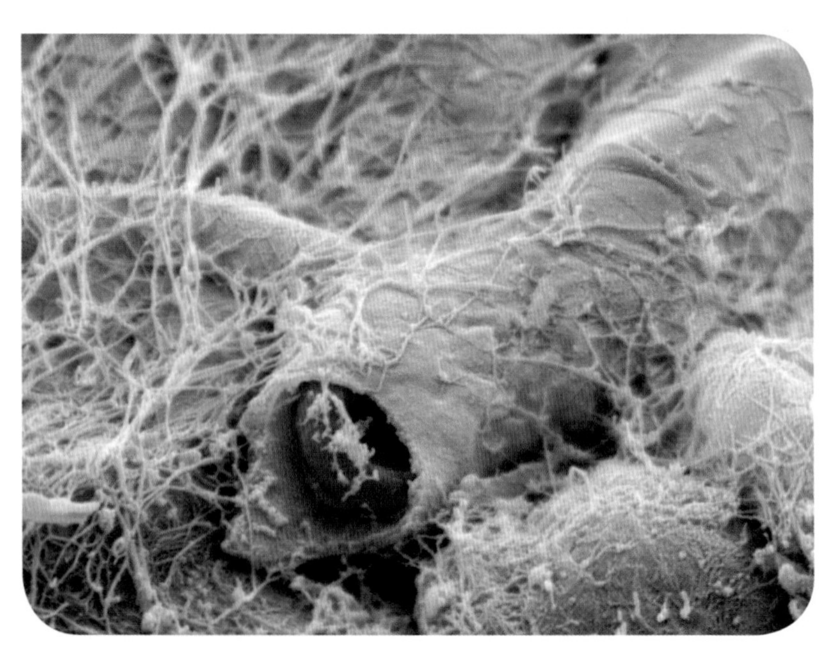

▲ False-coloured scanning electron micrograph of a capillary (pink) carrying blood to muscle fibres (grey) (×7000). Notice how just one red blood cell at a time can pass through.

A What are the functions of capillaries?

B Explain how the structure of capillaries enables them to carry out their functions.

What is exchanged?

Capillaries deliver blood to your tissues, carrying substances such as:

- oxygen for aerobic respiration
- glucose for respiration
- amino acids for making proteins for growth and repair
- other nutrients, such as fatty acids to be stored
- water to keep cells hydrated
- hormones to cause their target cells to respond.

At your tissues, some of the liquid blood plasma is forced out through the tiny gaps in the capillary walls. This liquid, now called tissue fluid, bathes your cells. Substances dissolved in tissue fluid can diffuse into your cells. Substances made in your cells can diffuse into your tissue fluid. This tissue fluid goes back into your capillaries and then into your veins and back to your heart.

Because there is a great network of capillaries at each organ, none of your cells is very far from a capillary. This is how capillaries deliver oxygen, nutrients, and other useful substances such as hormones, to your cells.

Your cells carry out all sorts of chemical reactions (your metabolism) and these reactions produce waste products.

Capillaries carry waste products away from cells and tissues, including:

- carbon dioxide from aerobic respiration
- lactic acid from anaerobic respiration
- urea from the breakdown of excess amino acids and haemoglobin from old red blood cells (in liver cells)
- spent hormones
- water, a by-product of respiration
- heat made by respiring muscles and liver cells
- hormones from the cells of glands where they are made.

Did you know...?

In the seventeenth century, William Harvey, a doctor at St Bartholomew's hospital in London, published a book showing how blood circulated in the body, from the heart and back, and to and from the lungs. He realised that the pulse in arteries was linked to the contraction of the left ventricle of the heart. He discovered that veins have valves to prevent backflow. He postulated that there were capillaries, but did not see them as he did not have a microscope.

Questions

1 List four substances that your capillaries deliver to your tissues.

E

2 If you gain a pound of fat tissue, one mile of new capillaries will be made to serve those fat cells. How can a mile of capillaries fit into such a small amount of tissue?

C

3 Where in your body do you think the heat that your blood carries away from your muscle cells is dissipated?

A*

4 Which organs do you think deal with the urea that your blood carries away from your liver?

Learning objectives

After studying this topic, you should be able to:

- ✔ understand the structure and functions of blood

Key words

tissue

Blood is a tissue

Blood is a **tissue**. It consists of red blood cells, white blood cells, and platelets suspended in a non-living, straw-coloured fluid (liquid) called plasma. Because blood is liquid, it can flow around your body. Blood is the only fluid tissue in your body.

Blood is made in the red bone marrow of your long bones. You have about five litres of blood in your body. This volume has to be maintained, otherwise your blood cannot circulate properly. Blood is slightly alkaline and its temperature is a little higher than your body temperature, at 38 °C.

Blood is for

- transport
- protection
- regulation.

Transport

Red blood cells carry oxygen from your lungs to your heart, and then to your tissues.

Plasma carries

- soluble products of digestion from your small intestine to other organs
- urea from the liver to the kidneys
- carbon dioxide from the organs to your lungs
- hormones from glands to target cells.

Protection

If you cut yourself, your blood clots and forms a scab over the wound. This stops blood loss and prevents pathogens entering.

If pathogens do get into your body, certain white blood cells deal with them.

Regulation

Your blood helps to maintain your

- body temperature – by distributing heat from respiring muscles and liver cells to other organs and to your skin
- pH in body tissues – some of the plasma proteins act as buffers, which means they resist changes in pH.

A What makes blood a unique tissue?

B Where in your body is blood made?

C State the three functions of blood.

Did you know...?

Normal blood has a pH of 7.2. If it changes to above 7.4 or below 6.8, you are in serious danger.

Plasma

Plasma contains about 90% water. There are many dissolved substances in it. Some are being carried to and from cells. These dissolved substances include:

- glucose
- amino acids
- fatty acids
- vitamins
- hormones
- cholesterol
- carbon dioxide and hydrogencarbonate ions
- mineral ions, such as sodium, calcium, potassium, and chloride
- fibrous proteins that are important for blood clotting when you cut yourself
- antibodies.

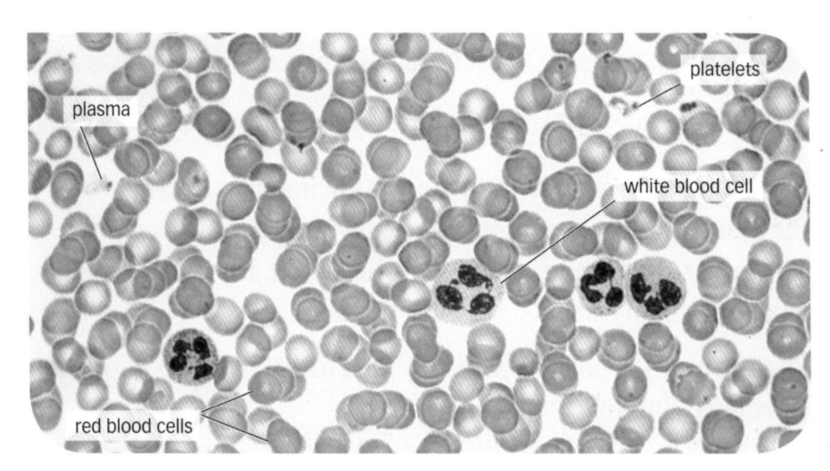

Human blood smear as seen with a light microscope

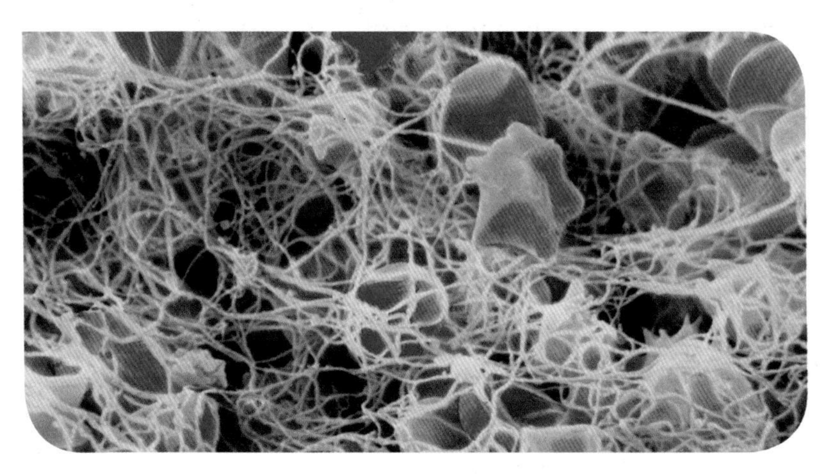

False colour scanning electron micrograph of a blood clot forming (×2000). Soluble blood proteins have come out of solution to form threads that make a mesh and trap red blood cells.

Exam tip AQA

✔ Remember that blood is a tissue – a collection of different types of cells that work together to perform certain functions.

Questions

1. Name four substances carried by the blood. For each one, state where it is being carried to and from.
2. List ten substances dissolved in your blood plasma.

E

3. How does your blood protect you?
4. How does your blood help regulate your body temperature?

C

5. Explain the importance of maintaining a constant blood pH. How does the blood resist changes in pH?

A*

Learning objectives

After studying this topic, you should be able to:

- ✔ recognise the structures and functions of red blood cells, white blood cells, and platelets

Key words

haemoglobin, oxyhaemoglobin

Did you know...?

You have about 5 billion red blood cells in each cm^3 of your blood. This means you have five thousand billion in each litre of blood, so you have 25 million million red blood cells in your body. This is about 30–40% of the total number of cells in your body. You have about 600 times more red blood cells than white blood cells. People who live in a high-altitude area have even more red blood cells.

A What is the function of red blood cells?

B Explain how the size and shape of red blood cells enables them to carry out their function.

Red blood cells

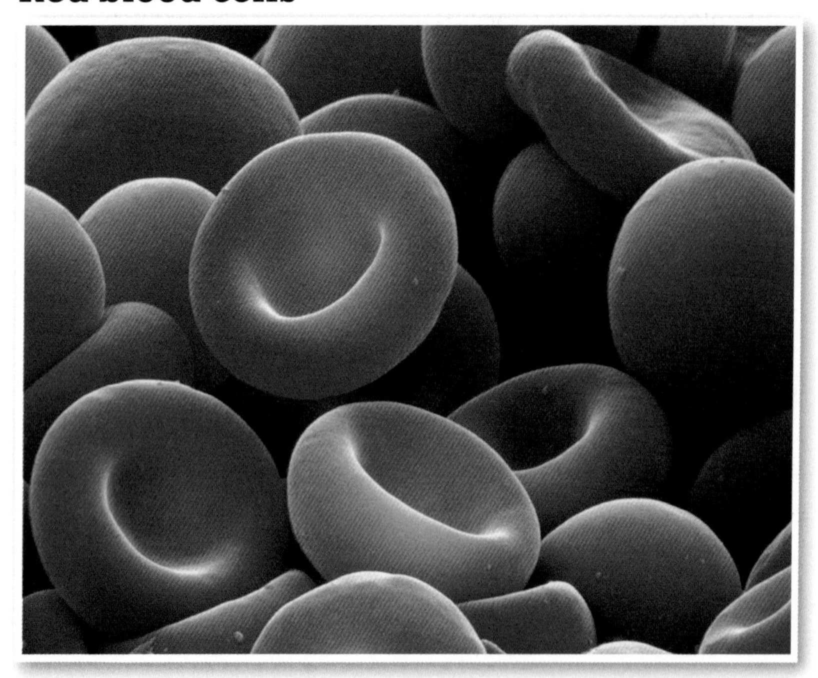

▲ Coloured scanning electron micrograph of human red blood cells (×7000)

Red blood cells are made in your bone marrow. Just before they get into your bloodstream, their nucleus breaks down. This gives them their characteristic shape. Each is a biconcave disc – each side caves inwards.

Each red cell is packed with **haemoglobin**. Haemoglobin is a protein that also contains iron.

In your lungs, oxygen diffuses from the alveoli into your blood, and then into red blood cells. Because red blood cells are very small and biconcave, they have a large surface area compared with their volume. Oxygen can easily diffuse through their cell membrane and reach all the haemoglobin molecules inside the cell. Oxygen combines with haemoglobin in red blood cells, to form **oxyhaemoglobin**.

The oxygenated blood returns from your lungs to your heart and then travels all over your body. At respiring tissues, oxyhaemoglobin splits into oxygen and haemoglobin. The oxygen diffuses into your cells to be used for aerobic respiration.

Red blood cells live for about four months. They cannot divide. Old ones are broken down in your liver. They are replaced by new ones being made in your bone marrow.

White blood cells

White blood cells have a nucleus. There are different types of white blood cell. They form part of your body's defence system against microorganisms. Some types ingest pathogens and other foreign particles. Some produce antibodies.

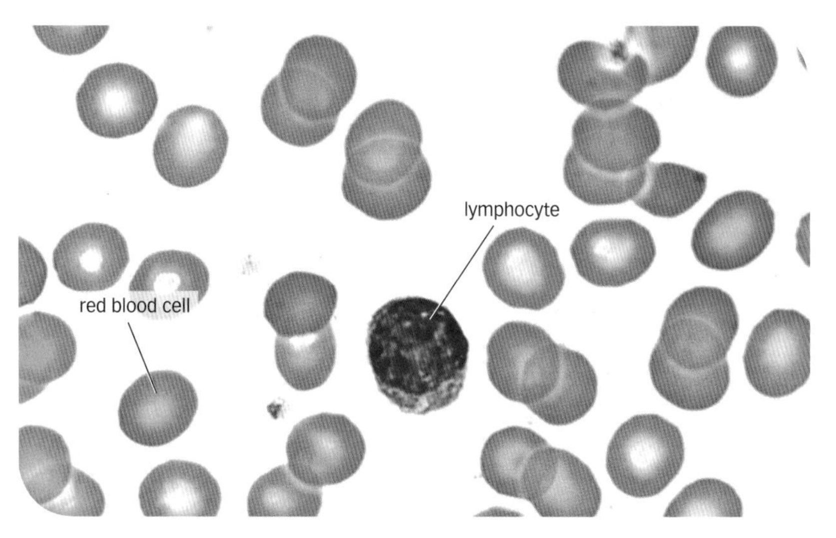

A purple-stained lymphocyte amongst red blood cells, seen under a light microscope. Most lymphocytes are the same size as red blood cells, about 7–8 μm in diameter. Before they produce antibodies they become bigger and may reach 16 μm in diameter. This one is 12 μm in diameter.

Platelets

Platelets are small cell fragments. Platelets do not have a nucleus. They are very important in helping blood to clot at a wound. Each platelet lasts for about a week, but new ones are being made all the time in your bone marrow. You should have about 300×10^9 platelets in each litre of your blood. Old platelets are destroyed by phagocytes (another type of white blood cell) in your liver.

Questions

1 State two functions of white blood cells. (E)

2 Why do you think red blood cells cannot divide?

3 If red blood cells are 7.5 μm in diameter, calculate the diameter of the platelets in the electron micrograph on the right. (C)

4 Red blood cells lack mitochondria. What type of respiration can they use? (A*)

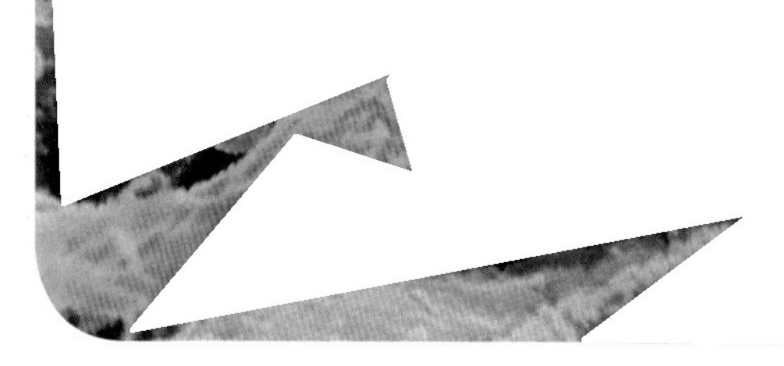

False-coloured scanning electron micrograph of two macrophages in the lungs. One is elongated to engulf (ingest) a round particle. Macrophages in your lungs remove dust particles, pollen, and pathogens.

C Knowing that the lymphocyte in the picture on the left is 12 μm in diameter, calculate the magnification of this light micrograph.

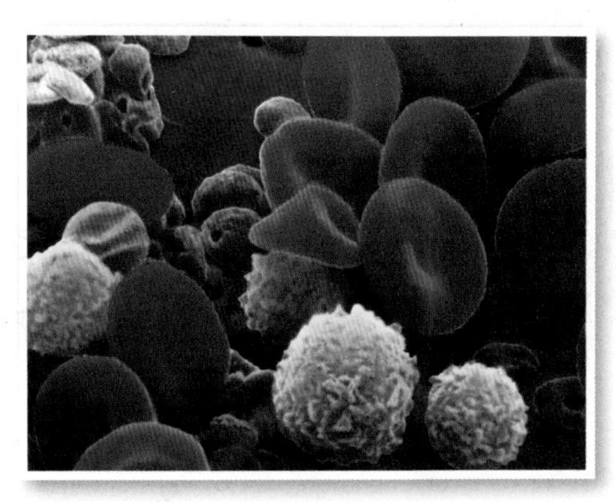

Coloured scanning electron micrograph of blood cells. The small, pink cell fragments are platelets.

Exam tip AQA

✔ If you are asked to 'suggest', or are asked 'why do you think?', you need to use your knowledge from other biology topics that you have studied and apply it to a new situation.

Learning objectives

After studying this topic, you should be able to:

- ✔ know the structure of xylem and phloem, and their role in transport
- ✔ describe the process of transpiration

Key words

xylem, phloem, translocation

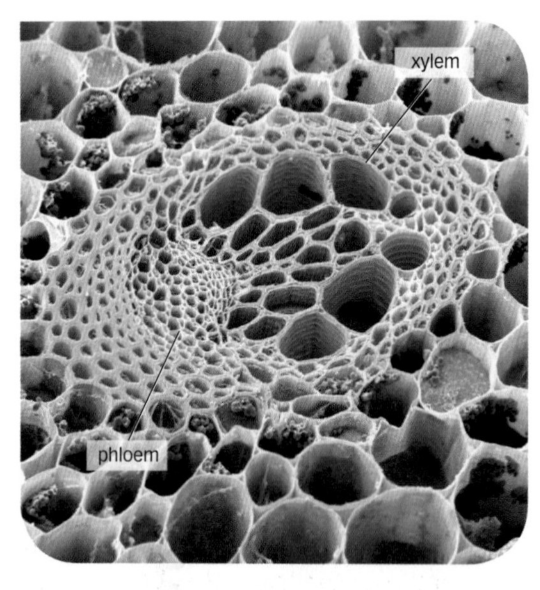

▲ A section through a buttercup stem to show the vascular bundles (× 150)

A Explain why it is important that xylem vessels are hollow.

B Describe the distribution of the vascular bundles in the stem.

Inside a plant

Inside a plant organ are tissues made up of similar cells working together. Two major tissues are **xylem** and **phloem**, which are found in the vascular bundles.

A closer look at vascular bundles

The vascular bundles form a continuous system from the roots, through the stem, and into the leaves. They carry out two major functions:

- transport
- support.

Both xylem and phloem are involved in the transport of water and dissolved substances through the plant.

- Xylem: these cells are dead and stacked on top of one another to form long, hollow, tube-like vessels. Xylem is involved in the transport of water and dissolved minerals from the roots to the shoots and leaves.
- Phloem: these cells are living and are also stacked on top of one another in tubes. They transport the food substances made in the leaf to all other parts of the plant.

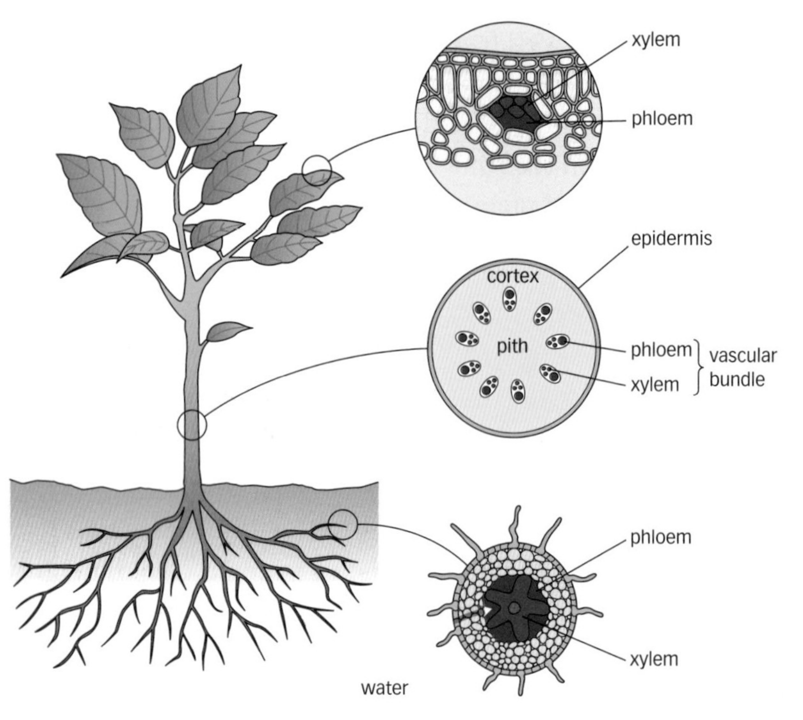

▲ This drawing shows sections cut through the root, stem, and leaf. It shows the different tissues involved in transport around the plant.

Moving substances through the plant

Plants can be very big. They need to move substances from one part of the plant to another. They need to move the water absorbed in the root, and the sugars made in photosynthesis in the leaves, throughout the plant to the parts that need them. Plants move substances in the vascular tissue – the xylem and phloem.

The transpiration stream

Xylem continually transports water and minerals up from the root to the leaf in the transpiration stream.

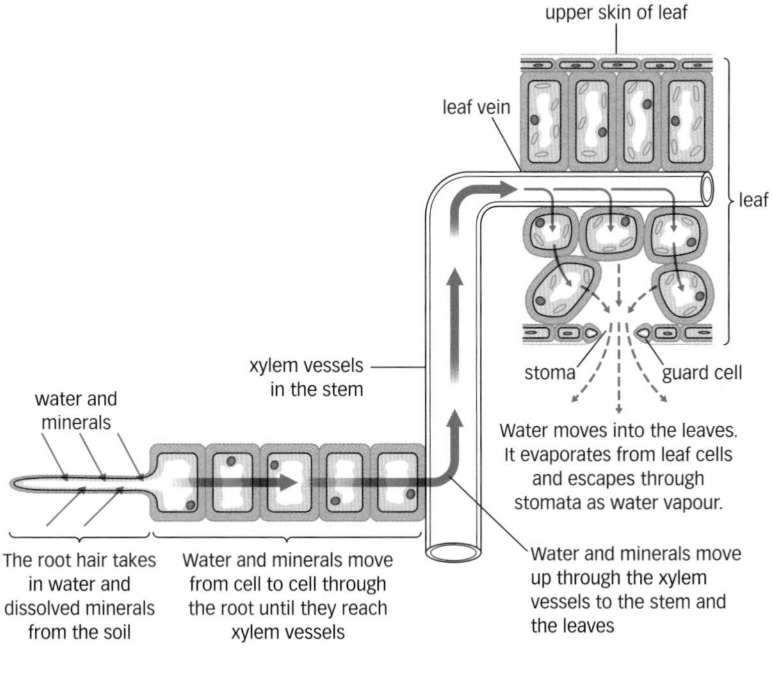

▲ The process of transpiration

Translocation

Phloem transports the sugars made by photosynthesis in the leaf (known as the source) to areas of the plant that are for storage or are still growing (known as the sink). This process is called **translocation**.

Glucose is made in the cells of the leaf by photosynthesis. The glucose is converted to sucrose, which dissolves easily. It is moved into the phloem by active transport. It is then transported to areas of the plant where it is needed, such as the growing tips, or storage areas like fruits.

C What is the name of the process by which water is transported?

D What is the difference between transpiration and translocation?

E Which plant tissue is responsible for the transport of dissolved sugars?

Questions

1. What is: (a) a source, and (b) a sink? (E)
2. Why is glucose converted to sucrose for transport in the phloem? (C)
3. Describe how the distribution of vascular bundles in the plant changes at ground level.
4. The water in the xylem is under tension. Explain how the walls of the xylem are adapted to cope with this. (A*)

Course catch-up

Revision checklist

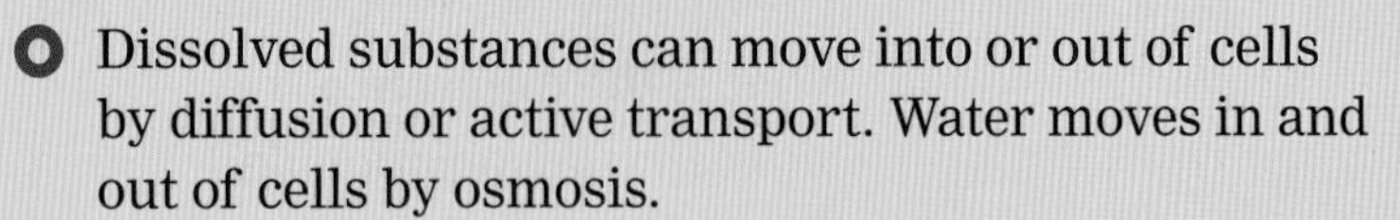

- Dissolved substances can move into or out of cells by diffusion or active transport. Water moves in and out of cells by osmosis.
- Water keeps cells and bodies hydrated. Exercise uses glucose and causes loss of water and ions in sweat. Sports drinks can replace these.
- Active transport moves substances across cell membranes against the concentration gradient, using protein carriers.
- As organisms get bigger, their surface area to volume ratio becomes smaller. They need special exchange surfaces such as lungs, gills, villi, leaves, and roots.
- Exchange surfaces are thin, have a large surface area, may be supplied with blood, and have a concentration gradient.
- Breathing brings air, with oxygen, into the lungs. Gaseous exchange occurs at the alveoli.
- Plants lose water by transpiration (evaporation from stomata).
- Factors such as air movement, light intensity, temperature, and humidity affect the rate of transpiration. Rate of transpiration can be measured with a potometer.
- Large, complex organisms need a circulatory system that transports substances around the body. The circulatory system can be medically assisted.
- Humans have a double circulatory system. Artificial hearts and valves may treat some heart problems.
- Arteries carry blood away from the heart and veins return blood to the heart.
- At body tissues, substances are exchanged between blood in capillaries and cells.
- Blood is a tissue, consisting of different cells suspended in liquid plasma. It is for transport, protection, and temperature regulation.
- Red blood cells contain no organelles and are full of haemoglobin. They carry oxygen. New ones are made in the bone marrow.
- White blood cells are for defence. Some ingest pathogens, some make antibodies.
- Platelets are for blood clotting.
- Plants have a transport system. Xylem carries water and dissolved minerals from roots to leaves; phloem carries food substances made in the leaves to other parts of the plant.

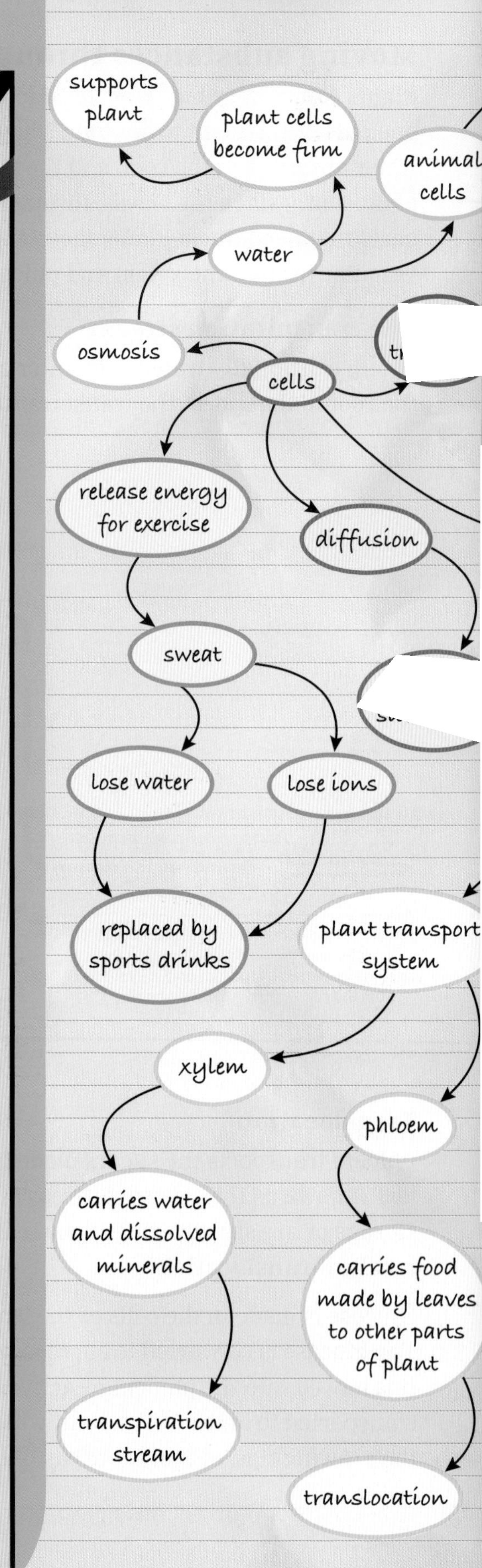

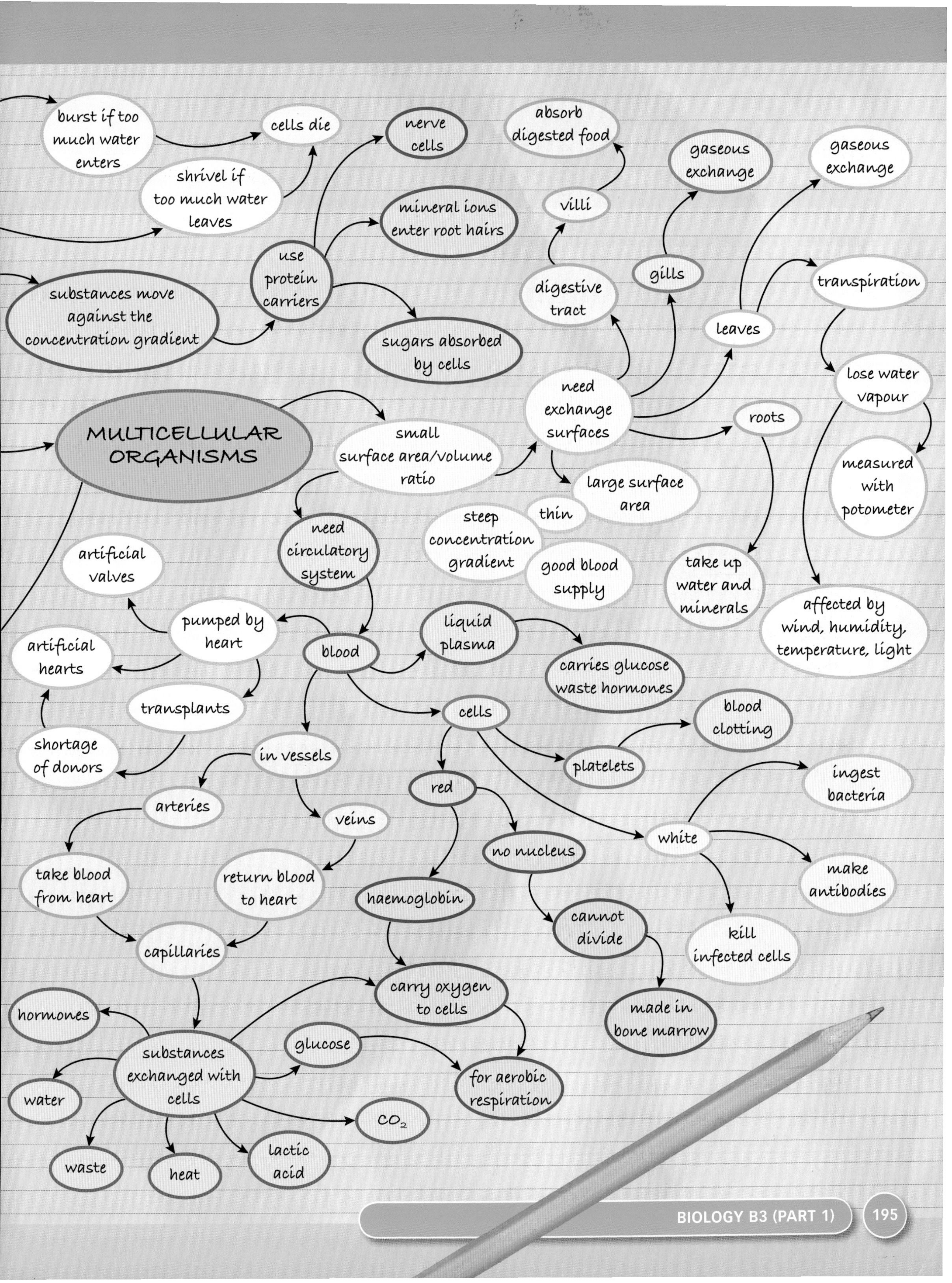
burst if too much water enters
cells die
nerve cells
absorb digested food
shrivel if too much water leaves
gaseous exchange
gaseous exchange
villi
mineral ions enter root hairs
use protein carriers
gills
transpiration
substances move against the concentration gradient
digestive tract
leaves
sugars absorbed by cells
need exchange surfaces
lose water vapour
roots
MULTICELLULAR ORGANISMS
small surface area/volume ratio
measured with potometer
large surface area
thin
steep concentration gradient
need circulatory system
artificial valves
good blood supply
take up water and minerals
affected by wind, humidity, temperature, light
pumped by heart
liquid plasma
artificial hearts
blood
carries glucose waste hormones
transplants
cells
blood clotting
shortage of donors
in vessels
platelets
red
ingest bacteria
arteries
veins
white
no nucleus
make antibodies
take blood from heart
return blood to heart
haemoglobin
cannot divide
kill infected cells
capillaries
carry oxygen to cells
made in bone marrow
hormones
substances exchanged with cells
glucose
for aerobic respiration
water
CO_2
waste
heat
lactic acid

Answering Extended Writing questions

QUESTION

Humans have a double circulatory system, which is more efficient than the single circulation that fish have. The human heart is a double pump. Explain what is meant by a double circulatory system. Explain how the heart acts as a double pump.

The quality of written communication will be assessed in your answer to this question.

G–E

The hart is made of mussel and pumps blood. It pumps blood to the body. Then the blood goes back to the hart. Then it goes to the lungs to get oxygen. Then it goes back to the hart and round the body again.

Examiner: This answer shows some understanding of double circulation. It describes the blood going through the heart twice. However, no chambers are named and it does not make clear that the left and right sides of the heart and separate. There are some spelling mistakes.

D–C

Blood goes from the heart to the body and back again. Then it gets pumped to the lungs to pick up oxygen. Then it goes back to the left atrium of the heart. It then goes through the valve and into the left ventricle and out to go all over the body. It takes oxygen to the cells.

Examiner: This answer demonstrates that the candidate understands what a double circulatory system is. It mentions the left atria and ventricle, but does not make clear that the blood returns from the body to the right atrium and leaves the right ventricle to go to the lungs.

B–A*

A double circulation means there are two circuits from the heart. The blood has to pass twice through the heart. The heart is made of muscle and as the ventricles contract it pumps the blood out.

Blood is pumped from the left ventricle into the aorta and to the body organs. It delivers oxygen. Then the blood goes back, in veins, to the right atrium. It then goes to the right ventricle and to lungs to get oxygen and back to the left atrium.

Examiner: This is an excellent answer. It clearly explains what a double circulatory system is. By saying that the blood passes through the heart twice and that the heart is made of muscle that contracts, it explains how the heart acts as a double pump. It adds more detail about double circulation by describing the route taken by the blood from heart to body to heart, and then to lungs and back to the heart.

Exam-style questions

1 The diagram shows a small part of a lung.

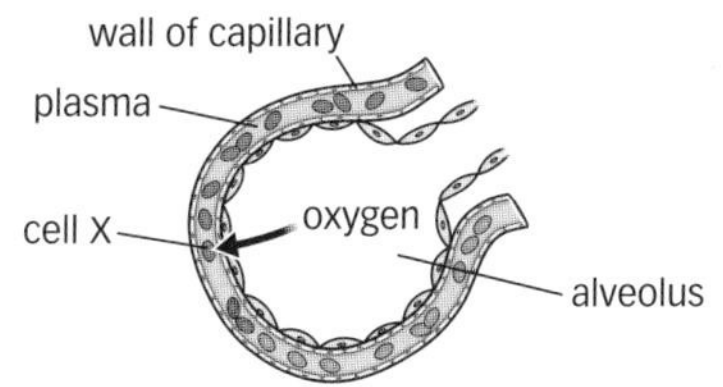

A01 **a** What type of cell is cell X?

A01 **b** By what process does oxygen move from the alveolus to cell X?

A01 **c** What substance in cell X does oxygen combine with?

A01 **d** Describe how the lungs are well adapted for gaseous exchange.

G–E

2 A group of students looked at stomata on the leaves of four plant species.

Plant species	Estimated number of stomata per cm² of leaf surface	
	upper surface	lower surface
A	4000	28000
B	0	800
C	8500	15 000
D	8000	26000

A03 **a** Which plant lives in a dry region?

A02 **b** Suggest why all four species of plant have more stomata on the underside of their leaves.

A01 **c** What environmental factors, besides water availability, affect the rate of transpiration in plants?

A01 **d** Name a piece of apparatus that can be used to measure the rate of transpiration.

D–C

3 The graph shows blood pressure measurements for a person at rest.

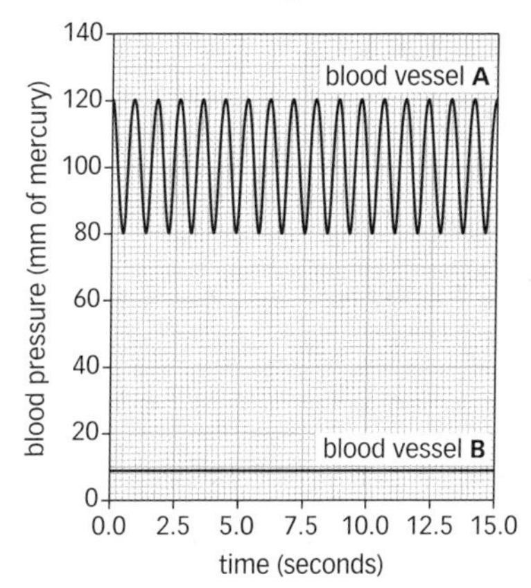

A03 **a** Which blood vessel, A or B, is an artery? Give two reasons.

A03 **b** How many times did the heart beat in: **i** 15 seconds, and **ii** a minute?

A02 **c** At tissues the blood vessels form a network of capillaries. Name two useful substances supplied to muscle cells by capillaries when a person runs.

A01 **d** Name two waste substances that pass from muscle to capillaries.

B–A*

Extended Writing

4 Explain why large multicellular organisms need a gaseous exchange system and a circulatory system. A01 (G–E)

5 Describe how materials are transported in plants. A01 (D–C)

6 Describe the structure and functions of blood. A01 (B–A*)

A01 Recall the science
A02 Apply your knowledge
A03 Evaluate and analyse the evidence

B3 Part 2

Regulating the human and natural environments

AQA specification match

Why study this unit?

In this unit you will find out about regulation of the internal conditions within the human body, as well as the impact humans have on their external environment. We all need to understand how our activities can disrupt the balance of the external environment, as these disruptions will affect other living organisms, and ultimately affect us.

In this unit you will learn how your body temperature and the amount of salt and water in your body are regulated, and what might happen if they are not properly regulated. You will also learn how we can grow enough food to feed our growing population, and how we can meet our increased demand for fuel as fossil fuels run out. These activities will inevitably affect the environment, and you will learn how we may minimise and repair any environmental damage.

You should remember

1. You are made of cells organised into tissues, organs, and systems – such as the reproductive system.
2. Plants make food by photosynthesis, and form the basis of nearly all food chains.
3. Consumers may be herbivores or carnivores. Humans are omnivores, as they eat plants and meat.
4. Farming alters the environment and may disrupt food chains.
5. Human activities may have significant effects on the environment.

Research on marine wood borers – has taken a new turn. Researchers hoping to find ways to reduce the damage done by these wood-eating organisms have instead discovered a means of producing biofuels more efficiently. They found that these creatures can produce an enzyme that digests cellulose in plant waste, such as agricultural waste products from crops like maize (pictured) more efficiently than the cellulase enzymes that are currently used. The gene for the wood borers' enzyme has been sequenced, and it is hoped that it can be mass-produced and used in the production of biofuels such as bioethanol from maize. This research is an example of serendipity – research into one topic has produced valuable information about a totally different aspect with an important application.

B3 15: Keeping internal conditions constant

Learning objectives

After studying this topic, you should be able to:

- ✔ understand the need to remove toxic waste products from the body
- ✔ understand why body temperature and blood glucose must be kept within narrow limits

Key words

excretion, **urea**

Exam tip 

- ✔ Do not confuse the terms secretion and excretion. Glands secrete useful substances. Excretory organs excrete toxic waste made in your body.

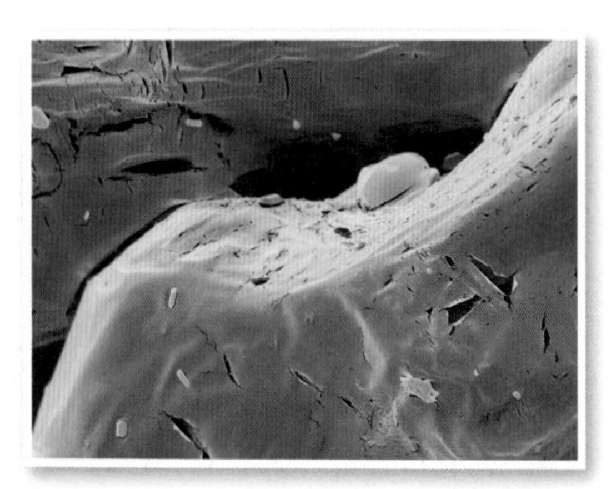

▲ Electron micrograph of urea, a waste product excreted by the kidneys in dissolved form

Staying in balance

In each of your cells there are always many chemical reactions going on. Many of these reactions produce toxic by-products. If these toxins were not removed from your body, you would die. Removal of toxic substances made in your body is called **excretion**.

Internal conditions that are controlled include:

- pH
- water content
- ion (salt) content
- temperature
- blood sugar levels.

Carbon dioxide

Respiring cells produce carbon dioxide. If carbon dioxide accumulated it would lower the pH of your cells, tissue fluid, and blood. The drop in pH would interfere with enzyme action and could also change the shapes of other protein molecules such as receptors on cell membranes. This would be very harmful.

The carbon dioxide produced in respiring cells diffuses into your blood to be carried to the lungs. It diffuses from your blood into your alveoli, from where you breathe it out.

A If your muscle cells respire anaerobically, what waste product do they make?

B How might this be harmful if it were not carried by the blood to the liver to be broken down?

Urea

Your liver cells break down excess amino acids and make ammonia. This is very alkaline and would raise the blood pH to a dangerous level. In liver cells, ammonia reacts with carbon dioxide to make **urea**. This is still alkaline, but is less toxic than ammonia and is carried in your blood to your kidneys in order to be removed.

Water and ions

You take water and ions into your body when you eat and drink. Your blood plasma is about 90% water. It also contains ions. If there were too much water in your blood, your blood cells would take in water by osmosis and would swell and burst. If there were too little water in your blood, your blood cells would shrivel and not function. Water would also leave your body cells to go into your blood. Your body cells would dehydrate and not be able to carry out their functions.

▲ Food and drinks contain water and ions

Temperature

Your body temperature may fluctuate slightly depending on your age, the time of day, and how active you are, but it is kept within narrow limits. Normally, human body temperatures are between 36 °C and 37.5 °C.

Most of your enzymes would still work at temperatures a bit higher than these, but other proteins in your body and your cell membranes would be damaged if your temperature rose above 40 °C. Humans, like other mammals, can regulate their body temperature so it stays fairly constant regardless of the outside temperature.

Blood sugar

Your blood should always have about 900 mg of sugar (glucose) in each litre of blood. Your cells need a continuous supply of glucose for respiration. This is delivered to them by your blood. If your blood sugar level dropped, you would feel faint and tired. If there was too much sugar in your blood, water would leave your body cells by osmosis. This would make your cells dehydrate.

Did you know...?

Some types of cells in your body can respire other substances as well as glucose, such as fatty acids. However, your brain cells can only respire glucose. Having the right amount of sugar in your blood is crucial for your brain to function properly.

Questions

1 Explain why your body temperature has to be regulated. (E)

2 Explain why your blood sugar level has to be regulated. (C)

3 How do you think raising the blood pH would be harmful to your body?

4 By what process would water leave your body cells? (A*)

16: The kidney

Learning objectives

After studying this topic, you should be able to:

- understand the role of the kidney in producing urine

Key words

filtration, **reabsorption**

Urea

Proteins are digested into amino acids, which are absorbed into the blood from the small intestine. If you eat more protein than you need for growth and repair, the excess amino acids are carried in the blood to your liver. Here, your liver cells convert them to ammonia. The ammonia then reacts with carbon dioxide to make urea. The urea passes into your blood and is carried to your kidneys to be removed from the body in the urine.

Urea is toxic because it is alkaline. If it accumulates in your blood, it will make the blood far too alkaline.

How your kidneys filter your blood

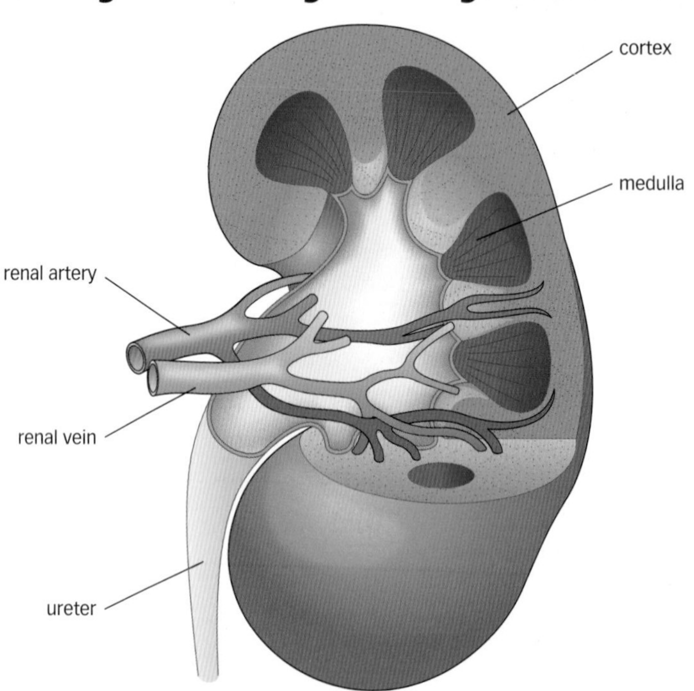

The structure of the kidney. Blood flows in through the renal artery, is filtered and leaves the kidney in the renal vein. The waste urine, containing water, salts, and urea, passes down the ureter to the bladder.

A Describe how urea is made in the body.

B Explain why urea has to be removed from the body.

C How is urea transported from where it is made to the kidneys?

Did you know...?

Your kidneys help regulate the pH of your blood. Besides removing alkaline urea, they also remove excess hydrogen ions that would otherwise make the blood more acidic. So the pH of urine can vary. Your kidneys also produce some hormones, one of which stimulates the production of red blood cells in your bone marrow.

Each of your kidneys has about one million **filtration** units. Blood enters your kidney in the renal artery. It is filtered under high pressure and lots of substances are filtered out of your blood:

- glucose
- salts (ions)
- water
- urea.

Then the useful substances:

- all the glucose
- the ions that are needed by your body
- as much water as your body needs

are **reabsorbed** into the blood. The remaining liquid, called urine, contains urea, excess ions, and water. It passes down the ureter to the bladder. It is stored in the bladder and passed out when convenient.

As well as removing urea, your kidney also regulates the amount of salt and water in your body.

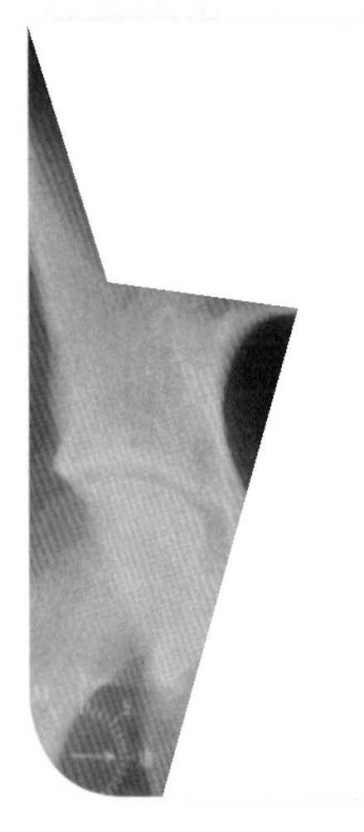
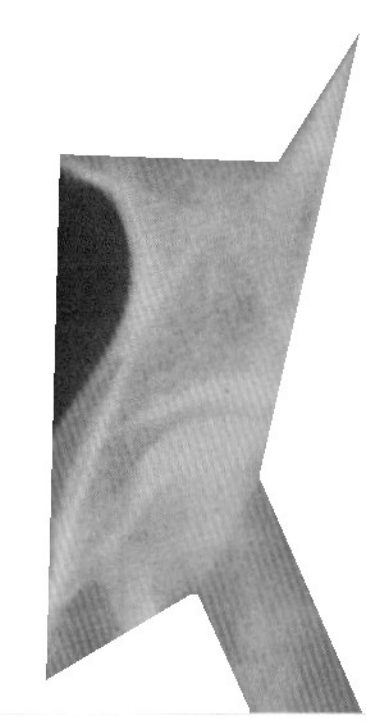

A coloured X-ray urogram of urine collecting in the bladder

Questions

1. Why is all the glucose that was filtered out of the blood by the kidney reabsorbed?

E

2. Why do you think it is harmful for your blood to become more alkaline?
3. Explain why it is important for the amount of salts (ions) and water in your body to be regulated.

C

4. How do you think the amount of urea in your urine would change if you started eating a high-protein diet? Explain your answer.
5. How do you think the water content of your urine would change if you drank a lot of tea and lemon squash? Explain your answer.
6. How do you think the water content of your urine would change on a hot day if you ran around and did not drink any extra water? Explain your answer.
7. How do you think your urine would change after you ate a very salty meal?

A*

Exam tip AQA

✔ You do not need to learn the structure of the kidney. The diagram on the opposite page is to help you understand how the kidney filters blood to remove urea, and reabsorbs the useful substances.

Learning objectives

After studying this topic, you should be able to:

- ✔ understand how renal dialysis filters the blood of patients with renal failure

Key words

renal, dialysis

Renal failure

Renal failure is a condition where your kidneys fail to function adequately. 'Renal' means relating to the kidney. There are many causes of renal failure.

Acute renal failure

Sometimes the condition is acute (sudden and short lived). Acute renal failure is usually caused by an infection or drugs, and the patient will recover with treatment. The patient may need dialysis whilst recovering.

Chronic renal failure

Renal failure may be chronic (long lasting), and the patient will not recover. Chronic renal failure may develop slowly. It may be caused by overuse of some medicines, by diabetes, or it may be genetic.

In cases of chronic renal failure the kidneys do not filter efficiently, and urea accumulates in the blood. The ion balance of the body is disrupted. There are many symptoms, including:

- feeling sick and losing weight
- muscle paralysis
- back pain
- anaemia
- swollen ankles, feet, face, or hands.

People with chronic renal failure need regular renal **dialysis**, which filters their blood artificially. They may have dialysis while waiting for a kidney transplant to replace the damaged kidney, which cannot be repaired or cured.

A What is renal failure?

B Explain the difference between acute and chronic renal failure.

Renal dialysis

The patient's blood is taken from their forearm and passed via tubes into a dialysis machine. An anticoagulant such as heparin is added to the blood to prevent clots forming whilst the blood flows through the machine.

Inside the machine, the blood flows on one side of a partially permeable membrane. On the other side of the membrane is the dialysis fluid, which contains

- water and ions of the same concentration as should be in your blood
- glucose in the same concentration as should be in your blood.

Exam tip

- ✔ Remember the difference between an acute illness and a chronic illness. An acute illness begins suddenly but can be cured. A chronic illness is of slow onset and is treatable but not curable.

The dialysis fluid and the membranes have to be sterile (very clean), so there is no risk of infection. The fluid is at body temperature, to prevent the blood losing or gaining heat as it passes through the machine.

The patient's blood passes through the machine several times over a period of about six hours, three to four times a week. During this time the patient cannot move around, so often the dialysis is done at night, whilst the patient sleeps.

If the patient's blood contains excess salts and water, these will diffuse across the partially permeable membrane into the dialysis fluid. Glucose will not pass across the membrane, as it is at the same concentration in the dialysis fluid as it is in the patient's blood.

Patients with renal failure have to regulate their diet carefully. They must limit their fluid, salt, and protein intake. However, whilst they are actually undergoing dialysis they can treat themselves to extra drinks and salty snacks.

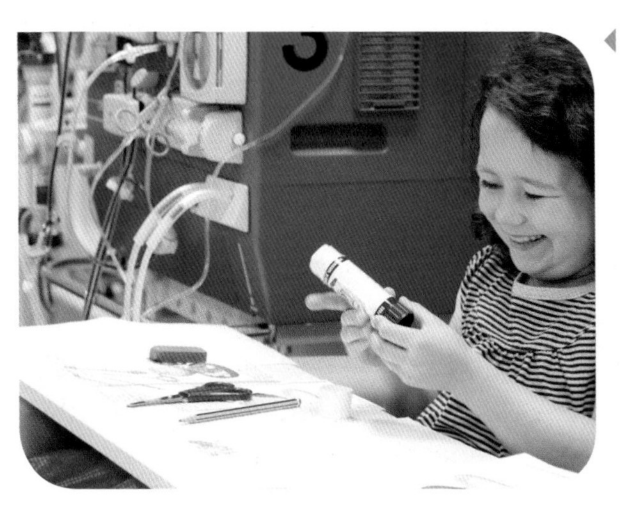

This girl is undergoing renal dialysis. Tubes take blood from her forearm, into the machine. Here the blood is filtered and then passes back into her arm.

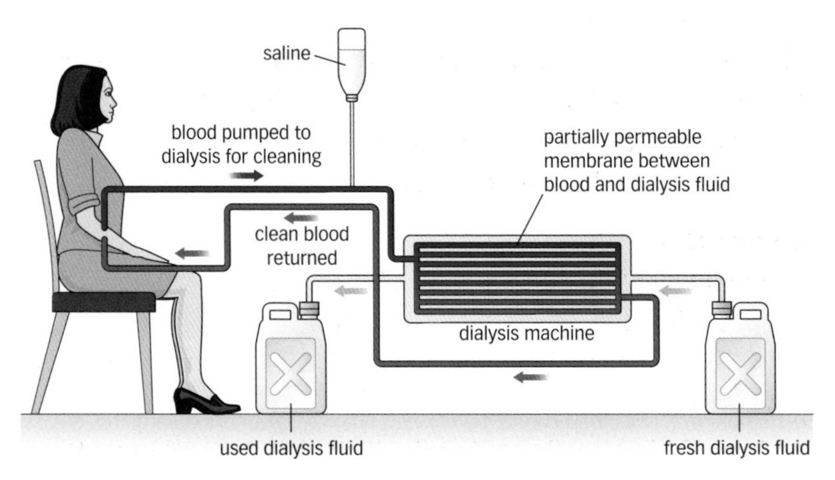

How a renal dialysis machine works

Did you know...?

A Dutch doctor made the first improvised dialysis machine in 1943. He used sausage skins, drinks cans, and a washing machine. He treated 16 patients, but none successfully. Then in 1945 he successfully treated a 67-year-old woman who survived acute renal failure and lived for a further seven years, before dying of an unrelated condition.

Questions

1 Why does the renal dialysis fluid and the rest of the machine have to be sterile?

2 Why is the dialysis fluid at body temperature?

3 Why is heparin added to the patient's blood before the blood enters the dialysis machine?

E

4 Explain why the dialysis fluid contains glucose in the same concentration as in healthy human blood.

5 Why do you think the renal dialysis fluid has to be changed after a dialysis session?

C

6 Why do you think people with renal failure get swollen ankles, feet, legs, and face?

7 Why do you think people with renal failure become anaemic?

A*

18: Kidney transplants

Learning objectives

After studying this topic, you should be able to:

- understand the advantages of a kidney transplant
- understand the problems of rejection and how they may be overcome

Key words

donor, antigen, tissue-typing, rejection, immunosuppressant

Advantages of kidney transplants

You can survive with just one kidney. So if someone has renal failure and needs a transplant, only one kidney needs to be transplanted. The donated kidney could be taken from a close living relative, or the **donor** may have recently died.

If the kidney is from an unrelated person, tests are done to see if the tissue types of donor and recipient are similar. The donor and recipient should also be of the same blood group, and the organs may need to be matched for size.

The recipient's kidneys may be left in place and the donated kidney implanted in the abdomen, with its blood vessels joined to the recipient's iliac artery and vein. If the recipient had kidney cancer, the diseased kidney would be removed.

Having a functioning kidney inside your body means that you can live a normal life, as you do not need to be connected to a dialysis machine several times a week. You can go on holiday, for example. It is also cheaper for the NHS to carry out an operation than to provide a dialysis machine for you in the long term. However, the problem of rejection has to be overcome, and enough suitable donors have to be found.

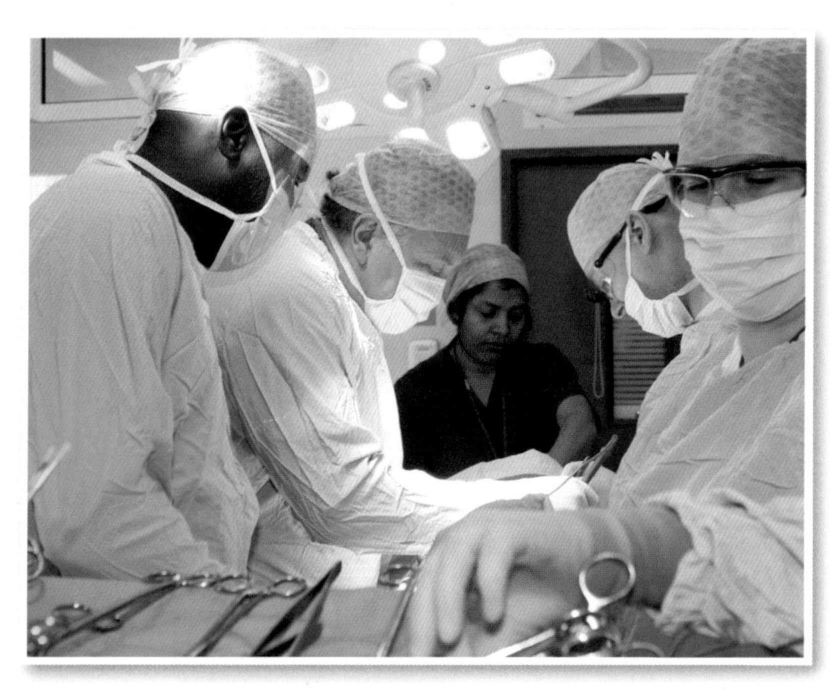

▲ A surgical team performing a kidney transplant operation. This is the safest and most commonly performed type of organ transplant. It has a high success rate, although patients have to take immunosuppressant drugs for the rest of their life to prevent rejection.

A What are the advantages of having a kidney transplant, compared with renal dialysis?

B What precautions does the medical team take to reduce the risk of rejection, when transplanting a kidney?

Rejection

You have proteins called **antigens** on the surface of all of your cells. Every one of us, except for identical twins, has particular antigens, unique to us. Your antigens may be similar to those of some of your close relatives. They may also, by chance, be similar to those of some unrelated people. **Tissue-typing** is about matching the antigens of donors and recipients.

Even with a close match, your immune system can recognise that the antigens on the transplanted kidney are not your own. Your white blood cells and antibodies may attack and destroy the cells of the transplanted kidney. The transplanted organ would cease to function and would die. This is called **rejection**.

Doctors give transplant patients **immunosuppressant** drugs. As their name suggests, these suppress the recipient's immune system to prevent rejection. If your immune system is suppressed, you are more at risk of infection. However, you can try and avoid contact with people who have colds and flu, and the benefits of having a successful kidney transplant outweigh the risks of infection.

Many people have had transplants and live healthy lives afterwards. Every year in the UK, a game of cricket is played between the England Transplant team and the Australian Transplant team. Every player has had at least one transplant (heart, kidney, liver, or heart–lung) and they are all fit and healthy.

Did you know...?

The most widely used immunosuppressant drug for transplant patients, cyclosporin, was discovered by 'accident'. In science this is called serendipity. It means making a useful discovery whilst looking for something else. In 1972 two Swiss doctors were testing cyclosporin as a treatment for cancer. They gave it to mice and found that it suppressed their immune systems. Cyclosporin is a short protein made by a fungus that lives in soil. It was approved for use as an immunosuppressant in 1983.

Questions

1 Why are transplant patients more at risk of infection? (E)

2 Explain why a recipient may reject a transplanted kidney. (C)

3 Some people carry a card and join the Organ Donor Register to say that they are willing to donate organs if they die. It is suggested that we should all be viewed as potential donors, and instead have to carry a card if we do *not* want to donate our organs. What are your views on this idea? (A*)

Exam tip

✔ For questions like Question 3, think of some pros and some cons. Then, if possible, reach a conclusion.

Learning objectives

After studying this topic, you should be able to:

- ✔ understand the role of the thermoregulatory centre in the brain and the temperature receptors in the skin
- ✔ describe the changes that bring the body temperature back to normal when it is too high

Key words

thermoregulation, **nerve impulses**, **evaporation**, **vasodilation**, **radiation**

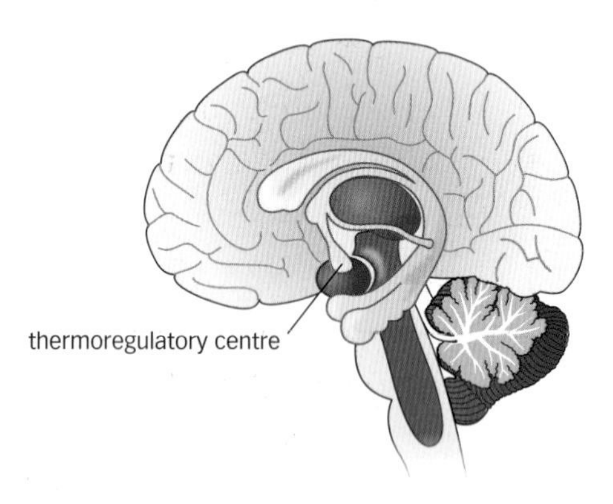

▲ The thermoregulatory centre (hypothalamus) in the brain

Did you know...?

If you overheat too much, your thermoregulatory centre shuts down. You get hotter and hotter and you would eventually die of heat stroke. You need to be immediately immersed in cool water and to drink lots of fluids, with some salt added.

Your body temperature

The temperature of the human body is normally kept at about 37 °C. This is the core temperature. At your fingers and toes it will be cooler. It is dangerous if your body temperature rises above 40 °C. **Thermoregulation** means keeping the body temperature within safe limits.

The thermoregulatory centre

A particular area of your brain, the thermoregulatory centre, monitors your blood temperature. It can do this because your blood flows through the brain every two or three minutes as it circulates your whole body. This area of the brain also receives information via nerves from your skin. You have special nerve endings in your skin that can detect changes in the external temperature. If your body temperature rises or falls, the thermoregulatory centre stimulates your body to make adjustments. This part of your brain acts as a thermostat. It consists of a heat-loss area and a heat-promoting area.

When you overheat

You may overheat when you

- exercise
- are dehydrated
- are exposed to very high external temperatures for too long.

Your blood carries heat away from respiring muscle tissue. As it flows through the brain your thermoregulatory centre detects the raised temperature. It then activates the heat-loss area, which causes more blood to flow to your skin. Your skin looks red because more blood flows through the capillaries, and more heat is lost.

How else does the thermoregulatory centre bring about heat loss?

Increased sweating

Nerve impulses stimulate the sweat glands in your skin to release large amount of sweat. This pours out onto your skin surface. The water in the sweat evaporates, which uses heat energy from your skin and blood. This **evaporation** cools you. It works best if the air is dry. In humid weather we feel hot and sticky because our sweat does not evaporate.

We can also seek shade or use a fan to move the air around us and promote evaporation and cooling. Removing some clothing may also help.

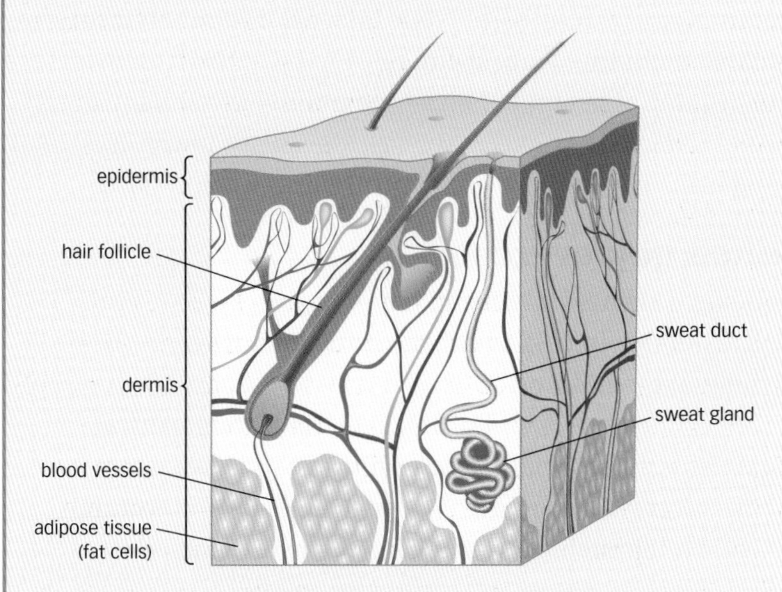

▲ Section through human skin showing sweat glands and sweat ducts. The sweat glands extract water and salts from the blood vessels. More sweat pours on to the surface of the skin when your body is overheating. Evaporation of the water in sweat takes heat from your body and cools you.

Vasodilation

Nerves stimulate the blood vessels (arterioles) that supply the capillaries in the skin to dilate (widen). This **vasodilation** allows more blood to flow though the skin capillaries. The excess heat from the blood is lost by **radiation**. Your skin looks flushed and feels hot because more blood is flowing just beneath it.

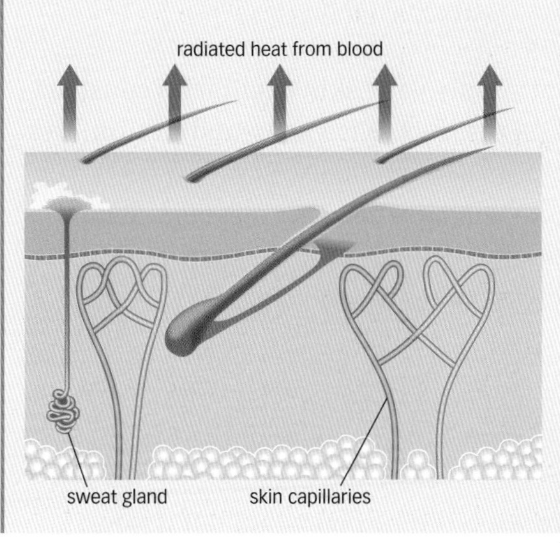

◀ Blood vessels in the skin dilate and heat is lost by radiation. Your skin looks flushed and feels hot.

A Explain why exercise can cause you to overheat.

B Why do you think being dehydrated could cause you to overheat?

Questions

1 Explain how increased sweating can cool you.

2 Explain why your skin looks red and feels hot when you are overheating.

3 Why do you think you should drink more water during hot weather?

E

4 How does the thermoregulatory centre in your brain stimulate the blood vessels in your skin to widen?

C

5 Why do you feel cold when getting out of a swimming pool, even on a hot day?

6 Dogs do not sweat. Instead they pant. How do you think panting cools them?

A*

7 Elephants and rabbits have large ears. When they are hot, more blood flows through their ears. How do you think this cools them?

Learning objectives

After studying this topic, you should be able to:

- ✔ describe the changes that bring the body temperature back to normal when it is too low

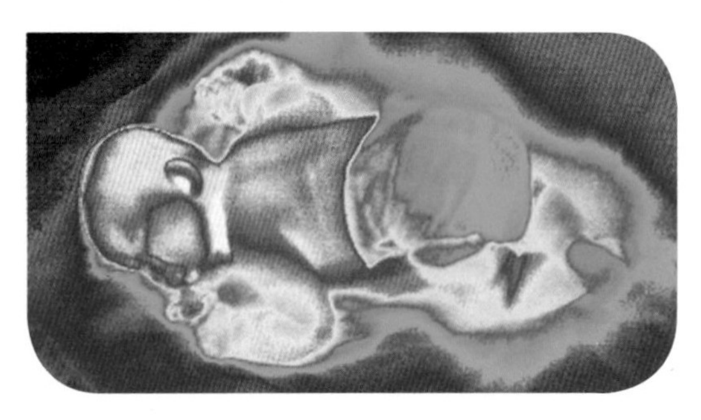

▲ Thermogram of a baby. The hottest areas are white. The temperature scale then goes red, yellow, green, blue, and purple (coldest).

A Why do you think it is dangerous if your core body temperature drops below 35°C?

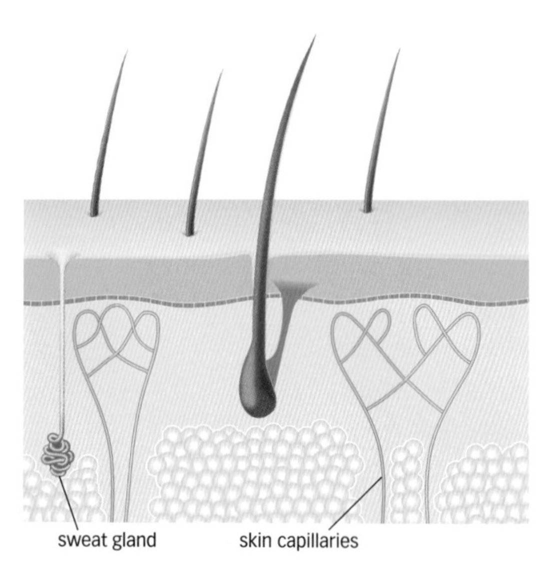

▲ Blood vessels supplying skin capillaries constrict so less blood flows near your skin surface. This reduces heat loss. Core body heat is conserved.

Core and extremities

When we talk about body temperature we are referring to the temperature of your core – your internal organs. Your internal organs, including your brain, need to have their temperature maintained at around 37°C all the time. The temperature at your extremities (fingers and toes) is always lower. It is dangerous if your core temperature drops below 35°C.

When you overcool

You may lose heat when you

- are outside in cold weather for too long without good insulating clothing
- are immersed in cold water, even for a short period
- are elderly, especially if your home is not adequately heated
- are a baby, as you have a large surface area to volume ratio and much heat is lost through your surface.

If you lose too much body heat, your thermoregulatory centre initiates mechanisms to adjust your body temperature back to normal. It will stimulate heat-promoting mechanisms.

Reduced sweating

If you are cold, your sweat glands produce very little sweat. This reduces heat loss by evaporation.

B Explain why reduced sweating prevents you from overcooling.

Vasoconstriction

The arterioles that supply your skin capillaries constrict (get narrower). As a result of this **vasoconstriction**, less blood flows into your skin capillaries. Less heat is lost by radiation. Your skin will look pale and feel cold. This conserves heat at your core and protects your internal organs from heat loss and damage.

However, in extreme cases of prolonged exposure to cold, you could lose fingers or toes through frostbite. The cells at your extremities would die, as the reduced blood supply means that less oxygen and glucose reaches them. You could still live with fewer fingers, but not if your heart, liver, or brain died.

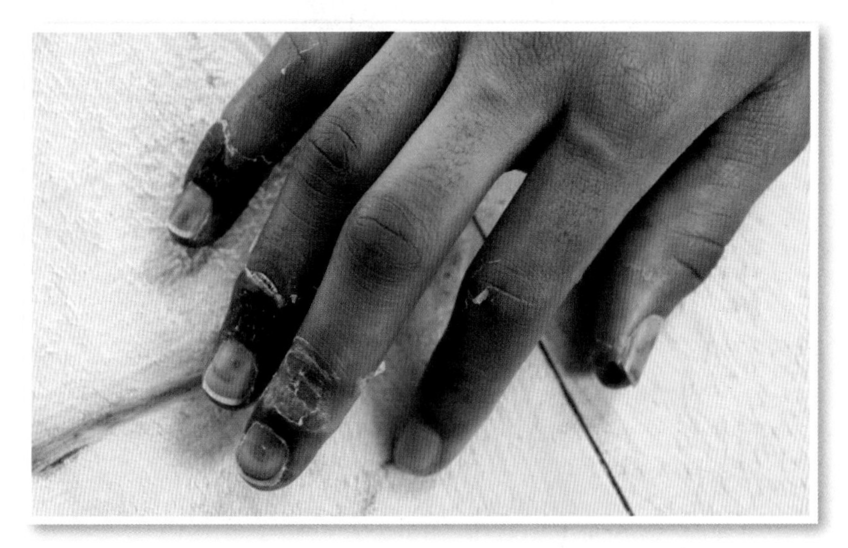

The fingers of a young man with severe frostbite after a blizzard

Shivering

Your thermoregulatory centre stimulates **shivering** in your skeletal muscles. They contract and relax quickly to give a shuddering motion that is involuntary – it is beyond your control. Muscles need energy to contract, so they respire more when shivering. Respiration always releases some of its energy as heat, so increased respiration releases more heat.

You can of course also alter your behaviour to avoid overcooling, by putting on more clothes, moving about more, eating hot food or drinks, and hunching up and placing your hands under your arms to keep them warm.

Questions

1 State two reasons why you may overheat. (E)

2 If your core body temperature drops below 35°C, you have hypothermia. Your heat-generating mechanisms do not get activated and you continue to lose heat. This would be fatal if not treated by being wrapped in an insulating blanket and gently warmed up. Why do you think untreated hypothermia would be fatal? (C)

3 Why do you think people with more brown fat are thinner than people with less brown fat? (A*)

Key words

vasoconstriction, shivering

Did you know...?

Babies do not shiver. Instead they have special fat called brown fat. The fat cells contain lots of mitochondria that contain iron, so this fat looks reddish/brown. When cold, babies respire this fat in a way that generates more heat than usual. Now scientists have found that adults also have some brown fat in the neck region. Thinner people have more brown fat than heavier people. Heavier people have more white fat that is not respired so readily.

Exam tip

- Remember, 'hypo' means below, as in a hypodermic (below the skin) needle. So hypothermia is a reduced body temperature.

21: Regulating blood glucose levels

Learning objectives

After studying this topic, you should be able to:

- ✔ understand the roles of the pancreas, insulin, and glucagon in controlling blood glucose levels

Key words

glucose, insulin, glucagon

▲ Testing blood glucose level

Why does your blood glucose need to be regulated?

All your body cells need to respire the sugar **glucose**, to release energy for metabolic reactions, such as:

- making proteins including enzymes, antibodies, and haemoglobin
- muscle contraction
- cell division
- replication of DNA
- active transport of substances into and out of cells
- nerve action.

Glucose is your cells' energy source. Your blood carries glucose, so it is vital that it always has enough to deliver to all cells. However, it should not contain too much glucose, as this would make the blood too concentrated, and water would leave your body cells by osmosis. Your body cells would dehydrate and their enzymes would not function. The cells would not be able to carry out their chemical reactions.

Your blood glucose level has to be kept within narrow limits.

Your blood glucose level rises after you have eaten a meal. The glucose from digested carbohydrates is absorbed from the small intestine into your blood.

A Explain why your blood glucose level rises after you have eaten a meal.

B Explain why your blood glucose level drops when you have not eaten for several hours.

C Explain why your blood glucose level drops after you have been swimming.

The pancreas

Your pancreas sits within the U bend, or loop, of your duodenum. It is a double gland – as well as making digestive enzymes, it also makes hormones. As your blood flows through it, the pancreas monitors the blood glucose level.

If your blood glucose level rises above normal

Special cells in your pancreas secrete (make) the hormone **insulin**. The insulin passes straight into your bloodstream and is carried to all cells, including its target cells. Most cells in your body respond to insulin. Insulin causes the cells to take up more glucose from your blood. In some cells the glucose is used for respiration. In other cells (in the liver and muscles) the glucose is stored as glycogen (a large carbohydrate molecule). This taking up of more glucose into the cells lowers the blood glucose level back to normal.

If your blood glucose level drops below normal

However, as your blood continues to deliver glucose to respiring cells, if you have not eaten for a while, your blood glucose level may drop below normal. As your blood flows through the pancreas, this lower level of glucose is detected. Some other special cells in your pancreas make another hormone called **glucagon**. This hormone passes straight into the blood and is carried to all your cells, including its target cells. It causes your liver cells to break down some of their stored glycogen and release glucose into your blood, to top it up to normal levels.

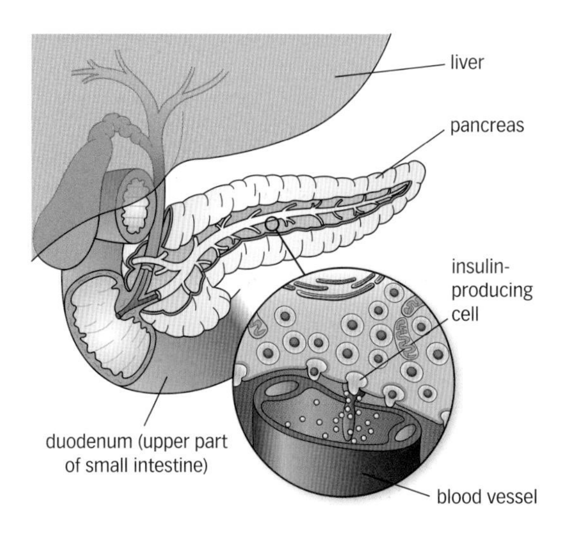

▲ The position of the human pancreas. Special cells secrete insulin straight into the bloodstream.

Did you know...?

Muscle cells also store glycogen, but it is only for their use. Their glycogen does not get released as glucose into the blood.

Questions

1 Which organ monitors your blood glucose level?

2 Which hormone is released into your blood when your blood glucose level rises above normal levels?

3 Which hormone is released into your blood when your blood glucose levels are lower than normal?

E

4 Explain how the hormone in Question 2 causes your blood glucose levels to drop back to normal.

5 Explain how the hormone in Question 3 causes your blood glucose level to rise back to normal.

C

6 Which of the following are carbohydrates and which are proteins?

glucagon   glucose   glycogen   insulin

7 Draw a flow diagram to show how your blood glucose level is regulated.

A*

Exam tip

- ✔ Remember that glucagon is made when your glucose is low – when your glucose has gone.
- ✔ If you talk about *glycogen* being broken down to glucose, do not confuse it with *glucagon*. You must get the spellings correct.

22: Type 1 diabetes

Learning objectives

After studying this topic, you should be able to:

- ✔ know the causes and treatments for type 1 diabetes

Key words

diabetes

Exam tip AQA

- ✔ Always refer to type 1 diabetes if that is what you are talking about. There are other types of diabetes: type 2 diabetes and diabetes of pregnancy.

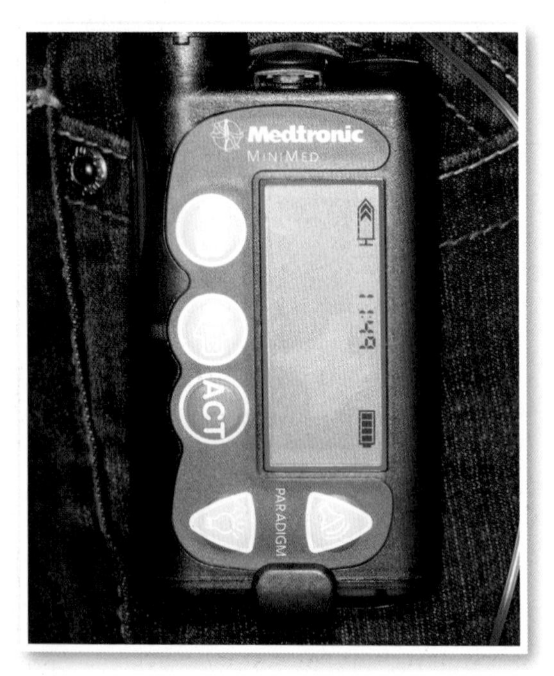

▲ A digital insulin pump. This can be attached to a belt and carried around by a person with diabetes. It delivers constant very small doses of insulin, and a larger dose before meals or if the blood glucose level is high.

What happens if you do not make enough insulin?

Some people do not make enough insulin in their pancreas. Their blood glucose level is not controlled. They suffer from a type of **diabetes** called type 1 diabetes.

If you do not make enough insulin, your blood glucose level will stay high after a meal, because there is no insulin to make your body cells take up the extra glucose.

When your blood is filtered by your kidneys, too much glucose is filtered out and the kidneys do not reabsorb it all.

Glucose passes out in your urine. A lot of water also goes with it, so you urinate more frequently and feel thirsty because your blood lacks water.

A Explain why not producing insulin will lead to a permanently high blood glucose level.

B What are the symptoms of type 1 diabetes?

How is type 1 diabetes treated?

People with type 1 diabetes have to pay careful attention to their diet. They should

- eat regular meals containing plenty of fibre and other complex carbohydrates, but not too much sugar. These complex carbohydrates are digested to sugar, but only slowly, and so after a meal the sugar is absorbed more steadily into the blood. More protein and unsaturated fats should replace sugary foods.
- take regular exercise
- regularly monitor their blood glucose level and inject insulin into their blood at mealtimes
- avoid alcohol.

Long-term consequences of diabetes

Careful treatment and management of type 1 diabetes is very important, as it can shorten the person's life expectancy. High blood glucose levels damage red blood cells and blood vessels.

Without proper control, diabetes can lead to

- blindness due to damage to the retina
- poor wound healing
- ulcers, which may lead to gangrene and loss of toes or feet
- increased risk of stroke
- increased risk of heart attack.

People with diabetes test their blood to see how much glucose is in it.

What causes type 1 diabetes?

In people with type 1 diabetes, the cells of their own immune system have destroyed the special cells in their pancreas that produce insulin.

Scientists are not certain why, in some people, the immune cells do this. There may be a number of factors.

There are environmental factors:

- They may have suffered from a particular viral infection, such as rubella. The white blood cells that deal with this virus as part of the immune system also destroy insulin-making cells.
- Some scientists think that giving babies cow's milk too early may increase the risk of type 1 diabetes in later life. It may cause them to make antibodies against proteins in the cow's milk, and these antibodies may also attack insulin-making cells.
- Certain antibiotics and anti-cancer drugs may destroy the insulin-making cells.
- Trauma injury and pancreatic tumours also destroy these cells.

There is also a genetic component, because these white blood cells and antibodies only attack insulin-making cells that have particular-shaped protein antigens on their surface.

Genes govern the way you make proteins. People with type 1 diabetes do not make enough of a certain protein that kills the white cells that destroy insulin-making cells.

There are cases of identical twins, one with type 1 diabetes and one without, so for type 1 diabetes to develop it seems you need particular genes plus an environmental trigger.

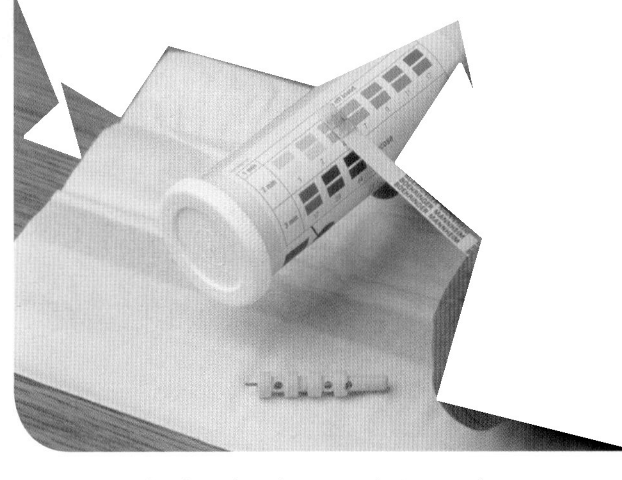

▲ A test strip placed against a scale, to test the blood glucose level. A yellow lancet was used to prick the finger and place a drop of blood on the test strip. The blood reacts with two coloured squares at the tip of the test strip. Depending on the blood glucose level, the squares change colour. The scale shows that this blood glucose level is normal.

Did you know...?

People with type 1 diabetes do not make enough of a certain protein, called TNF-α, that kills the white cells that destroy insulin-making cells. One scientist has successfully treated diabetic mice with this protein. Other scientists are researching the use of stem cells to replace destroyed insulin-making cells.

Questions

1. How do people with type 1 diabetes manage their condition?
2. What are the long-term consequences of type 1 diabetes?
3. Why do you think people with type 1 diabetes are advised to take regular exercise?

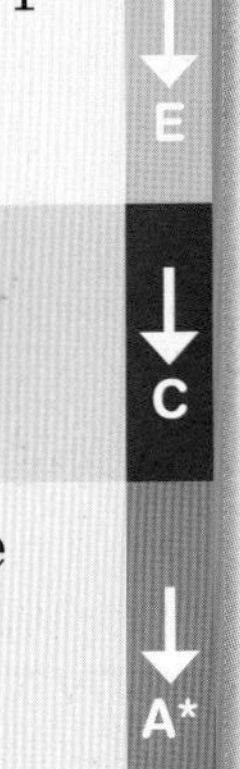

23: Human populations

Learning objectives

After studying this topic, you should be able to:

- ✔ know that the human population is increasing
- ✔ understand that this increase is unsustainable and leads to pollution

Key words

population, pollution, sustainable

A Families had large numbers of children before 1900. Suggest why.

B The death rate decreased steadily after 1800. Suggest why this might have happened.

C What must happen to the birth rate and death rate to keep the population fairly constant?

The human population

The number of humans on the planet has been increasing. But the increase is not steady. Biologists have studied the growth of the human **population** and have found some interesting points.

1000 years ago, the population of the UK was stable:

- There was very little increase.
- The reason for this was that there were not many people to reproduce, and the food supply was limited.

Between 1600 and 1900, there was a steady increase.

- This was due to better farming methods and improving hygiene.

After 1900, there was a dramatic rise in the population:

- Diet improved.
- Hygiene improved.
- Healthcare improved.
- Infant mortality (death rate) fell.

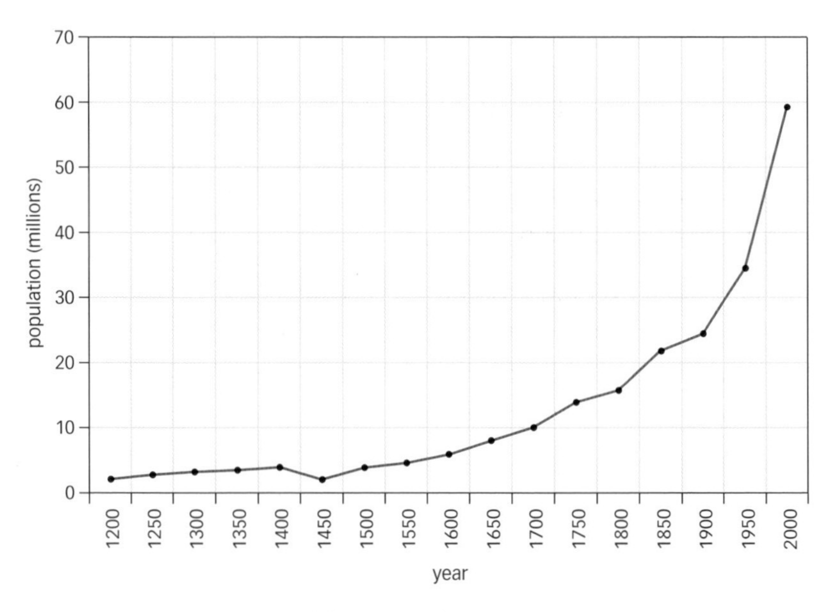

▲ The human population of the UK has been rising fast

The population explosion

The massive increase in human population seen in the UK since 1900 is repeated in most countries in the world. At the same time, the standard of living has improved in many countries. So there is much greater demand for material products and for the resources needed to make them.

These changes have a number of effects:

- There is a shortage of food in some countries.
- More land is being used for building and farming.
- More **pollution** is being produced.
- The world's resources are being used up more quickly than they can be replaced.

This growth is not **sustainable**. The impact of so many humans on the environment is harmful.

▲ Our modern way of life uses many more resources than lifestyles in the past

D State why the human population explosion is not sustainable.

E Biologists say that the use of resources must be sustainable. Describe what this might mean.

Sustainable development

Sustainable development is a term that is used increasingly often. It means using resources for human needs without harming the environment.

Humans use a lot of resources to live modern lifestyles. If these resources are taken from the environment without being replaced, so that they are being used up, this use is unsustainable. This was often the case in the past, and unsustainable use of resources has harmed both the environment and the organisms in it.

Humans need to manage their use of resources. There are two major approaches to this:

1. Replace resources where possible. For example, we use wood for many things, but new trees can be planted after felling.
2. Avoid overuse where resources cannot be replaced as quickly as they are used. For example, quotas are now used to stop overfishing in the oceans – limits are set on how many fish can be caught.

These methods require planning. As the population increases, our demand for resources grows. Resources, habitats, and species must all be considered. Planning for sustainable development needs to become increasingly global.

▲ Commercial replanting of conifers in a managed woodland

Questions

1 Recycling helps increase sustainability. Explain why this is the case. (E)

2 Explain why an increase in the human population might lead to an increase in pollution. (C)

3 Explain the shape of the human population curve shown on the previous page. (A*)

Learning objectives

After studying this topic, you should be able to:

- ✔ know that the actions of humans release pollutants
- ✔ be aware of the range of pollutants released and their sources
- ✔ recognise the impact of various pollutants

Key words

pollutant

Did you know...?

Finland is a Green Champion. It was the first country to introduce a carbon tax. The government imposes a tax on companies that release carbon dioxide into the air.

Human influences on the environment

Humans have an impact on the environment by reducing the amount of land available for other animals and plants, and by producing pollution. These influences can be grouped into two major areas:

Agriculture

- use of fertiliser
- use of pesticides
- loss of habitat
- deforestation
- monoculture
- animal waste.

Towns and industries

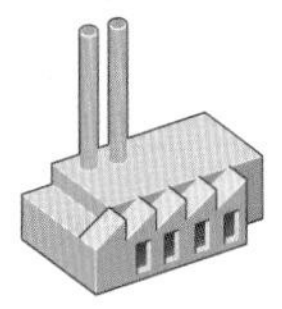

- loss of habitat
- quarrying and extraction of raw materials
- dumping of wastes
- production of toxic chemicals
- sewage.

The development of towns, industries, and farms has a number of effects on the natural environment. Building projects remove large areas of Britain's natural woodlands. Agriculture and industry also produce a number of **pollutants** which affect the water, air, and land. Many land pollutants have their effect when they are washed from the land into waterways.

Air pollution

Pollutant	Source	Effect on environment
Sulfur dioxide and nitrogen oxides	Burning (combustion) of fossil fuels	Dissolves in rain to form acid rain: damages plant leaves acidifies lakes changes minerals available in water supplies causes bronchitis.
Carbon dioxide	Burning (combustion) of fossil fuels	Dissolves in rain to form acid rain (see above). A greenhouse gas which keeps more heat in the atmosphere, leading to global warming.
Methane	Cows and rice fields	Another greenhouse gas which contributes to global warming.
Smoke	Burning (combustion) of fuels	Releases particles into the air which cause: bronchitis reduction in photosynthesis.
CFCs	Aerosols and refrigerator coolants	Destroy the ozone layer in the upper atmosphere, allowing more ultraviolet radiation through and contributing to skin cancers.

Smoke and steam from a petrochemical plant

A List three ways in which farming can adversely affect the environment.

B How does sulfur dioxide produce acid rain?

Water pollution

Pollutant	Source	Effect on environment
Untreated sewage	Sewage works	Bacteria can cause disease, eg typhoid and cholera. Nitrates in the water lead to the death of fish by a process called eutrophication.
Fertilisers	Farms	Nitrates in the water lead to the death of fish by a process called eutrophication.
Pesticides and herbicides	Farms	These chemicals are washed into waterways and may build up in food chains to toxic levels.
Toxic chemicals	Factories and mining	Poison organisms directly, or may build up in food chains to toxic levels.
Detergents	Domestic use	Kill valuable microbes.
Oil	Tankers and pipelines	Kills birds and fish. Pollutes the sea bed.

A biologist collecting water samples for testing

Questions

1. Give: (a) the source, (b) the effects of smoke in the atmosphere. (E)
2. Use information on these pages to explain why the increase in the human population is leading to more pollution of the air.
3. Explain why people in most towns are now only allowed to burn smokeless fuel. (C)
4. Everybody wants the latest mobile phone. However, there is a cost to the environment in buying such products. Identify effects on the environment that producing mobile phones may have. (A*)

Learning objectives

After studying this topic, you should be able to:

- ✔ understand what deforestation is
- ✔ know the consequences of deforestation

Key words

deforestation, peat

This truck is carrying logs from felled rainforest in Gabon, west central Africa

Can't see the woods or the trees

Wood is a highly commercial product, and the land that trees grow on is sometimes prized even more. The large-scale felling of trees, called **deforestation**, can be a lucrative business. Deforestation is carried out for two major reasons:

- to provide timber for furniture, building, and fuel
- to use the forest land for farming, towns, and industries.

In many developing regions of the world, felling trees and selling the timber provides a quick income. What is more, generating farmland allows cash crops to be grown, which again generates a rapid income. However, while this has a short-term benefit to the local economy, the trees are often not replaced. This is happening on a massive scale, leading to a net reduction in the area of forest in the world. Deforestation has a lasting harmful impact worldwide.

A State two major reasons for deforestation.

B Why do many people in developing countries carry out deforestation?

The cost of deforestation

1. Slash and burn 	When forests are cleared for farms, the tree litter and stumps are often burnt to remove them. This releases carbon dioxide into the air.
2. Effect on global gases 	Deforestation leads to a rise in carbon dioxide levels in the atmosphere by • the release of carbon dioxide during burning • the release of carbon dioxide during the decomposition of felled trees by microorganisms • a reduction in photosynthesis, so less carbon dioxide is taken up by trees and locked up in wood for many years.

3. Land for cattle and rice 	Forest land is often cleared for cattle farms or rice fields. Both of these release large quantities of methane gas into the atmosphere. Methane is a major greenhouse gas.
4. Land for biofuel crops 	Land can be cleared to grow crops to make biofuels based on ethanol. These are a low-polluting fuel. However, the loss of forest land still reduces the amount of carbon dioxide taken up.
5. Loss of biodiversity 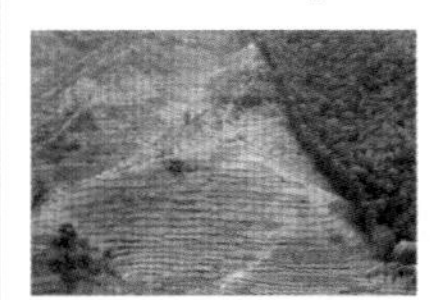	A diverse forest community is removed and replaced with a single crop. This provides few habitats and removes large amounts of the same mineral from the soil.
6. Loss of future resources 	With this loss of biodiversity, many species may become extinct, some of them as yet unknown to humans. These species, particularly tree species, may hold valuable potential uses for us, as cures for diseases, for example.
7. Soil erosion 	The removal of trees may lead to soil erosion, as there are no tree roots to hold the soil together. This allows heavy rain to wash the soil away.

Destruction of peat bogs

Peat is produced over thousands of years from sphagnum moss in very wet acidic boggy areas. The acid preserves the dead moss for thousands of years. The peat is very nutrient rich, and has been dug extensively, mainly to produce compost for the garden market. This gradually led to the removal of vast areas of peat bog. More than 90% of the bogs in England have been damaged or destroyed. Not only are we losing a diverse habitat, but as soon as the peat is aerated in gardens, the decay process begins, releasing carbon dioxide back into the atmosphere. Many gardeners now use peat-free composts.

Exam tip 

✔ Link the two major gases mentioned here, carbon dioxide and methane, to the process of global warming described in the next spread.

C Why is the drugs industry concerned about deforestation?

D Explain how trees prevent soil erosion.

Did you know...?

Peat bogs once covered 15% of Ireland. Much of this has been removed for burning as well as for gardens.

Questions

1 Explain how deforestation can lead to a rise in methane levels in the atmosphere.

E

2 Explain three ways in which deforestation results in a rise in carbon dioxide in the atmosphere.

C

3 Explain why peat-free composts are important for the environment.

4 Suggest how the human population can use wood in a sustainable way.

A*

Learning objectives

After studying this topic, you should be able to:

- ✔ understand the causes, process, and effects of global warming

Key words

global warming, greenhouse gas

Too hot to handle

Global warming is an overall rise in average global temperatures. Most scientists think this is due to more heat being trapped in the Earth's atmosphere. The atmosphere naturally contains gases called **greenhouse gases**, which trap heat and make Earth warmer than it would otherwise be. However, we are producing more and more greenhouse gases, which increase the atmosphere's natural 'greenhouse effect'.

Biologists have become increasingly worried that global warming is happening at an alarming rate, caused by humans altering the balance of gases in the atmosphere.

Gases that trap heat in

The two major greenhouse gases that are increasing in the atmosphere and contributing to global warming are

- carbon dioxide – from increased combustion of fossil fuels to supply us with energy, and as a result of deforestation
- methane – from cattle, rice fields, and decaying waste.

The greenhouse effect

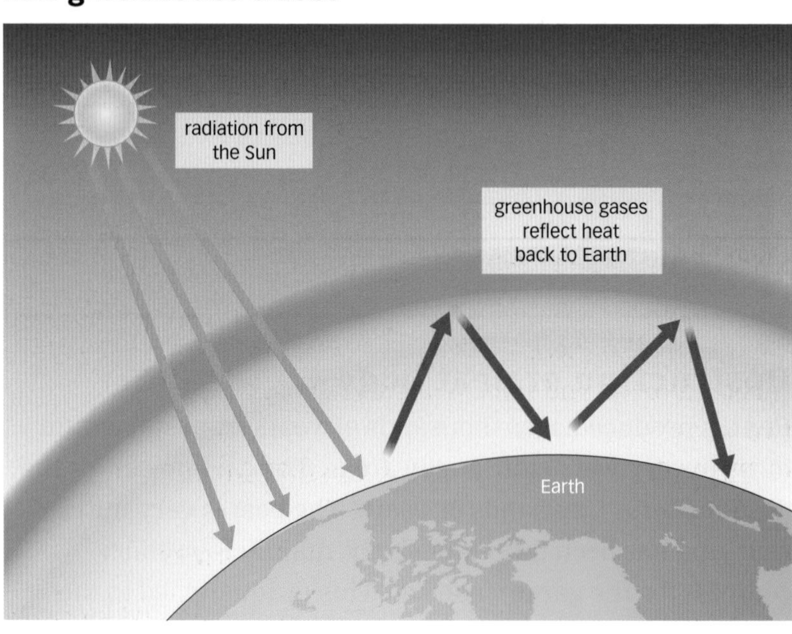

▲ Greenhouse gases reflect heat back to Earth, keeping the temperature higher

Any activity that increases the amount of greenhouse gases in the atmosphere, such as combustion and deforestation, will lead to an increased greenhouse effect and global warming.

A Which two gases are the major greenhouse gases?

B What are the major sources of these gases?

Exam tip AQA

- ✔ When answering questions about global warming, you need to focus on the two major greenhouse gases, carbon dioxide and methane. You need to know how they cause global warming, and the effects of global warming.

The effects of global warming

A small rise in global temperatures may not feel significant to us. However, the global effects of a rise of just one or two degrees could be catastrophic:

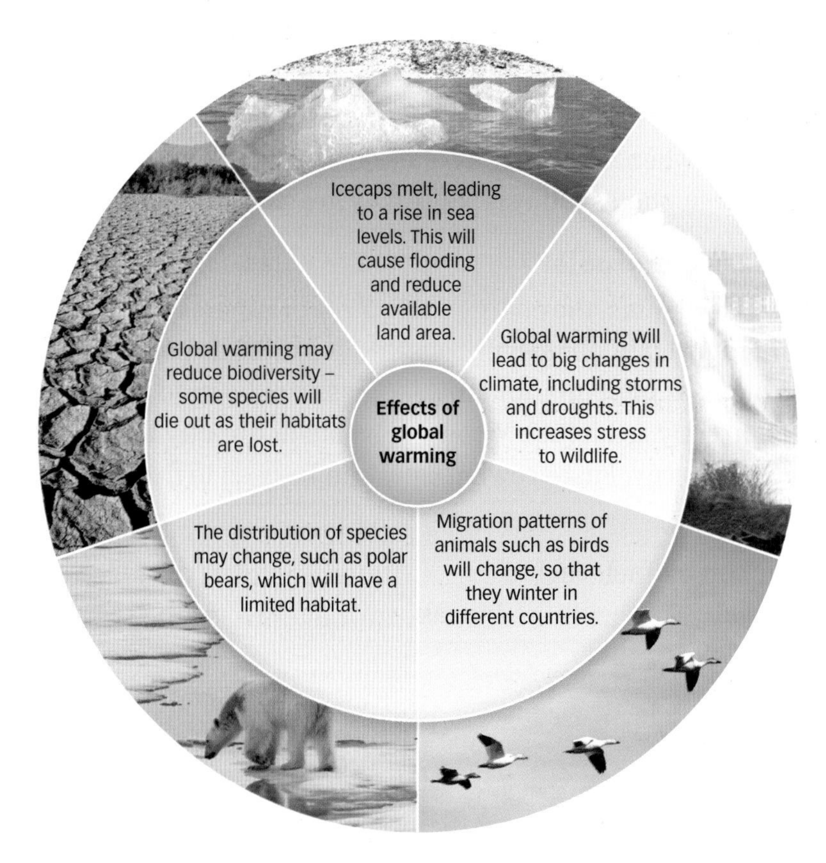

▲ Global warming may have many harmful effects on the environment

Oceans to the rescue

Large bodies of water absorb large amounts of carbon dioxide. So the oceans, lakes, and ponds of the world store carbon dioxide and remove it from the atmosphere in two ways:

- Phytoplankton absorb carbon dioxide during photosynthesis.
- Carbon dioxide dissolves directly in the water.

Both these processes increase as the carbon dioxide levels in the atmosphere rise. So the oceans are acting as a counterbalance to our increased release of carbon dioxide. However, they cannot counteract an excessive rise in atmospheric carbon dioxide levels.

Questions

1 Explain why there is an increase in the level of the two major greenhouse gases in the atmosphere at present.

E

2 Explain why global warming is affecting the distribution of the polar bear.

3 How might droughts affect biodiversity?

C

4 Draw a flow chart which starts with burning fossil fuels, and ends in the melting of icecaps.

5 Explain how carbon dioxide is taken out of the atmosphere.

A*

27: Biofuels

Learning objectives

After studying this topic, you should be able to:

- know that biological materials can be used as a fuel source
- give examples of biofuels
- understand the production, uses, and composition of biogas
- know that fuel can be made from alcohol

Key words

biofuel

A What is the source of energy for making biofuels?

B Name three types of biofuel.

C State one advantage and one disadvantage of using biofuels.

This researcher is culturing algae as part of research into biofuels

Producing wood chips to be burnt in a wood-fired power station

Greener fuels

The burning of fossil fuels harms the environment, as it produces waste gases including carbon dioxide, which leads to global warming. A variety of fuels from biological materials can be used as an alternative. These are called **biofuels**. They are better for the environment because the carbon dioxide produced when they burn is balanced by the carbon dioxide they use in photosynthesis while they are growing.

What are biofuels?

As in fossil fuels, the energy in biofuels originates from sunlight used in photosynthesis. Photosynthesis produces the biomass in plants, and this biomass can be used directly or indirectly as biofuel. Wood can be burnt directly to release energy. Fast-growing trees can be used to fire power stations.

Common biofuels include:

- wood
- biogas
- alcohol.

Advantages of using biofuels	Disadvantages of using biofuels
Reduce fossil fuel consumption by providing an alternative fuel.	Cause habitat loss because large areas of land are needed to grow the plants.
No overall increase in levels of greenhouse gases, as the plants take in carbon dioxide to grow, and release it when burnt.	Habitat loss can lead to extinction of species.
Burning biogas and alcohol, produces no particulates (smoke).	

Balancing the books

To burn fuels while maintaining no overall increase in greenhouse gases is a difficult balancing act. When we burn biofuels we have grown, the carbon dioxide taken in during photosynthesis is then released during combustion.

However, land is needed to grow these crops. In some areas forests are cleared for the cash crop. This leads to a loss of plants to absorb carbon dioxide, and an increase in carbon dioxide released by decaying wood. It also causes a loss of habitat.

Biogas

Biogas is made by the fermentation of carbohydrates in plant material and sewage by bacteria. This fermentation occurs naturally, for example in marshes, septic tanks, and even inside animals' guts. Biogas is also produced at some landfill sites, where the gas can be burnt. Sometimes the biogas can explode, making the landfill site unusable for many years.

Biogas is a mixture of gases:

- methane (50–75%)
- carbon dioxide (25–50%)
- hydrogen, nitrogen, and hydrogen sulfide (less than 10%).

This mixture will burn in oxygen, so forms a useful fuel.

There are a few technical issues with the production of biogas. First, since many different waste materials are used, a large range of bacteria are needed to digest the waste.

Biogas is a cleaner fuel than petrol or diesel, as fewer particulates are released. However, burning biogas releases 4.5–8.5 kWh/m^3 of energy compared with natural gas, which releases 9.8 kWh/m^3. This is because biogas contains less methane than natural gas.

Biogas production on a larger scale

The gas is generated commercially in large anaerobic tanks. Wet plant waste or animal manure is constantly added, and the gas produced is removed. Gas production is fastest at a temperature of 32–35 °C, because the fermenting bacteria grow best at this temperature. The remaining solids need to be removed from the tanks and can be used as a fertiliser in some cases.

Biogas has a number of uses:

- as vehicle fuel
- to generate electricity
- for heating systems.

Bioethanol

Alcohol is produced from plant material by yeasts in brewing. On a larger scale this alcohol can be used as a fuel. Mixed with petrol it produces gasohol, which can fuel cars. This is a particularly economic fuel in countries that produce large amounts of plant waste, such as Brazil. Brazil has no oil reserves and plenty of sugar cane waste to make the alcohol.

▲ A bioethanol fuel pump

Questions

1. Explain why biogas from landfill sites is particularly dangerous.
2. Give two reasons why gasohol is used in Brazil.

↓ E

3. Why must a biogas digester be kept airtight?

↓ C

4. Explain why using biofuels should not contribute to any net increase in greenhouse gases, in contrast to using fossil fuels.

↓ A*

Learning objectives

After studying this topic, you should be able to:

- ✓ know that fungi are used to produce foods
- ✓ understand how mycoprotein is produced

Key words

mycoprotein, **fermenter**

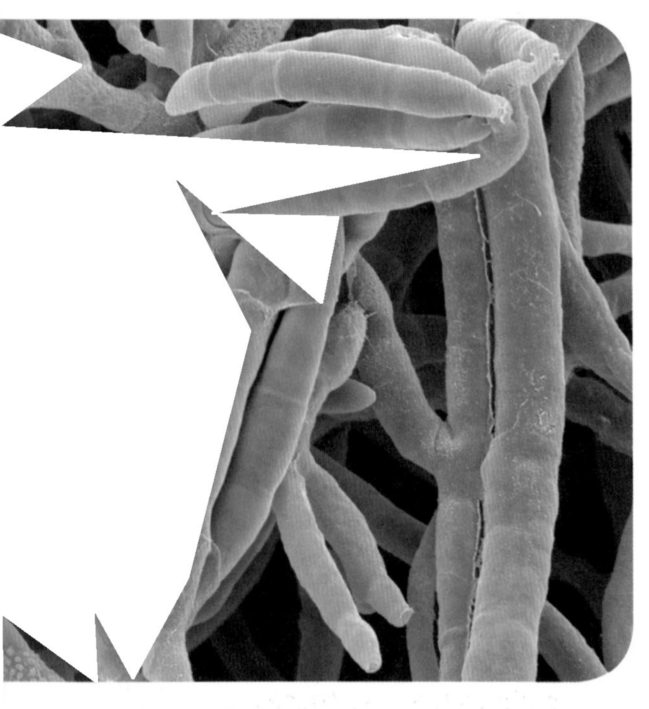

▲ Fusarium venenatum

> **A** Name some fungi that are eaten directly.
>
> **B** Name two foods produced by yeasts.
>
> **C** Which fungus is used to produce mycoprotein?

Fungal foods

A number of foods are produced with the help of fungi.

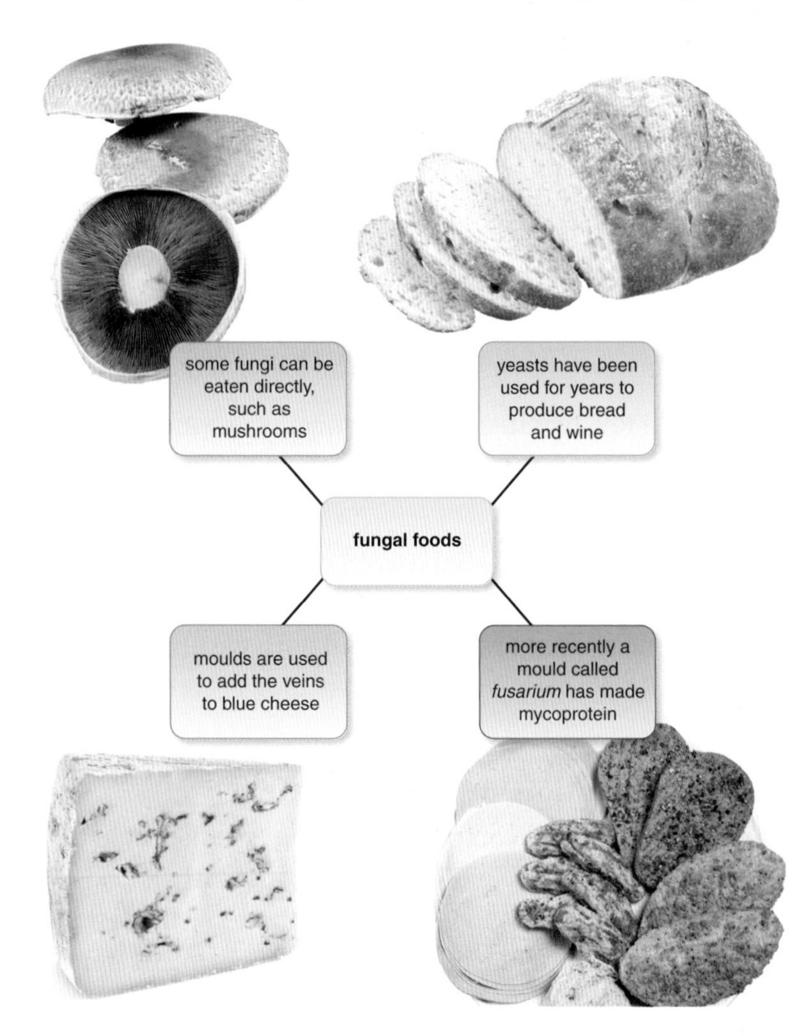

▲ Some fungi are eaten directly in their natural state; others are used to produce foods

Mycoprotein

Why make mycoprotein?

In the 1960s, scientists became concerned about possibilities of food shortages as the world population increased. They were particularly concerned about producing enough protein to feed the world. Biologists looked for an organism that could convert waste plant materials into a protein-based food. In 1967 a fungus was discovered that could do this – *Fusarium venenatum*. After about ten years of research and development, **mycoprotein** was produced. This is a high-protein food product suitable for vegetarians.

Mycoprotein production

Mycoprotein is produced in large tanks called **fermenters**. The two largest fermenters in the world are used to produce mycoprotein. It is a continuous process. The fungi are placed in the tank filled with water and glucose syrup.

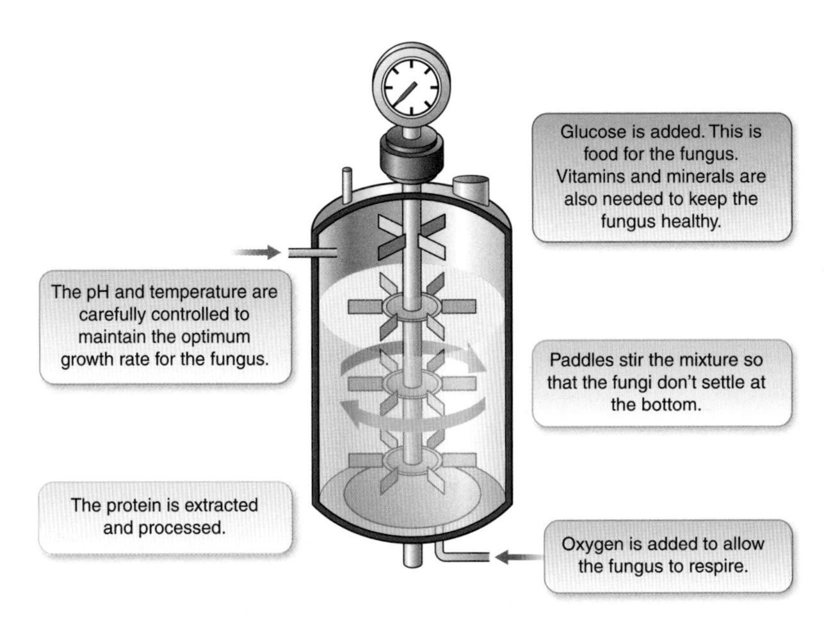

▲ A fermenter for producing mycoprotein

Nutrient	Amount (g) per 100 g of mycoprotein
protein	11.0
carbohydrate	3.0
all fat	2.9
saturated fat	0.7
fibre	6.0

▲ The nutrients in mycoprotein

Processing to give the finished product

The mycoprotein is removed from the fermenter. Next, a number of processing steps are carried out:

- The crude product is heated to 65 °C to remove excess nucleic acids. High levels of nucleic acid in the protein might cause gout.
- It is centrifuged – spun at high speed – to remove water.
- Egg is added to bind the product together.
- The mycoprotein is frozen to create fibres, making it resemble meat.

The nutritional value of mycoprotein

Mycoprotein is an excellent source of protein in the diet. As well as providing protein, it is a low-fat, high-fibre food, which many people think is healthier than meat.

Questions

1 What conditions must be kept constant in the mycoprotein fermenter?

2 Why is the product centrifuged?

E

3 What type of respiration occurs in a mycoprotein fermenter?

C

4 When mycoprotein was developed, some people didn't like the thought of eating a food produced by a microbe. Make a nutritional argument for eating mycoprotein.

A*

Learning objectives

After studying this topic, you should be able to:

- know how energy is transferred through a food chain
- understand the methods of maximising efficiency of production from farm animals

Key words

battery farming

Energy transfer in food chains

You may remember that at each link in a food chain, the amount of energy and biomass becomes less. Some energy is transferred to the environment at each stage. So, as you move along a food chain, there is less and less of the original energy and biomass available. The longer the food chain, the less food is available to organisms at the end.

This affects our place in food chains. If you think about the food humans eat, you will realise that most of our food is either a producer or a herbivore. We seldom eat carnivores. Farming produces either plant material or herbivores for our consumption. This makes our food chains as efficient as possible. They are short, with fewer links at which energy and biomass can be transferred out to the environment.

The food web below shows some examples of human food chains:

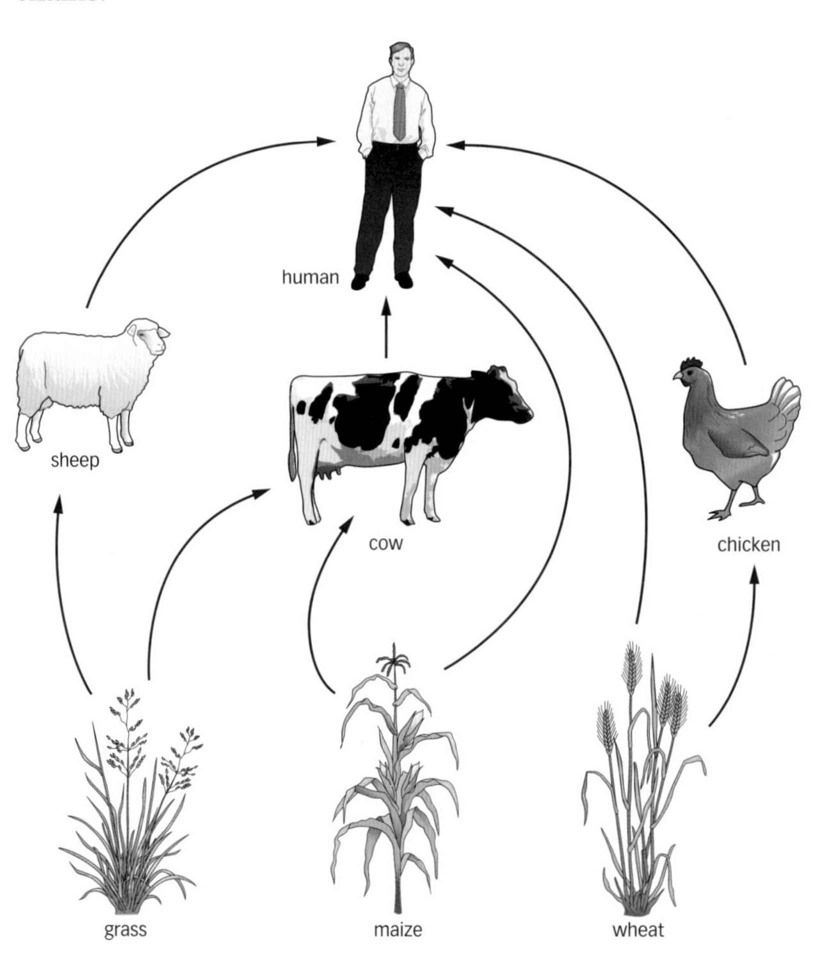

▲ Farming food web

A What two things become less available as you move along a food chain?

B List ten foods that a human might eat, and identify which level these foods occupy in a food chain.

C Why is it more efficient for humans to eat only herbivores and producers, rather than carnivores?

Maximising energy transfer

As biologists began to understand the ideas of energy transfer in food chains, they were able to look at farming methods and maximise the energy efficiency of their processes. For example, look at these two methods of rearing chickens – free range farming and intensive **battery farming**, in which large numbers of chickens are reared closely together indoors.

These principles are not just used for chickens – many other animals, such as pigs and calves, are reared intensively.

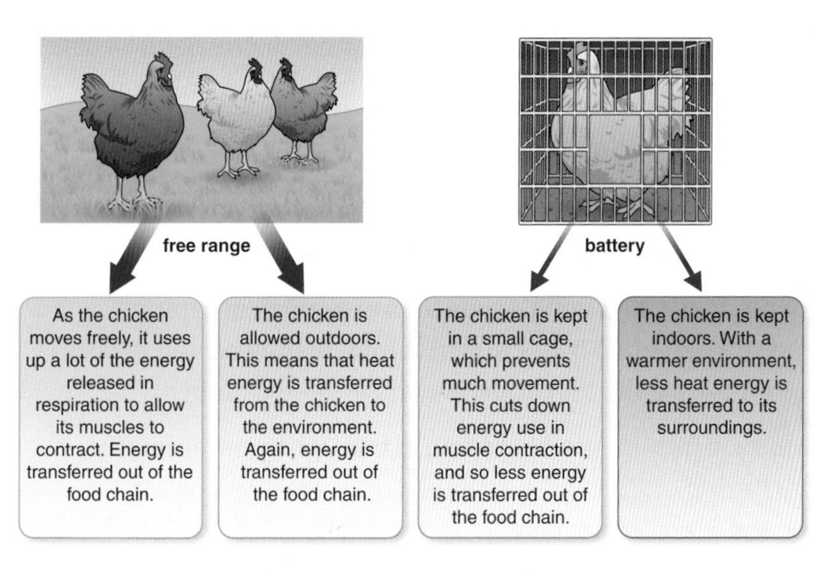

▲ Intensive farming maximises efficiency and retains more energy in the food chain

▲ An intensive dairy farm

Advantages and disadvantages of energy-efficient farming

Advantages	Disadvantages
Less energy is transferred out of the food chain, so more energy is available for human consumption.	There is a greater risk of disease spreading through the animals as they are in close contact.
It is less labour intensive, as the animals are all contained in a limited area.	Some people feel that the technique is inhumane or cruel to the animals.
There is less risk of attack from predators such as foxes.	Some people believe that the quality of the product is poorer.
Production costs are cheaper.	

Questions

1 Give two ways in which energy is transferred out of the food chain when rearing animals. E

2 How do battery farming methods maximise energy transfer? C

3 Discuss why some people do not like to buy intensively reared meats. A*

30: Fishing

Learning objectives

After studying this topic, you should be able to:

- know that overfishing will reduce fish stocks
- know that government quotas and net size rules have helped fish stocks recover

▲ Dragnet fishing in the North Sea

▲ Cod has always been a popular fish with consumers, leading to a dramatic fall in cod populations

Gone fishing

When human populations were small, little farming occurred and most food was either hunted or gathered. As the population increased, more food was needed. This led to the development of farms. Some hunting still takes place, but the only hunting carried out on a more commercial basis today is open-ocean fishing.

Trawlers and nets

Commercial fishing features large ocean-going vessels, which lower large nets into the sea and catch shoals of fish. Modern developments in fishing fleets have resulted in greater numbers of fish being caught.

Modern fishing fleets benefit from technological advances such as:

- sonar to locate fish
- sophisticated net designs to prevent fish escaping
- well-designed boats to travel long distances and process and store the fish after they have been caught.

Many different species of fish are caught commercially, including:

- cod
- herring
- mackerel
- haddock.

Too efficient

Developments in fishing technology have provided bigger and bigger catches, but this has led to problems. Fishing fleets have overfished many seas, and the populations of some of our more popular fish have been seriously reduced. By the end of the 1960s, fish like cod were no longer common. The situation became so critical that international governments had to intervene.

> **A** Explain why more fish can now be caught than 100 years ago.

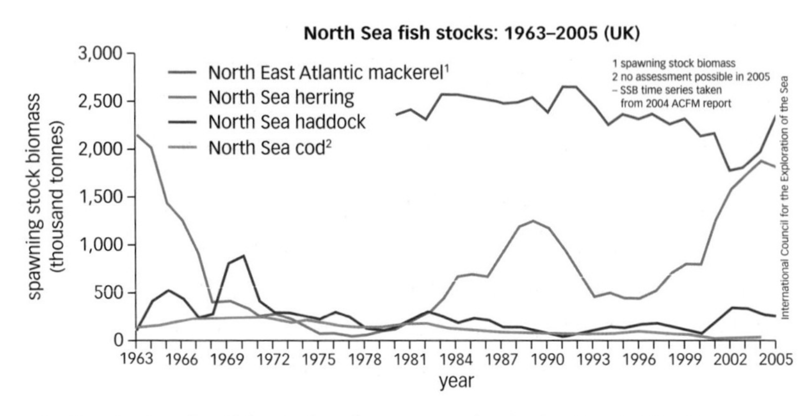

▲ Graph showing fish stocks of some species in the North Atlantic

B List some fish species caught in UK waters.

The main fishing waters for British trawlers are the seas to the north of the UK. Figures have been recorded by the Department for Environment, Food and Rural Affairs (Defra), which have tracked **fish stocks** in the North Sea for several years. They show several key points:

- A general increase in fishing in response to increased demand by a growing human population.
- Stocks of many fish have fallen as a result.
- During the 1970s the levels of fish in the sea were very low.
- This level of fishing was unsustainable.
- The European Community (EC) intervened to protect the fish.
- This has allowed the recovery of some fish, such as herring.
- There is now a sustainable level of fishing.
- Some species, such as cod, are still dangerously low in numbers, and have been low for a long time.

Protecting fish stocks

National governments realised that they had to act to protect fish stocks in their own waters. Both the British Government and the EC looked for ways to allow the fish stocks to recover. The two major conservation methods they employed were:

- Net size – the size of the holes in the nets was increased. This would allow the smaller, younger fish to escape, as they would pass through the holes. These younger fish would then survive, breed, and help the population recover.
- Fishing **quotas** – each government set a limit on the number of fish of each species that could be caught. This meant that fewer fish from the species most in danger were removed, so the populations could recover.

These measures have enabled fishing to be sustainable in the North Sea. Fishing has been reduced to a level at which the fish populations can slowly recover. However, the disadvantage has been that many fishing communities have suffered unemployment as a result of reduced fishing.

Key words

fish stocks, quotas

C Which fish seems most likely to disappear from the North Sea?

D Why were the fishing methods of the 1960s unsustainable?

Questions

1. What actions have international governments taken to conserve fish populations? (E)
2. Look at the graph on the previous page. Which fish species has shown the greatest recovery in recent times? (C)
3. How effective have these government measures been, and how do we know this? (A*)

Course catch-up

Revision checklist

- Living organisms produce toxic wastes, such as carbon dioxide, urea, and heat, which must be removed.
- Body temperature and blood glucose must be kept within narrow limits.
- The water and ion content of the body has to be regulated.
- Kidneys filter urea, excess salts, and water from the blood.
- Renal (kidney) failure can be treated by dialysis or a kidney transplant.
- Recipient and donor have to be matched for blood group and tissue type. The recipient must take immunosuppressant drugs to prevent rejection of the transplanted organ.
- Sweating and vasodilation prevent overheating.
- Reduced sweating, vasoconstriction and shivering prevent overcooling.
- Insulin lowers the blood glucose level after meals.
- Glucagon increases the blood glucose level after fasting.
- Some people do not make enough insulin and they suffer from type 1 diabetes. Type 1 diabetes is treated with insulin injections.
- The human population is increasing. More land will be needed to grow more food and to house the extra people.
- This leads to depletion of resources and more pollution. Humans are developing sustainable development methods.
- Agriculture and industry produce many pollutants that contaminate air and water.
- Deforestation (removal of forests) increases the carbon dioxide and methane content of the air; reduces resources and biodiversity, and leads to soil erosion.
- Increased methane and carbon dioxide in the air cause the greenhouse effect that leads to global warming.
- Oceans can absorb some of the excess carbon dioxide.
- Biofuels could replace use of fossil fuels and reduce the greenhouse effect. However, producing them takes up a lot of land that could be used to grow food.
- Knowledge of how energy passes along food chains enables energy-efficient farming.
- Modern fishing has been so efficient that fish stocks are dwindling. Some governments are acting to protect fish stocks.

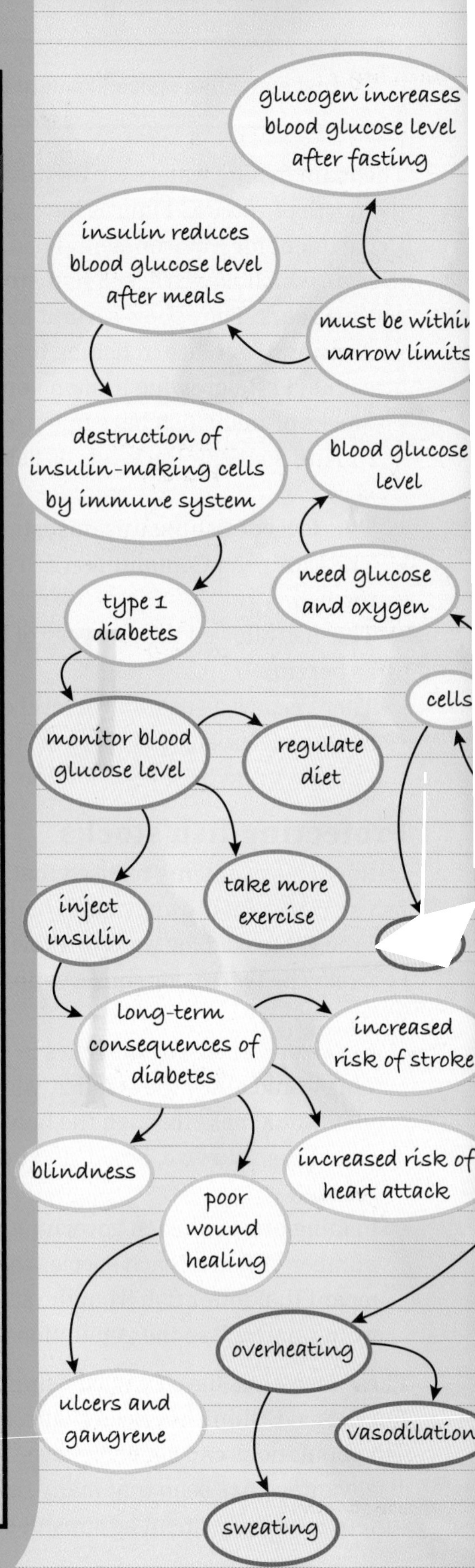

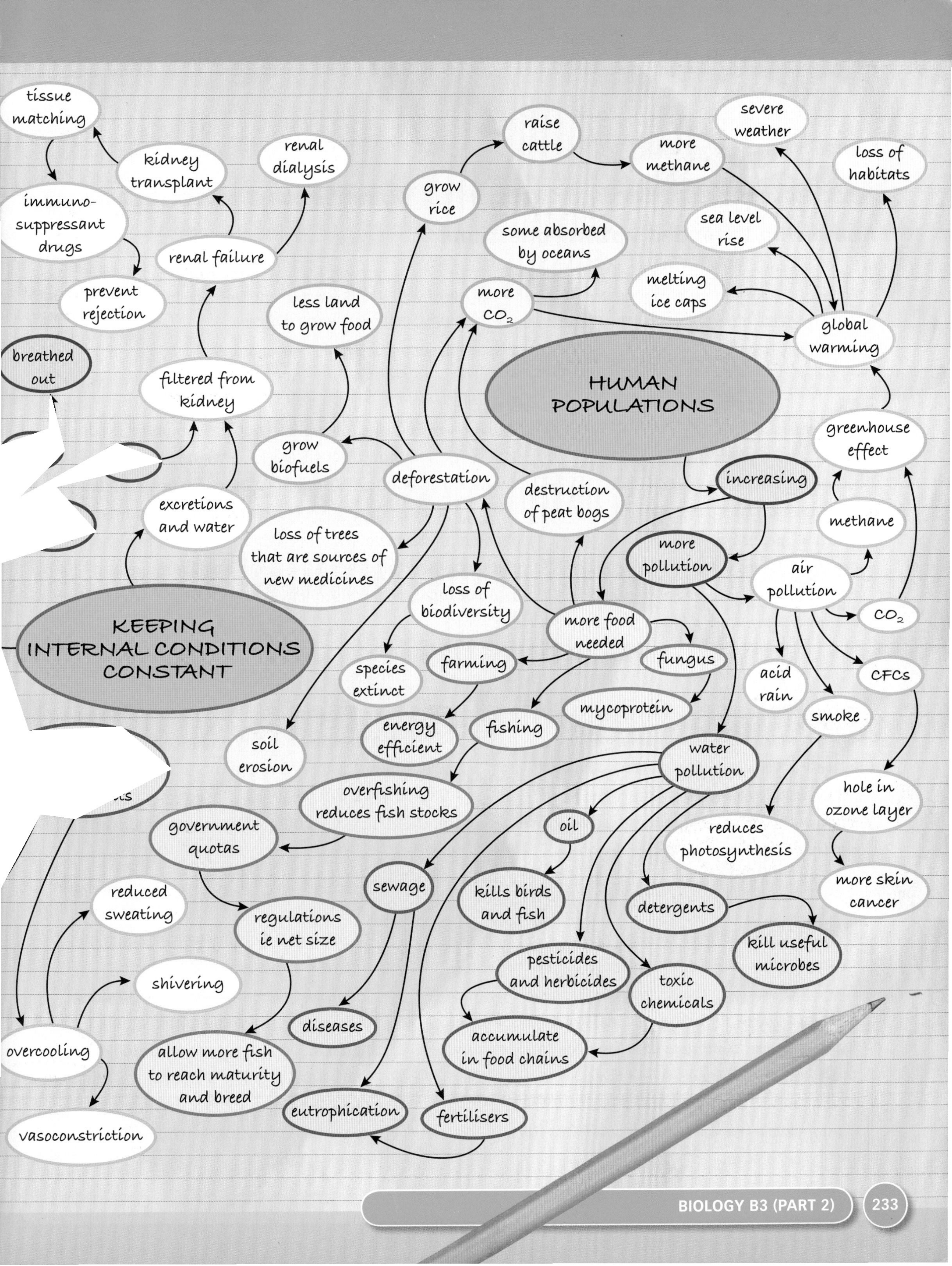
tissue matching
kidney transplant
renal dialysis
immuno-suppressant drugs
renal failure
prevent rejection
breathed out
filtered from kidney
excretions and water
KEEPING INTERNAL CONDITIONS CONSTANT
reduced sweating
shivering
overcooling
vasoconstriction
grow rice
raise cattle
more methane
severe weather
loss of habitats
sea level rise
some absorbed by oceans
melting ice caps
more CO_2
global warming
less land to grow food
HUMAN POPULATIONS
grow biofuels
deforestation
destruction of peat bogs
greenhouse effect
increasing
methane
more pollution
air pollution
CO_2
loss of trees that are sources of new medicines
loss of biodiversity
more food needed
fungus
mycoprotein
species extinct
farming
energy efficient
fishing
acid rain
CFCs
smoke
soil erosion
water pollution
hole in ozone layer
overfishing reduces fish stocks
government quotas
oil
reduces photosynthesis
more skin cancer
sewage
kills birds and fish
detergents
regulations ie net size
kill useful microbes
pesticides and herbicides
toxic chemicals
diseases
accumulate in food chains
allow more fish to reach maturity and breed
eutrophication
fertilisers

Answering Extended Writing questions

QUESTION

Patients with kidney failure may be treated by renal dialysis or with a kidney transplant. Describe how renal dialysis works. State one advantage of receiving a kidney transplant over having dialysis.

The quality of written communication will be assessed in your answer to this question.

G–E

Dialysis is using a kidney machine. it cleans the blood. it takes out salts and waste. if you have a transplant you have to have drugs to supress your ammune system for ever. With a kidney transplant you can eat a normal diet and have more drinks.

Examiner: Significant spelling and grammatical errors. The section on how dialysis works is vague and gives no detail. One advantage of having a kidney transplant is stated, but a disadvantage is also stated. The candidate does not make clear which is which. You need to clearly state whether you are talking about an advantage or a disadvantage, rather than leave the examiner to do the work.

D–C

The person has to use a kidney machine a few times a week. There blood goes into the machine. Poisonus waste, acess salts and water go acros a membrane into the fluid. Having this treatment is ecspensive. Kidney transplants are cheaper in the long run.

Examiner: This answer describes how dialysis works, but does not use any technical terms. An advantage of kidney transplants is clearly stated, although it is not clear that the cost savings are to the NHS, not the patient. There are some spelling errors.

B–A*

The patient's blood goes through a dialysis machine. The blood is on one side of a membrane. On the other side is the dialysis fluid. Waste urea, and unwanted salts and water diffuse across the membrane into the fluid. Then the blood goes back into the patient's arm. This takes several hours 3 or 4 times a week and the patient can't go anywhere and has to watch their diet. If they have a transplant they can live a normal life and go on holiday.

Examiner: This answer explains how dialysis works and uses technical terms such as 'diffusion' and 'urea'. It does not mention the anticoagulant heparin, or that the fluid has to be sterile and at body temperature, but there is enough detail here. An advantage of a kidney transplant is clearly stated.

Exam-style questions

1 The European Commission makes annual proposals for cod fishing quotas in EU waters, based on scientific data. Fishermen say that scientists are exaggerating the danger to cod stocks. Scientists say that fishermen are ignoring warnings about low cod populations, and that because they only fish in areas where there are lots of cod, fishermen get the wrong impression of the total population size.

A03 **a** Explain why scientists and fishermen have different ideas about the size of the cod population.

A03 **b** Suggest two reasons why the size of the catch allowed may not depend entirely on scientific data.

2 The table gives the composition of two drinks.

Drink	sugar (g per litre)	sodium (mmol per litre)	chloride (mmol per litre)
isotonic	73	24	12
cola	105	3	1

A03 Explain why the isotonic drink would be best for a runner on a hot day.

3 A01 **a** Give three reasons why deforestation increases the carbon dioxide level in the atmosphere.

A01 **b** Deforestation also leads to loss of biodiversity. What is biodiversity?

A02 **c** Why is it important to prevent the extinction of species of trees?

G–E D–C

4 Describe the harmful effects of the following water pollutants:

A01 **a** sewage **b** fertilisers **c** oil.

5 Governments are encouraging businesses to reduce CO_2 emissions.

A01 **a** Explain the link between CO_2 and the greenhouse effect.

A01 **b** Describe two possible outcomes of the greenhouse effect.

6 The kidneys excrete water and urea. The graph shows the amount of sweat and urine produced at different temperatures.

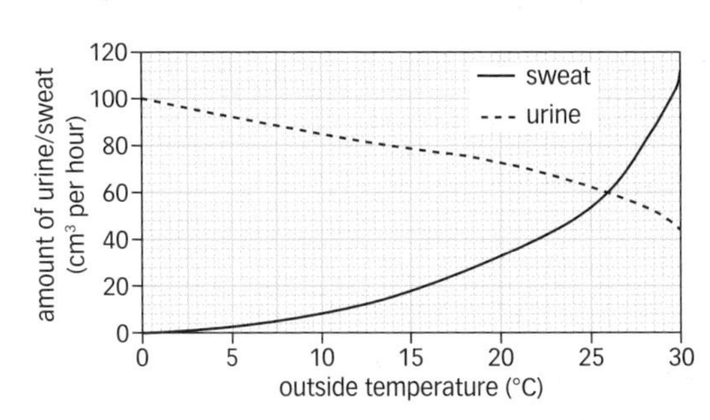

A03 **a** Describe how the amount of urine changes as temperature increases.

A02 **b** Explain what will happen to the amount of water in the urine as temperature increases.

D–C B–A*

Extended Writing

7 A01 Describe how air pollution may harm the environment.

8 A01 Explain how the body prevents overheating and overcooling.

9 A01 Explain how insulin and glucagon regulate the blood glucose level, and why this is necessary.

G–E D–C B–A*

A01 Recall the science
A02 Apply your knowledge
A03 Evaluate and analyse the evidence

Glossary

active transport Process that can move substances across cell membranes from low concentrations to high concentrations (against the concentration gradient). Active transport uses energy and is carried out by protein carriers in the cell membrane.
adaptation Feature of the body that helps an animal to survive.
addiction The body is dependent on a drug and will not function properly without it.
adult cell cloning Removing the nucleus from an unfertilised egg and replacing it with the nucleus from another cell taken from an adult organism. The resulting cell can develop into a new individual.
aerobic Using/in the presence of oxygen.
allele Version of a gene.
alveolus Structure in the lung that provides a large surface area with an extensive network of capillaries to carry out gaseous exchange.
amino acids Molecules containing carbon, hydrogen, nitrogen and oxygen. Many amino acids join together to make a protein.
amylase Enzyme that catalyses the breakdown of starch to sugar molecules.
anaerobic Without using/not in the presence of oxygen.
antibiotic Chemical, usually made by fungi or bacteria, that can be used as medicine to kill other fungi or bacteria in an infected person or animal.
antibodies Special proteins in the body that can bind to a particular antigen and destroy a particular pathogen.
antigen Special protein on the surface of a cell such as pathogen. The body makes antibodies to an antigen.
asexual reproduction Reproduction without gametes/sex cells, using mitosis.
atria Upper chambers of the heart that receive blood from veins.
auxin Plant hormone that causes shoots to bend.
bacteria Single-celled microorganism, 1–5 μm in length. The DNA is not enclosed in a nucleus. Bacterial cells have cytoplasm, a cell membrane, and a cell wall.
bacteria Single-celled microorganisms, 1–5 μm in length. The DNA is not enclosed in a nucleus. Bacterial cells have cytoplasm, a cell membrane, and a cell wall.
balanced diet Diet that has the right amount of proteins, fats, carbohydrates, vitamins, minerals, water, and fibre and gives you enough energy.
battery farming Rearing large numbers of animals in small spaces, to maximise efficiency of production.
biofuel Fuel such as wood or ethanol, derived from biological materials that absorb carbon dioxide while they are growing, so their use is less harmful to the environment than burning fossil fuels.
blood Tissue specialised for transport of substances including oxygen, carbon dioxide, food molecules, and wastes in larger animals. Blood is made up of plasma, red and white blood cells, and platelets.
breathing Movements of rib cage and diaphragm that cause air to enter and leave the lungs.
camouflage The body or colouring of an animal that allows it to blend in with its surroundings.
capillaries Small blood vessel with a very thin wall and narrow diameter. Capillaries allow exchange of substances between cells and blood.
carbohydrase Enzyme that catalyses the breakdown of large carbohydrate molecules such as starch to smaller sugar molecules.
carbon cycle Process by which carbon moves between the living and non-living world in a cyclic flow.
catalyst Substance that speeds up a reaction without being used up in the reaction.
cell Building block of living things.
cell membrane A thin layer around a cell that controls the movement of substances into and out of the cell.
cell wall Rigid cellulose layer outside the cell membrane in a plant or bacterial cell.
central nervous system The brain and spinal cord.
chlorophyll Green substance found in chloroplasts, where light energy is trapped for photosynthesis.
chloroplast Small disc in the cytoplasm of plant cells, containing chlorophyll. Photosynthesis occurs in chloroplasts.
chromosome Structure in a cell nucleus that consists of one molecule of DNA that has condensed and coiled into a linear structure.
classification Sorting organisms into groups according to their characteristics.
clinical trial Testing a drug to make sure it works and has no harmful side-effects.
clones Genetically identical individuals.
community All the populations of organisms that live together and interact in the same area.
competition The struggle between organisms to get sufficient resources to survive.
concentration gradient Difference in concentration of a substance from one region to another.
contraceptive Substance that prevents conception.
cuttings Taking part of a plant, eg the stem, and making new plants by asexual reproduction.

cytoplasm Jelly-like contents of a cell, inside the cell membrane. Cellular structures are suspended in the cytoplasm and this is where many of the cell's reactions occur.
decay Process by which microbes break down dead bodies or waste.
deficiency disease Disease caused by not eating enough of a particular nutrient.
deforestation Large-scale felling of trees to allow use of the land for building or agriculture.
dehydrated Not containing enough water to function properly.
denatured Describes the state of a protein when its shape has altered and it can no longer carry out its function.
detergent Substance added to water to improve its cleaning properties.
diabetes Condition in which blood glucose levels rise because the pancreas does not produce enough insulin.
dialysis Mechanical means of filtering the blood and removing harmful substances, used when the kidneys do not work properly.
differentiation Development of cells into types that are specialised for a particular function.
diffusion The spreading of the particles of a gas or a substance in solution, resulting in a net movement from a region of high concentration to a region of lower concentration. The bigger the difference in concentration, the faster the diffusion happens.
digestion Breaking down of large molecules into smaller, soluble particles that can be absorbed.
digestive system Organ system that digests food into smaller particles and absorbs these into the body.
digit Finger or toe.
diploid Describes a cell that has a nucleus with two sets of chromosomes; a body cell.
distribution Detail of where species are found over the total area where they occur. For example, woodlice may have a high distribution under a log.
DNA (deoxyribonucleic acid) Chemical that carries the genetic code.
DNA bases Molecules arranged in pairs within each molecule of DNA. A pairs with T and C pairs with G.
DNA fingerprinting Technique that analyses parts of the DNA of an individual and compares it with that of other individuals/DNA samples to find out whether someone committed a crime, or to establish whether individuals are related.
dominant Visible characteristic present in an organism even when only one allele of the gene is present.
donor Someone who donates (gives) something, such as an organ in an organ transplant.
double blind trial Clinical trial in which neither the patients nor the doctors know whether they are getting the real treatment or a placebo.
drug Chemical that alters the way your body or brain works.
effector Organ such as a gland or muscle that responds to a stimulus.
efficiency Carrying out a process, such as producing food, with the minimum loss of energy.
elastic Stretchy.
embryo transplant Method of cloning for animals. A fertilised embryo is split into four and each can develop into a new individual.
endangered Describes a species that has low numbers and is in danger of becoming extinct.
energy The ability to do work in the body to maintain life.
enzyme Biological catalyst made of protein. Enzymes catalyse chemical reactions in living organisms.
epidemic Sudden outbreak of a disease that affects many people in a country.
epidermal tissue Tissue one cell thick on the surface of plant roots, stems, and leaves, that protects the organs.
evaporation Liquid turning to gas. Evaporation transfers heat energy away so it cools the surface from which the liquid evaporates.
evolution Gradual change in an organism over time.
exchange system Organ system that exchanges materials, such as oxygen, carbon dioxide, and other wastes, between the organism and its surroundings.
excretion Removal of waste materials produced by the reactions of the body.
extinction End of a species, when all its members have died out.
extremophiles Organisms that can withstand extreme environmental conditions.
F_1 generation First filial (daughter) generation: the offspring from a genetic cross between two true-breeding parents.
fatigue Build-up of lactic acid in muscles that stops them contracting efficiently.
fermenter Large container used for growing large numbers of microorganisms.
fertilisation The joining of the male and female gametes to make a new individual.
fertiliser Chemical added to soil to increase the mineral content and promote plant growth.
filtration Removing particles from a liquid by passing it through a medium with small holes that do not allow the particles to pass through. In the kidney, filtration allows water and dissolved substances to pass into the kidney tubule, while blood cells and proteins remain in the capillaries.
fish stocks The numbers of fish in waters that are fished. Overfishing reduces fish stocks.
food chain A way of showing what organisms eat, showing the flow of food and energy from one organism to the next.
fossil Preserved remains of ancient living things.
fructose Type of sugar, which tastes sweeter than glucose.

FSH Female sex hormone, involved in making eggs in the ovary mature.

fungi Organisms with cells containing a membrane, cytoplasm, a nucleus, and a cell wall. The fungal cell wall is made of chitin rather than the cellulose of a plant cell wall.

gametes Sex cells. Male gametes are sperm. Female gametes are eggs. Gametes have half the normal number of chromosomes.

gene Length of DNA that codes for a characteristic/protein.

genetic engineering Changing an organism's genes by inserting a gene from another organism.

genetic modification Changing an organism's genes to give it desirable characteristics.

geotropic Describes a response of plants to the direction of the pull of gravity. Roots grow towards the pull of gravity and shoots grow away from it.

gland Structure that makes a useful substance for the living organism.

global warming Change in climate caused by the emission of more greenhouse gases.

glucagon Hormone that causes the liver to break down glycogen to glucose, and release more glucose into the blood.

glucose Type of sugar.

glycogen Large insoluble carbohydrate molecule; similar to starch but found only in animal cells, some bacterial cells, and some fungi.

greenhouse Structure where plants can be grown under controlled conditions.

greenhouse gas Gas that produces a greenhouse effect in the atmosphere, preventing heat energy being radiated away from the Earth.

growth Increase in size, usually with an increase in cell numbers.

haemoglobin Soluble protein that also contains an iron atom. Found in red blood cells. It carries oxygen from lungs to respiring tissues.

haploid Describes a cell that has a nucleus with only one set of chromosomes; a sex cell.

heart rate Number of times per minute a heart beats.

hectare Unit of area, equivalent to 2.4 acres.

herbicide Weedkiller.

hormone Chemical made by a gland and carried in the blood to its target organ(s).

hormone Chemical made by a gland and carried in the blood to its target organ(s).

immune system Your body's system that fights infections, involving white blood cells and antibodies.

immunisation Medical procedure to make people immune to a particular disease. It involves introducing dead or inactivated pathogens into the body to stimulate the immune system to make antibodies.

immunity You have immunity when you have made antibodies to a pathogen, so that next time the body can make them again quickly, and the pathogen does not make you ill.

immunosuppressant Describes drugs that suppress the immune system, given to recipients of a transplanted organ to reduce the risk of rejection.

in vitro fertilisation Fertilisation carried out outside the body, so the egg and sperm join in a glass dish. When the embryo begins to form from this fertilised egg it is put into the womanís uterus (womb).

indicator species Species that survive best at a certain level of pollution, and give an idea of the level of pollution.

inheritance factor Term used by Mendel. We now call this a gene.

insulin Hormone produced by the pancreas that causes body cells to take up glucose from the blood.

ion Charged particle, such as $Na+$, $K+$, $Mg+$.

isolation Separation of two populations of a species so that they cannot interbreed, for example by a geographical barrier such as an ocean or mountain range.

isomerase Type of enzyme that rearranges the atoms in a molecule.

isotonic Having equal concentrations of dissolved substances.

kingdom Major subdivision in the classification of living organisms, eg the plant kingdom.

lactic acid Chemical made from the incomplete breakdown of glucose during anaerobic respiration.

leaf Plant organ specialised for photosynthesis.

LH Female sex hormone, involved in triggering ovulation.

limiting factor Factor such as carbon dioxide level, light, or temperature, that will affect the rate of photosynthesis if it is in short supply. Increasing the limiting factor will increase the rate of photosynthesis.

lipase Enzyme that catalyses the breakdown of fats (lipids) to fatty acids and glycerol.

lumen Space in a blood vessel through which blood flows.

lung Organ of many land animals, specialised to exchange oxygen and carbon dioxide.

mean Average of a collection of data.

median Middle value of a collection of data.

meiosis Type of cell division that occurs to form sex cells, resulting in four genetically different cells.

menstrual cycle Monthly cycle in adult females. The first day of the cycle is the first day of the period. After that a new uterus lining forms and an egg is released in the middle of the cycle. If the egg is not fertilised the uterus lining passes out as the next period.

metabolic rate How quickly all the reactions are going on in cells.

metabolism All the chemical reactions going on in cells.

microbe Microscopic organism such as bacteria or fungi.

microorganism Small organism able to be seen with a microscope. Bacteria and viruses are microorganisms.

microscope Instrument used to view small objects. A light microscope magnifies up to about 1500 times; an electron microscope gives higher magnifications.

mitochondria Structures in animal and plant cells where aerobic respiration takes place.

mitosis Type of cell division that occurs in body cells, resulting in two genetically identical cells.

mode Most popular value of a collection of data.

monohybrid inheritance Inheritance pattern of a single characteristic, determined by one gene (that may have two alleles).

mucus Slimy substance made in special cells/glands.

multicellular Being built of many cells, all working together as an organism.

muscle Tissue made up of muscle fibres, specialised to contract and bring about movement in the organism.

mutation Change in the structure of a gene. A mutation may result in the gene coding for a different protein/characteristic.

mycoprotein High-protein food produced from a fungus.

natural selection The survival of better adapted organisms.

needles Reduced leaves with a small surface area.

nerve impulses Electrical signals that carry information along nerves.

neurone Cell that carries electrical impulses, sometimes called nerve cell.

nucleus Structure inside a cell that controls the cell's activities. It contains chromosomes made of DNA.

obese Describes someone who is very overweight.

oestrogen Female sex hormone, involved in regulating the menstrual cycle.

optimum Best.

organ Collection of different tissues working together to perform a function within an organism; examples include the stomach in an animal and the leaf in a plant.

organ system Collection of different organs working together to perform a major function within an organism; an example is the digestive system in an animal.

osmosis Diffusion of water through a partially permeable membrane, from a dilute solution to a more concentrated solution.

ovulation Release of a mature egg from an ovary.

oxygen debt Lack of oxygen in muscle cells. Oxygen is needed to oxidise lactic acid in the muscle to carbon dioxide and water.

oxyhaemoglobin Haemoglobin with oxygen atoms attached.

painkiller Drug that stops you feeling pain.

palisade layer A layer of tall columnar cells containing chloroplasts, where the majority of photosynthesis occurs in a leaf.

palisade mesophyll cells Tall columnar cells containing chloroplasts, where the majority of photosynthesis occurs in a leaf.

pandemic Epidemic that sweeps across continents or across the whole world.

partially permeable membrane Membrane that has small pores through which small molecules such as water can pass, but not larger molecules such as proteins.

pathogen Organism that can cause infectious disease.

peat Type of soil formed from rotting vegetation that is rich in carbon, and is burned as a fuel.

peripheral nervous system Nerves carrying information from sense organs in your body to the central nervous system, and from the central nervous system to effectors.

permanent vacuole Fluid-filled cavity in plant cells, containing sap.

pesticide Chemical sprayed onto crop plants to kill pests such as insects.

phloem Plant tissue made up of living cells that has the function of transporting food substances through the plant.

photosynthesis Process by which plants build carbohydrates from carbon dioxide and water, using sunlight energy.

phototropic Describes a response of plants to the direction of light. Shoots grow towards light and roots grow away from light.

placebo Dummy pill or treatment.

plasma Fluid in blood in which the cells are suspended. Plasma transports carbon dioxide, digested food molecules, and urea.

pollutant Substance put into the environment by human activity, which is not normally there.

pollution Contamination of the environment by harmful substances.

population The number of organisms of a species in a given area.

pore Small opening. Pores on the surface of a leaf allow water and gases to move in and out of the leaf.

potometer Apparatus used to measure the rate of transpiration by measuring the uptake of water by a plant.

progesterone Female sex hormone, involved in maintaining the uterus lining.

protease Enzyme that catalyses the breakdown of proteins to amino acids.

protein carrier Protein that carries something; for example, some protein carriers transport sodium ions out of nerve cells by active transport.

proteins Large molecules (polymers) made of many amino acids joined together. Proteins have many functions, including structural (as in muscle), hormones, antibodies, and enzymes.

pyramid of biomass A way of showing the biomass of organisms at each link in a food chain.

quotas Limits to the numbers of fish that can be caught in a particular area, set to prevent damage to fish stocks.

radiation Transfer of heat energy by infrared radiation.

rate of photosynthesis How quickly a plant is photosynthesising. The rate is affected by factors including carbon dioxide levels, light, and temperature.

rate of transpiration How quickly a plant is losing water by transpiration (evaporation from its leaves). The rate is affected by factors including humidity, air movement, and temperature.

reabsorption Process of reabsorbing (taking back) useful substances including glucose, ions, and water into the blood from the kidney tubule so they are not excreted. Takes place in the kidney.

receptor Cell or sense organ that detects stimuli.

recessive Visible characteristic that is only present in an organism if two alleles of the gene are present.

reflex action Fast automatic response of the body to a potentially dangerous stimulus, coordinated by the spinal cord.

rejection Attack on a transplanted organ by the immune system within a recipient's body.

relationship Interaction between different species living together in the same area, such that one species affects another. An example is a predator–prey relationship.

renal Relating to the kidney.

resistance The ability of a microorganism to withstand the effects of antibiotics and not be killed by them.

resource Something that an organism needs to survive, such as food, space, or oxygen.

respiration Process by which living things release energy from carbohydrates, also producing water and carbon dioxide.

sampling Counting a small number of a large total population in order to study its distribution.

secrete Produce a hormone in special cells of a gland.

sex chromosomes Chromosomes involved with determining the sex of an individual. In humans if you have two X chromosomes you are female; if you have one X and one Y chromosome you are male.

sex hormones Hormones produced in the sex organs (in the ovaries and testes).

sexual reproduction Reproduction involving the joining of gametes from two (usually unrelated) parents.

shivering Automatic response to cold, where muscles under the skin contract repeatedly to release more heat by respiration.

specialised Specialised cells have a structure well suited to their function.

speciation Separate evolution of two populations of the same species, to form two separate species.

species Group of organisms that are similar and capable of interbreeding.

specific Enzymes are specific; they act on only one substrate.

spines Leaves with reduced surface area and a pointed end.

starch Large insoluble carbohydrate molecule, stored in plant cells.

statins Drugs that reduce the amount of cholesterol made by the body.

stem cell Undifferentiated cell that can divide by mitosis and is capable of differentiating into any of the cell types found in that organism.

stent Narrow mesh tube that is inserted into a blocked artery.

stimulus Change in the environment, such as a temperature change, that you respond to.

stomata Pores on the surface of a leaf that allow water, carbon dioxide, and oxygen to move in and out of the leaf.

substrate Substance acted upon by an enzyme in a chemical reaction. The substrate molecules are changed into product molecules.

surface area The total amount of the surface of an organism or part of an organism.

sustainable Describes the use of resources for/by humans without harming the environment.

symptoms How you feel when you have a disease, eg headache, feeling sick. Do not confuse symptoms with signs – signs are the measurable changes when you have a disease, such as increased temperature or a rash.

synapse Small gap between one neurone and another, or between a neurone and a muscle cell.

target organ Organ or part of the body that responds to a particular hormone.

Thalidomide Drug used as a sleeping pill and for morning sickness, that caused birth defects when pregnant women took it.

thermoregulation Regulating body temperature.

tissue Group of cells of similar structure and function working together, such as muscle tissue in an animal and xylem in a plant.

tissue culture Growing whole new plants from small groups of cells taken from one plant.

tissue-typing Matching the tissues of donor and recipient in a transplant operation, so that their antigens are similar and there is less chance of rejection.

toxin Poison.

tract A system of tubes in the body, such as the digestive tract or respiratory tract.

translocation Transport of sugars made in photosynthesis in the leaf to areas of the plant that store them or use them.

transpiration Movement of water through the xylem of a plant, from the roots up to the leaves.

transpiration stream Continuous flow of water through the xylem of a plant, from the roots up to the leaves where it evaporates.

trauma Serious injury, for example resulting from an accident.

tropism Growth response to a stimulus. The direction of the stimulus determines the direction of the growth response.

urea Substance produced in the liver from the breakdown of amino acids, and removed in the urine by the kidneys.

valves Structures that allow one-way movement of a fluid and prevent backflow.

vasoconstriction Narrowing of blood vessels, so that less blood can flow through.

vasodilation Widening of blood vessels, so that bood can flow through more easily.

ventilation Movement of air into and out of the lungs, brought about by inhaling and exhaling.

ventricles Lower chambers of the heart that contract and force blood out of the heart into the arteries.

villus Structure in the small intestine that provides a large surface area with an extensive network of capillaries to absorb the products of digestion by diffusion and active transport.

virus Extremely small infectious agent that can only be seen with an electron microscope and can only live or reproduce inside a host cell.

withdrawal symptoms If you are addicted to (dependent on) a drug, when you stop taking it you get unpleasant symptoms such as pain and tremors.

xylem Plant tissue made up of dead cells that has the function of transporting water and dissolved substances through the plant.

yield The amount of useful product, eg from crop plants.

zygote (Usually) diploid cell resulting from the fusion of an egg and a sperm.

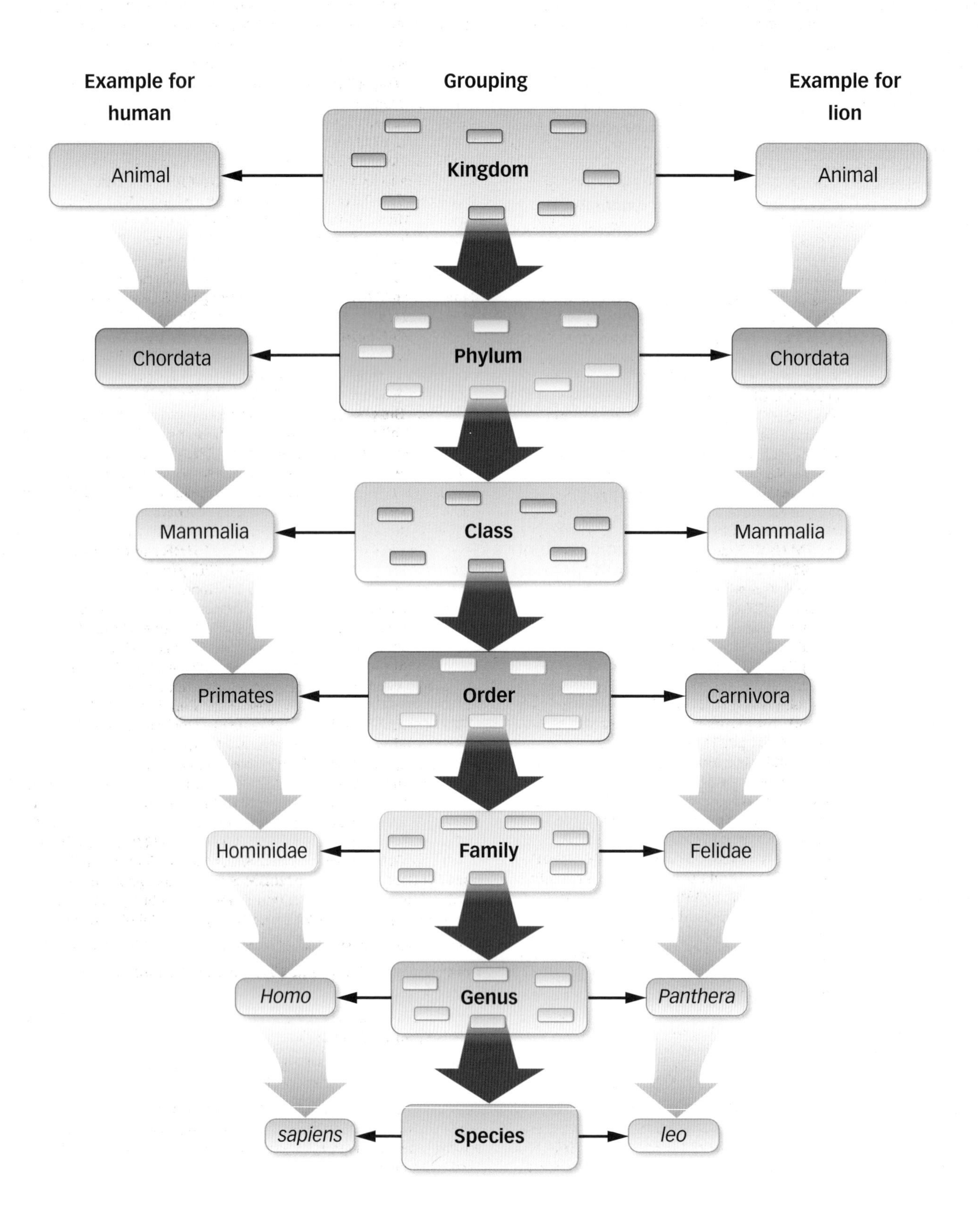

Example for human
Grouping
Example for lion
Animal
Kingdom
Animal
Chordata
Phylum
Chordata
Mammalia
Class
Mammalia
Primates
Order
Carnivora
Hominidae
Family
Felidae
Homo
Genus
Panthera
sapiens
Species
leo

Acknowledgements

The publisher and authors would like to thank the following for their permission to reproduce photographs and other copyright material:

p8T Martyn F. Chillmaid/SPL; **p8B** Chris Pearsall/Alamy; **p13** www.frankhuelsboemer.de; **p14T** Jochen Tack/Photolibrary; **p14B** Bjorn Svensson/SPL; **p15** Monkey Business Images/Dreamstime; **p16** Elena Schweitzer/Shutterstock; **p17** Life In View/SPL; **p18T** Mark Thomas/SPL; **p18B** Medical RF.Com/SPL; **p19** Steve Gschmeissner/SPL; **p20** GustoImages/SPL; **p21** Mike Devlin/SPL; **p22M** SPL; **p22R** Dr Gladden Willis, Visuals Unlimited/SPL; **p22L** Thierry Berrod, Mona Lisa Production/SPL; **p23** Samuel Ashfield/SPL; **p24** Medical RF.Com/SPL; **p25** Iofoto/Shutterstock; **p27** Tony McConnell/SPL; **p29R** AJ Photo/SPL; **p29L** Medical RF.Com/SPL; **p30L** Corbin O'Grady Studio/SPL; **p30R** Zephyr/SPL; **p31** Pascal Goetgheluck/SPL; **p32** Jerome Wexler/SPL; **p33L** Dan Suzio/SPL; **p33M** John Kaprielian/SPL; **p33R** Peter Anderson/Dorling Kindersley/Getty Images; **p34** E.M. Welch/Rex Features; **p35R** Ria Novosti/SPL; **p35L** Bob Gibbons/SPL; **p36** SPL; **p37R** Cordelia Molloy/SPL; **p37L** Colin Cuthbert/SPL; **p43** Thomas Nilsen/SPL; **p47** Gregory Dimijian/SPL; **p48L** Martin Shields/SPL; **p48R** Reinhard Dirscherl, Visuals Unlimited/SPL; **p49T** Bernhard Edmaier/SPL; **p49B** Wolfgang Baumeister/SPL; **p53B** Power and Syred/SPL; **p53T** Robert Brook/SPL; **p54** Brian Bell/SPL; **p55R** Ken Harris; **p55TL** Massimo Brega/The Lighthouse/SPL; **p55M** Dr J. Bloemen/SPL; **p55BL** Duncan Shaw/SPL; **p59** Curt Maas/AGStockUSA/SPL; **p60L** Sinclair Stammers/SPL; **p60R** Bob Gibbons/SPL; **p61** GustoImages/SPL; **p63** Tony Craddock/SPL; **p64T** Justyna Furmanczyk/Shutterstock; **p64BL** Mark Burnett/SPL; **p64BR** Mark Burnett/SPL; **p65L** Dr Gopal Murti/SPL; **p65R** Patrick Landmann/SPL; **p66B** Science Pictures Limited/SPL; **p66T** Eye Of Science/SPL; **p67L** Geoff Kidd/SPL; **p67M** Geoff Kidd/SPL; **p67BR** Royal Botanic Garden Edinburgh/SPL; **p67TR** Mark Thomas/SPL; **p68T** Rosenfeld Images Ltd/SPL; **p68BL** Geoff Tompkinson/SPL; **p68BR** Dr Najeeb Layyous/SPL; **p69** GustoImages/SPL; **p72T** Steve Percival/SPL; **p72B** Golden Rice; **p73T** Subbotina Anna/Shutterstock; **p73M** Martin Bond/SPL; **p73B** Jon Naustdalslid/Shutterstock; **p74R** Robert Brook/SPL; **p74L** Chris Knapton/SPL; **p76BL** Maximilian Stock Ltd/SPL; **p76BM** S.R. Maglione/SPL; **p76T** Dr Keith Wheeler/SPL; **p76BR** Steve Gschmeissner/SPL; **p77L** John Beatty/SPL; **p77R** Mark Phillips/SPL; **p78** William Ervin/SPL; **p80L** Tony Camacho/SPL; **p80R** K. Jayaram/SPL; **p81** Michael W. Tweedie/SPL; **p82TL** National Library Of Medicine/SPL; **p82BL** SPL; **p82R** Gary Hincks/SPL; **p83** SPL; **p89** Carmen Martinez Banus/Istockphoto; **p90M** Photo Researchers/SPL; **p90R** Omikron/SPL; **p90L** Dr Gopal Murti/SPL; **p91** John Durham/SPL; **p92** Dr Terry Beveridge, Visuals Unlimited/SPL; **p93** Science VU, Visuals Unlimited/SPL; **p94T** Power and Syred/SPL; **p94MT** Pasieka/SPL; **p94M** Michael Abbey/SPL; **p94MB** Eye of Science/SPL; **p94B** Steve Gschmeissner/SPL; **p95T** Dr David Furness, Keele University/SPL; **p95MT** Steve Gschmeissner/SPL; **p95MB** Dr David Furness, Keele University/SPL; **p95B** J.C. Revy, Ism/SPL; **p96L** Andrew Lambert Photography/SPL; **p96R** Andrew Lambert Photography/SPL; **p98T** Science VU, Visuals Unlimited/SPL; **p98M** Dr Richard Kessel & Dr Gene Shih, Visuals Unlimited/SPL; **p98B** CNRI/SPL; **p99** Dr Richard Kessel & Dr Gene Shih, Visuals Unlimited/SPL; **p101** Steve Gschmeissner/SPL; **p102** Gavin Kingcome/SPL; **p103** Power And Syred/SPL; **p106R** Veronique Leplat/SPL; **p106L** Power and Syred/SPL; **p107** Carlos Munoz-Yague/Eurelios/SPL; **p108** Scott Sinklier/AGStockUSA/SPL; **p110** Jennifer Fry/SPL; **p112T** Paul Harcourt Davies/SPL; **p112BL** Martyn F. Chillmaid/SPL; **p112BR** Martyn F. Chillmaid/SPL; **p114** Colin Cuthbert/Newcastle University/SPL; **p115** George Steinmetz/SPL; **p121** PH. Plailly/Eurelios/SPL; **p122R** Tim Vernon/SPL; **p122L** John Bavosi/SPL; **p123** Laguna Design/SPL; **p126** Roger Harris/SPL; **p128R** Cordelia Molloy/SPL; **p128BL** Power and Syred/SPL; **p128TL** Maximilian Stock Ltd/SPL; **p129** Manfred Kage/SPL; **p130** AJ Photo/SPL; **p131T** Ashley Cooper, Visuals Unlimited/SPL; **p131B** Jonathan Hordle/Rex Features; **p132T** Tony Craddock/SPL; **p132B** Joe McDonald, Visuals Unlimited/SPL; **p133** Manfred Danegger/Okapia/SPL; **p134T** Dr Fred Hossler, Visuals Unlimited/SPL; **p134B** GustoImages/SPL; **p135** CNRI/SPL; **p136** Friedrich Saurer/SPL; **p137** Stigur Karlsson/Istockphoto; **p138** Pr. G. Gimenez-Martin/SPL; **p139** Herve Conge, ISM/SPL; **p141B** Eye of Science/SPL; **p141T** Adrian T. Sumner/SPL; **p142B** Bruno Petriglia/SPL; **p142T** James King-Holmes/SPL; **p146L** SPL; **p146R** Gary Parker/SPL; **p148** Jacopin/SPL; **p150L** Larry Dunstan/SPL; **p150R** CNRI/SPL; **p152** Professor Miodrag Stojkovic/SPL; **p153** Martin Shields/SPL; **p154TL** Frederick R. Mcconnaughey/SPL; **p154MT** Sinclair Stammers/SPL; **p154MB** Pasieka/SPL; **p154BR** Olivier Darmon/Jacana/SPL; **p156** Christian Darkin/SPL; **p157TL** Photo Researchers/SPL; **p157R** Dan Sams/SPL; **p157BL** Simon Fraser/SPL; **p158** Science Source/SPL; **p159** David Gifford/SPL; **p165** AJ Photo/SPL; **p166** Biophoto Associates/SPL; **p167BR** Michael Abbey/SPL; **p167TR** Michael Abbey/SPL; **p167L** Steve Gschmeissner/SPL; **p168** Mike Manzano/Istockphoto; **p169** Francesco Ridolfi/Istockphoto; **p172** Claude Nuridsany & Marie Perennou/SPL; **p173** Eye of Science/SPL; **p177T** Dr Jeremy Burgess/SPL; **p177B** Dr Jeremy Burgess/SPL; **p178L** Martyn F. Chillmaid/SPL; **p178R** Cindy Hughes/Shutterstock; **p180** Eye of Science/SPL; **p181** Jim Varney/SPL; **p183R** Hank Morgan/SPL; **p183L** Steve Allen/SPL; **p184** Dr Gladden Willis, Visuals Unlimited/SPL; **p186L** Pasieka/SPL; **p186R** Eye of Science/SPL; **p189T** Dr Gopal Murti/SPL; **p189B** Prof. P. Motta/Dept. of Anatomy/University 'La Sapienza', Rome/SPL; **p190** Power and Syred/SPL; **p191TR** Dr Arnold Brody/SPL; **p191L** Michael Ross/SPL; **p191BR** National Cancer Institute/SPL; **p192** Power and Syred/SPL; **p199** Alex Bartel/SPL; **p200** Martin Oeggerli/SPL; **p201L** Ian Hooton/SPL; **p201R** Sheila Terry/SPL; **p203** BSIP VEM/SPL; **p205** Life in View/SPL; **p206** Life in View/Science Photo Library; **p210** Tony McConnell/SPL; **p211** Annie Griffiths Belt/National Geographic/Getty Images; **p212** AJ Photo/SPL; **p214** Dr P. Marazzi/SPL; **p215** Saturn Stills/SPL; **p217B** Bjorn Svensson/SPL; **p217T** Paul Bradbury/OJO Images/Getty Images; **p219L** Jeremy Walker/SPL; **p219R** Brian Bell/SPL; **p220L** Gary Cook, Visuals Unlimited/SPL; **p220TR** Brasil2/Istockphoto; **p220BR** Jacques Jangoux/SPL; **p221T** Guenter Guni/Istockphoto; **p221MT** Chris Hellier/SPL; **p221M** Prill Mediendesign & Fotografie/Istockphoto; **p221MB** Adrian Brockwell/Istockphoto; **p221B** Chris Sattlberger/SPL; **p223T** Michael Szoenyi/SPL; **p223TR** Simon Fraser/SPL; **p223BR** S.R. Maglione/SPL; **p223BL** John Shaw/SPL; **p223TL** Jerry Schad/SPL; **p224L** Hubert Raguet/Look At Sciences/SPL; **p224R** Simon Fraser/SPL; **p225** Victor de Schwanberg/SPL; **p226TR** Geoff Kidd/SPL; **p226TM** Geoff Kidd/SPL; **p226BM** Subjug/Istockphoto; **p226BR** Cordelia Molloy/SPL; **p226L** Dennis Inc/Photolibrary; **p229** Perets/Istockphoto; **p230T** Dirk Wiersma/SPL; **p230B** HelleM/Istockphoto.

Cover image courtesy of GUSTO IMAGES/SCIENCE PHOTO LIBRARY

Illustrations by Wearset Ltd, HL Studios, Peter Bull Art Studio, Tristan Mitchell.

The publisher and authors are grateful for permission to reprint the following copyright material:

p230 Fish stocks, used with the kind permission of the International Council for the Exploration of the Sea.

Although we have made every effort to trace and contact all copyright holders before publication this has not been possible in all cases. If notified, the publisher will rectify any errors or omissions at the earliest opportunity.

Great Clarendon Street, Oxford OX2 6DP

Oxford University Press is a department of the University of Oxford.
It furthers the University's objective of excellence in research,
scholarship, and education by publishing worldwide in

Oxford New York

Auckland Cape Town Dar es Salaam Hong Kong Karachi
Kuala Lumpur Madrid Melbourne Mexico City Nairobi
New Delhi Shanghai Taipei Toronto

With offices in
Argentina Austria Brazil Chile Czech Republic France Greece
Guatemala Hungary Italy Japan Poland Portugal Singapore
South Korea Switzerland Thailand Turkey Ukraine Vietnam

First published 2011

British Library Cataloguing in Publication Data

Data available

ISBN 978-0-19-913598-1

10 9 8 7 6 5 4 3 2

Printed in Great Britain by Bell and Bain, Glasgow

Paper used in the production of this book is a natural, recyclable product
made from wood grown in sustainable forests. The manufacturing process
conforms to the environmental regulations of the country of origin.